ADVANCES IN CHEMICAL PHYSICS
VOLUME XVI

EDITORIAL BOARD

ADVANCES IN CHEMICAL PHYSICS

Edited by I. PRIGOGINE
University of Brussels, Brussels, Belgium
and S. A. RICE
University of Chicago, Chicago, U.S.A.

VOLUME XVI

INTERSCIENCE PUBLISHERS
a division of
John Wiley & Sons   London-New York-Sydney-Toronto

FIRST PUBLISHED 1969 BY JOHN WILEY & SONS LTD. ALL RIGHTS RESERVED

LIBRARY OF CONGRESS CATALOG CARD NUMBER 58-9935

SBN 471 69921 7

PRINTED IN GREAT BRITAIN BY JOHN WRIGHT & SONS LTD., AT THE STONEBBRIDGE PRESS, BRISTOL

INTRODUCTION

In the last decades, chemical physics has attracted an ever-increasing amount of interest. The variety of problems, such as those of chemical kinetics, molecular physics, molecular spectroscopy, transport processes, thermodynamics, the study of the state of matter, and the variety of experimental methods used, makes the great development of this field understandable. But the consequence of this breadth of subject matter has been the scattering of the relevant literature in a great number of publications.

Despite this variety and the implicit difficulty of exactly defining the topic of chemical physics, there are a certain number of basic problems that concern the properties of individual molecules and atoms as well as the behaviour of statistical ensembles of molecules and atoms. This new series is devoted to this group of problems which are characteristic of modern chemical physics.

As a consequence of the enormous growth in the amount of information to be transmitted, the original papers, as published in the leading scientific journals, have of necessity been made as short as is compatible with a minimum of scientific clarity. They have, therefore, become increasingly difficult to follow for anyone who is not an expert in this specific field. In order to alleviate this situation, numerous publications have recently appeared which are devoted to review articles and which contain a more or less critical survey of the literature in a specific field.

An alternative way to improve the situation, however, is to ask an expert to write a comprehensive article in which he explains his view on a subject freely and without limitation of space. The emphasis in this case would be on the personal ideas of the author. This is the approach that has been attempted in this new series. We hope that as a consequence of this approach, the series may become especially stimulating for new research.

Finally, we hope that the style of this series will develop into something more personal and less academic than what has become the standard scientific style. Such a hope, however, is not likely to be completely realized until a certain degree of maturity has been attained—a process which normally requires a few years.

At present, we intend to publish one volume a year, but this schedule may be revised in the future.

In order to proceed to a more effective coverage of the different aspects of chemical physics, it has seemed appropriate to form an editorial board. I want to express to them my thanks for their cooperation.

I. PRIGOGINE

CONTRIBUTORS TO VOLUME XVI

B. BARANOWSKI, Institute of Physical Chemistry, Polish Academy of Sciences, Warsaw, Poland

R. BERSOHN, Department of Chemistry, Columbia University, New York, N.Y. 10027

R. DAGONNIER, Université Libre de Bruxelles, Brussels, Belgium

A. HARING, FOM-Institute for Atomic and Molecular Physics, Amsterdam, The Netherlands

A. HAZI, Department of Chemistry and the James Franck Institute, University of Chicago, Chicago, Illinois 60637

J. HIJMANS, Koninklijke/Shell-Laboratorium, Amsterdam (Shell Research N.V.), The Netherlands

TH. HOLLEMAN, Koninklijke/Shell-Laboratorium, Amsterdam (Shell Research N.V.), The Netherlands

J. KARLE, Laboratory for the Structure of Matter, U.S. Naval Research Laboratory, Washington D.C., 20390

S. H. LIN, Department of Chemistry, Arizona State University, Tempe, Arizona 85281

L. MEYER, Department of Chemistry and the James Franck Institute, University of Chicago, Chicago, Illinois 60637

R. PAUL, Department of General Physics and X-rays, Indian Association for the Cultivation of Science, Calcutta 32

S. A. RICE, Department of Chemistry and the James Franck Institute, University of Chicago, Chicago, Illinois 60637

A. E. DE VRIES, FOM-Institute for Atomic and Molecular Physics, Amsterdam, The Netherlands

J. D. WEEKS, Department of Chemistry and the James Franck Institute, University of Chicago, Chicago, Illinois 60637

CONTENTS

THEORY OF QUANTUM BROWNIAN MOTION

R. DAGONNIER,* *Université Libre de Bruxelles, Brussels, Belgium*

CONTENTS

* Present address: Faculté des Sciences, Centre Universitaire de l'Etat,
Mons, Belgium.

1. INTRODUCTION

For several years, much interest has been raised by the "microscopic" justification of the Brownian motion theory. This theory was developed initially by Einstein and Smoluchowsky in terms of semi-phenomenological arguments and later elaborated by Langevin and others.[1] These authors describe the effects of the fluid particles (of mass m) on Brownian particles (of mass $M \gg m$) in a stochastic fashion. Their results are summarized in the well-known Fokker–Planck equation satisfied by the distribution function of the heavy particle in its position and momentum space. All descriptions of Brownian motion still relied, up to very recently, on stochastic assumptions because the dynamics of the motion of the fluid atoms, which cause the Brownian motion of the heavier particles, was not introduced explicitly.

In recent years, various authors[2-4] have developed a microscopic theory of Brownian motion using the latest developments of non-equilibrium statistical mechanics. Since the laws of mechanics alone describe the irreversible behaviour of large systems (provided suitable initial conditions are chosen) no extramechanical assumptions are needed for a theory of Brownian motion.

Starting from the Liouville equation for the total distribution function, these authors derive a general transport equation for the one particle distribution function of the Brownian particle by integrating over the variables of the fluid molecules in certain limits involving the time scale and the size of the system. As the fluid is taken at thermal equilibrium the motion of the Brownian particle is studied under the *quasi-equilibrium condition*:

$$(\langle p_1^2 \rangle / \langle P^2 \rangle)^{\frac{1}{2}} \simeq (m/M)^{\frac{1}{2}} \equiv \gamma \ll 1 \qquad (1)$$

which is obviously realized when both average momenta are in the thermal range (where P and p_f denote respectively the momentum of the Brownian particle and of the fluid molecule). The Fokker–Planck equation is then readily obtained as the limiting form of the Brownian particle generalized transport equation when γ tends to zero.

Very recently, some authors[5-10] have extended to quantum systems this method for obtaining a Fokker–Planck-like equation. The aim of the present paper is to give a general report of the study of the quantum-mechanical Brownian motion.

The quantum theory of Brownian motion appears to be useful in understanding various physical situations, for example theoretical interpretations of ionic mobilities in quantum liquids[8, 9, 14, 15] and predictions about non-classical isotope effects that should appear in diffusion experiments with heavy particles in quantum fluids at very low temperatures.[7] It should be therefore emphasized that limitations concerning the domain of validity of a Fokker–Planck-like description are necessary, chiefly in the very low temperature range where the zero-point motion of the particles plays an important part.

Here we shall apply the general theory of irreversible processes due to Prigogine and coworkers[11-13] by using for the description of quantum Brownian motion the method developed by Resibois and the author.[8, 9] Section 2 deals with the general formulation of the problem; we consider the model of one heavy charged Brownian particle (mass M, charge e) immersed in a fluid of N light quantum particles (for instance, bosons of mass m) and influenced by a weak electrostatic field E. Starting from the von Neumann equation for the $(N+1)$ particle density matrix, we obtain a generalized transport equation for the Brownian motion. From this equation we derive, in Section 3, the usual Fokker–Planck equation by expanding the dynamic properties of the Brownian particle in the mass ratio γ^2; the Fokker–Planck equation appearing then as the limiting form of the general transport equation when $\gamma \to 0$ [see Eq. (1)]. This requirement is obviously realized for a weakly coupled surrounding fluid such as a nearly perfect Bose gas. This model is then considered to interpret the experimental data obtained from measurements of ionic mobilities in liquid ^{4}He. In the next Section we discuss the problem of Brownian motion in a Fermi fluid where the convergence of the γ expansion appears to depend crucially on the temperature range considered; indeed it is shown that the condition of validity of the Fokker–Planck equation [Eq. (1)] should be replaced by the much more restrictive requirement:

$$(\langle p_1^2 \rangle / \langle P^2 \rangle)^{\frac{1}{2}} \simeq (m\varepsilon_{\mathrm{F}}/MkT)^{\frac{1}{2}} \equiv \gamma\xi \ll 1 \qquad (2)$$

where ε_{F} denotes the Fermi energy and ξ is defined by $(\varepsilon_{\mathrm{F}}/kT)^{\frac{1}{2}}$. To illustrate this case we give a theoretical interpretation of the behaviour of heavy ions in liquid ^{3}He.

Section 5 is devoted to the analysis of the next term of the

γ series appearing in the basic transport equation for Brownian motion. This examination of the first "correction" to the usual Fokker–Planck equation shows that condition (1) is no longer valid for strongly coupled systems (dense fluids) at very low temperatures, the zero-point motion of both kinds of particles starting to play a decisive part. Consequently when the localization effects are such that the convergence of the general transport equation is no longer ensured the whole framework of the usual Brownian motion theory breaks down.

Finally some lengthy calculations are reported in the Appendices.

The work reported in this paper has been carried out in the department of Professor I. Prigogine, at Brussels University. We wish to thank him for his continuous encouragement during its progress. We are also most indebted to P. Resibois, who suggested and collaborated in the major part of this work, and to H. T. Davis, who played an essential part in developing the results concerning the Fermi systems. Dr. J. Lekner read the manuscript and we thank him for his aid.

2. THE GENERALIZED TRANSPORT EQUATION FOR QUANTUM BROWNIAN MOTION

A. The Von Neumann–Liouville Equation

We consider a system enclosed in a box of volume Ω, made of one heavy charged particle (mass M, charge e) immersed in a fluid of N bosons of mass m $(m \ll M)$, submitted to the influence of a weak external electrostatic field E. As in the equivalent classical situation[2-4] the motion of the heavy ion is studied under the quasi-equilibrium condition (1). This requirement can obviously be realized at sufficiently low temperatures for a *weakly coupled system* (at $T = 0$, $\langle p_f^2 \rangle = 0$!). However for a system with *strong interactions*, the zero-point motion of both kinds of particles starts to play a part; it is difficult then to make general assertions and Eq. (1) will in this case be considered as a sufficient condition for the validity of our proof. This point will be considered in more detail in Section 5. Let us stress also, that although the theory is formulated for bosons, the method is quite general provided that (1) is valid.

If we use a second quantization representation for free particles, we can cast the Hamiltonian of our system into the well-known form:

$$H' = H + H^e; \quad H = H_0 + \lambda W \tag{3}$$

where the "internal" Hamiltonian H contains the kinetic energy term:*

$$H_0 = H_0^B + H_0^f = \frac{\hat{K}^2}{2M} + \sum_k \frac{k^2}{2} a_k^+ a_k \tag{4}$$

and the interaction energy between the particles:

$$\lambda W = \lambda V^f + \lambda V^{fB}$$

$$= \frac{\lambda}{2\Omega} \sum_{klpr} v(klpr) a_k^+ a_l^+ a_p a_r \, \delta^{Kr}(k+l-p-r)$$

$$+ \frac{\lambda}{\Omega} \sum_{kl} u(k-l) \exp\left[-i(k-l) R\right] a_k^+ a_l \tag{5}$$

where $(R, \hat{K})$ denote the coordinate and momentum operators of the Brownian particle; a_k^+ and a_k are the usual creation and destruction operators for bosons. Moreover $v(klpr)$ represents the fluid–fluid interaction while $u(k-l)$ corresponds to the potential energy between the fluid and the Brownian particle (both potentials are scaled with the dimensionless parameter λ). The eigenstates $|Kn\rangle$ of the unperturbed Hamiltonian in definition (3) are then given by

$$H_0 | Kn \rangle = \left(\frac{\hat{K}^2}{2M} + \sum_k \frac{k^2}{2} \hat{n}_k \right) | Kn \rangle \tag{6}$$

where $\mathbf{n}$ denotes the occupation numbers $(n_k, n_l, n_p, \ldots, n_r, \ldots)$.†
Moreover H^e represents the interaction between the charged Brownian particle with the external electrostatic field

$$H^e = -eER \tag{7}$$

We start from the von Neumann equation for the total density matrix ρ, i.e. in the $|Kn\rangle$ representation:

$$i\partial_t \langle Kn | \rho | K'n' \rangle = \langle Kn | [H', \rho] | K'n' \rangle \tag{8}$$

---

* Except when explicitly noticed, we drop the vector notation and set $\mathbf{h} = m = 1$.

† As usual, we have for bosons:

$$a_k^+ | n_k \rangle = (n_k + 1)^{\frac{1}{2}} | n_k + 1 \rangle, \quad a_k | n_k \rangle = n_k^{\frac{1}{2}} | n_k - 1 \rangle$$

We now rewrite this equation using the following definition valid for any operator A:[11, 16]

$$\langle K\mathbf{n}|A|K'\mathbf{n}'\rangle \equiv A_{K-K',\mathbf{n}-\mathbf{n}'}\left(\frac{K+K'}{2},\frac{\mathbf{n}+\mathbf{n}'}{2}\right) \equiv A_{\kappa,\nu}(P,\mathbf{N}) \quad (9)$$

Here

$$\kappa = K - K', \quad P = \tfrac{1}{2}(K+K') \quad (10)$$

define a new set of variables for the Brownian particle, while for the fluid molecules we use

$$\mathbf{\nu} = \mathbf{n} - \mathbf{n}', \quad \mathbf{N} = \tfrac{1}{2}(\mathbf{n}+\mathbf{n}') \quad (11)$$

with $\mathbf{\nu} = (\nu_k, \nu_l, \nu_p, ..., \nu_r, ...)$ and $\mathbf{N} = (N_k, N_l, N_p, ..., N_r, ...)$. It is then easy to express Eq. (8) in terms of these variables; we obtain

$$i\partial_t \rho_{\kappa,\nu}(P,\mathbf{N};t)$$

$$= \sum_{\kappa',\nu'} \left[H'_{\kappa-\kappa',\nu-\nu'}\left(P+\frac{\kappa'}{2},\mathbf{N}+\frac{\nu'}{2}\right) \rho_{\nu'}\left(P+\frac{\kappa'-\kappa}{2},\mathbf{N}+\frac{\nu-\nu'}{2};t\right) \right.$$

$$\left. - H'_{\kappa-\kappa',\nu-\nu'}\left(P-\frac{\kappa'}{2},\mathbf{N}-\frac{\nu'}{2}\right) \rho_{\nu'}\left(P-\frac{\kappa'-\kappa}{2},\mathbf{N}-\frac{\nu'-\nu}{2};t\right) \right]$$

$$(12)$$

where the summation over ν' runs over all possible values from $-N$ to N [that is for bosons: $\nu_k = n_k - n'_k$; ν_k in the range $(N, N-1, ..., -N+1, -N)$].

If we now introduce displacement operators $\zeta^{\pm\kappa}$, $\eta^{\pm\nu}$ such that for any function of P or $\mathbf{N}$:

$$\zeta^{\pm\kappa} f(P) = \exp\left[\pm\frac{\kappa}{2}\frac{\partial}{\partial P}\right] f(P) = f\left(P\pm\frac{\kappa}{2}\right) \quad (13)$$

$$\eta^{\pm\nu} f(\mathbf{N}) = f\left(\mathbf{N}\pm\frac{\nu}{2}\right) \quad (14)$$

we can cast Eq. (8) into the following form:

$$i\partial_t \rho_{\kappa,\nu}(P,\mathbf{N};t) = \sum_{\kappa',\nu'} \langle \kappa\nu | \mathcal{H}'(P,\mathbf{N}) | \kappa'\nu' \rangle \rho_{\kappa',\nu'}(P,\mathbf{N};t) \quad (15)$$

where the "von Neumann–Liouville operator"* $\mathcal{H}'$ is defined by

$$\langle \kappa\nu | \mathcal{H}'(P,\mathbf{N}) | \kappa'\nu' \rangle = [\zeta^{\kappa'}\eta^{\nu'} H'_{\kappa-\kappa',\nu-\nu'}(P,\mathbf{N}) \zeta^{-\kappa}\eta^{-\nu}$$

$$- \zeta^{-\kappa'}\eta^{-\nu'} H'_{\kappa-\kappa',\nu-\nu'}(P,\mathbf{N}) \zeta^{\kappa}\eta^{\nu}] \quad (16)$$

* Operators of this kind will be represented by script letters.

The $\rho_{\kappa\nu}(P,\mathbf{N})$ are the Fourier components of the Wigner distribution function of the whole system. The states $\kappa\nu \neq 0$ correspond to the existence of spatial correlations between the Brownian particle and the fluid; the dynamic evolution of these correlations is described by the von Neumann–Liouville equation (15). As in the case of quantum gases[16] Eq. (15) is the strict analogue of the classical Liouville equation. This great formal similarity will allow us to apply to this problem the general technique developed by Prigogine and coworkers[11-13,16] to deal with non-equilibrium situations.

Before going to our derivation of a general transport equation for the Brownian particle momentum distribution function let us give the definitions of the script operators which correspond to the various terms of our starting Hamiltonian. Using definitions (9) and (16) with Eqs. (4), (5), and (7) we get after some elementary algebra:

$$\mathscr{H}' = \mathscr{H} + \mathscr{H}^{\mathrm{e}}; \quad \mathscr{H} = \mathscr{H}_0 + \lambda\mathscr{W} \tag{17}$$

where

$$\langle\kappa\nu|\mathscr{H}_0|\kappa'\nu'\rangle = \langle\kappa\nu|\mathscr{H}_0^{\mathrm{B}} + \mathscr{H}_0^{\mathrm{f}}|\kappa'\nu'\rangle = [\kappa P/M + \mathbf{\epsilon}\cdot\nu]\,\delta_{\kappa,\kappa'}^{\mathrm{Kr}}\,\delta_{\nu,\nu'}^{\mathrm{Kr}} \tag{18}$$

represents the Brownian particle kinetic energy term plus the fluid unperturbed Hamiltonian (notice that $\mathbf{\epsilon}\cdot\nu$ means $\sum_k \varepsilon_k \nu_k$). On the other hand, the potential energy corresponds to

$$\langle\kappa\nu|\lambda\mathscr{W}|\kappa'\nu'\rangle = \langle\kappa\nu|\lambda\mathscr{V} + \lambda\mathscr{U}|\kappa'\nu'\rangle \tag{19}$$

where the "fluid–fluid" interaction term is given by

$$\langle\kappa\nu|\lambda\mathscr{V}|\kappa'\nu'\rangle = \frac{\lambda}{2\Omega}\sum_{klpr} v(klpr)\,\delta^{\mathrm{Kr}}(k+l-p-r)$$

$$\times\left[\prod_{klpr}\left(N_k + \frac{\nu_k'+1}{2}\right)^{\frac{1}{2}}\eta^{-1_k-1_l+1_p+1_r}\right.$$

$$\left.-\prod_{klpr}\left(N_k - \frac{\nu_k'-1}{2}\right)^{\frac{1}{2}}\eta^{+1_k+1_l-1_p-1_r}\right]$$

$$\times\prod_{k,l}\delta_{\nu_k',\nu_k-1}^{\mathrm{Kr}}\prod_{p,r}\delta_{\nu_p',\nu_p+1}^{\mathrm{Kr}}\,\delta_{\{\nu\}',\{\nu\}'}^{\mathrm{Kr}}\,\delta_{\kappa,\kappa'}^{\mathrm{Kr}} \tag{20}$$

and the "Brownian particle–fluid" interaction operator, defined by

$$\langle \kappa \mathbf{v} | \lambda \mathscr{U} | \kappa' \mathbf{v}' \rangle = \frac{\lambda}{\Omega} \sum_{kl} u(k-l)\, \delta^{\mathrm{Kr}}(k-l+\kappa-\kappa')$$

$$\times \left[\prod_{k,l} \left(N_k + \frac{v'_k+1}{2} \right)^{\frac{1}{2}} \eta^{-1_k+1_l} \exp \left(\frac{\kappa'-\kappa}{2}\, \frac{\partial}{\partial P} \right) \right.$$

$$\left. - \prod_{k,l} \left(N_k - \frac{v'_k-1}{2} \right)^{\frac{1}{2}} \eta^{+1_k-1_l} \exp \left(-\frac{\kappa'-\kappa}{2}\, \frac{\partial}{\partial P} \right) \right]$$

$$\times \delta^{\mathrm{Kr}}_{v_{k'},v_k-1}\, \delta^{\mathrm{Kr}}_{v_{l'},v_l+1}\, \delta^{\mathrm{Kr}}_{\{v\}',\{v\}'} \tag{21}$$

Finally the "external" Hamiltonian $\mathscr{H}^{\mathrm{e}}$ has the following form:

$$\langle \kappa \mathbf{v} | \mathscr{H}^{\mathrm{e}} | \kappa' \mathbf{v}' \rangle = \mathrm{ie}\, E \left[\frac{\partial}{\partial \left(P + \dfrac{\kappa}{2} \right)}\, \delta^{\mathrm{Kr}}_{\kappa,\kappa'} \exp \left(\frac{\kappa'-\kappa}{2}\, \frac{\partial}{\partial P} \right) \right.$$

$$\left. - \frac{\partial}{\partial \left(P + \dfrac{\kappa}{2} - \kappa' \right)}\, \delta^{\mathrm{Kr}}_{\kappa,\kappa'} \exp \left(-\frac{\kappa'-\kappa}{2}\, \frac{\partial}{\partial P} \right) \right] \delta^{\mathrm{Kr}}_{v,v'} \tag{22}$$

B. The Master Equation in an External Electrostatic Field

We now analyse the evolution equation of the particle density matrix [Eq. (15)] in the special case where:

(1) the system is supposed to be uniform; we thus consider the evolution of the total momentum distribution function $\rho_0(P, \mathbf{N}; t)$;*

(2) the external electrostatic field E is switched on at the initial time. For $t \leqslant 0$, the system is at equilibrium:

$$\rho_{\kappa' v}(P, \mathbf{N}; 0) = \rho^{\mathrm{eq}}_{\kappa' v}(P, \mathbf{N})$$

(3) we suppose that this external field is weak and we limit ourselves to a *linear theory* in this smallness parameter.

We thus set:

$$\rho_0(P, \mathbf{N}; t) = \rho^{\mathrm{eq}}_0(P, \mathbf{N}) + \Delta\rho_0(P, \mathbf{N}; t)$$

* We shall often use the term "distribution function" in lieu of density matrix (recall how they are simply related in the definition of the Wigner distribution function).

the system being just off its equilibrium position under the influence of the weak external field.

In this case the von Neumann–Liouville equation [Eq. (15)] becomes

$$i\partial_t \rho_0(P, \mathbf{N}; t) = \sum_{\kappa'\nu'} \langle 0 | \mathscr{H}'(P, \mathbf{N}) | \kappa'\nu' \rangle \rho_{\kappa'\nu'}(P, \mathbf{N}; t) \qquad (23)$$

and its formal solution has the following form:

$$\rho_0(P, \mathbf{N}; t) = \sum_{\kappa'\nu'} \langle 0 | \exp[-i\mathscr{H}'(P, \mathbf{N}) t] | \kappa'\nu' \rangle \rho_{\kappa'\nu'}^{\mathrm{eq}}(P, \mathbf{N}) \qquad (24)$$

as may be checked by direct derivation. However, this expression is of no great help until we know how to operate explicitly with the very complicated exponential operator. In order to circumvent this difficulty, we shall use a resolvent technique: we define a resolvent operator $(\mathscr{H} - z)^{-1}$, a function of the complex variable z and write (see Résibois[13]):

$$\exp(-i\mathscr{H}'t) = -\frac{1}{2\pi i} \oint_C \frac{\exp(-izt)}{\mathscr{H}' - z} \, dz \qquad (25)$$

The contour C will always be chosen as a straight line parallel to the real axis in the upper half-plane and a large semi-circle in the lower half-plane: since $\mathscr{H}$ is Hermitian, all the singularities of the resolvant are on the real axis and are thus included in the contour C.

The resolvant technique furnishes a very elegant perturbation method for calculating Eq. (24), i.e.:

$$\rho_0(P, \mathbf{N}; t) = -\frac{1}{2\pi i} \oint_C dz \exp(-izt)$$

$$\times \sum_{\kappa'\nu'} \langle 0 | (\mathscr{H}' - z)^{-1} | \kappa'\nu' \rangle \rho_{\kappa'\nu'}^{\mathrm{eq}}(P, \mathbf{N}) \qquad (26)$$

Indeed, using the following formal expansion in the external force:

$$(\mathscr{H}' - z)^{-1} = (\mathscr{H} + \mathscr{H}^{\mathrm{e}} - z)^{-1}$$

$$= \sum_{n=0}^{8} (\mathscr{H} - z)^{-1} [-\mathscr{H}^{\mathrm{e}}(\mathscr{H} - z)^{-1}]^n \qquad (27)$$

one sees immediately that, in a linear theory in the external perturbation,* Eq. (26) may be cast into

$$\Delta\rho_0(P,\mathbf{N};t) = -\frac{1}{2\pi i}\oint_C \frac{dz}{z}\exp(-izt)$$
$$\times \sum_{\kappa'\mathbf{\nu}'}\langle 0|(\mathcal{H}-z)^{-1}\mathcal{H}^{\mathrm{e}}|\kappa'\mathbf{\nu}'\rangle \rho^{\mathrm{eq}}_{\kappa'\mathbf{\nu}'}(P,\mathbf{N}) \quad (28)$$

if one uses the obvious identity:

$$\mathcal{H}\rho^{\mathrm{eq}} \equiv [H,\rho^{\mathrm{eq}}] = 0 \qquad (29)$$

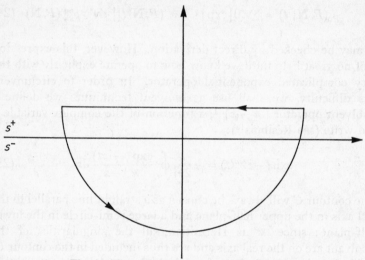

Fig. 1. The complex contour C in Eq. (25).

Moreover we may expand the right-hand side of Eq. (28) in the coupling constant λ, i.e. using the formal series (which is assumed to be convergent):

$$(\mathcal{H}-z)^{-1} = (\mathcal{H}_0-z)^{-1}\sum_{n=0}^{\infty}[-\lambda\mathcal{W}(\mathcal{H}_0-z)^{-1}]^n$$

Then we get

$$\Delta\rho_0(P,\mathbf{N};t)$$
$$= -\frac{1}{2\pi i}\oint_C \frac{dz}{z}\exp(-izt)$$
$$\times \sum_{\kappa'\mathbf{\nu}'}\sum_{n=0}^{\infty}\langle 0|(1/-z)[-\lambda\mathcal{W}(\mathcal{H}_0-z)^{-1}]^n\mathcal{H}^{\mathrm{e}}|\kappa'\mathbf{\nu}'\rangle \rho^{\mathrm{eq}}_{\kappa'\mathbf{\nu}'}(P,\mathbf{N}) \quad (30)$$

* The external field is supposed weak and the series (27), convergent.

This expression describes the departure of the system from its equilibrium position under the influence of an external perturbation supposed weak. Here this deviation is expressed in terms of the spatial correlations existing between the Brownian particle and the fluid. In order to isolate in Eq. (30) the various contributions of physical interest let us introduce the following operators whose physical meaning will become clear.

We define the "diagonal fragment" $\Psi'_{00}(z)$ as the sum of all "irreducible" transitions leading from the state $|0\rangle$ to the same final state; by irreducible, we mean that all intermediate states $|\kappa\mathbf{v}\rangle$ are such that $\kappa\mathbf{v} \neq 0$ and this condition will be indicated by a dash $(')$ in all subsequent formulae. We thus have

$$\Psi'_{00}(z) = \sum_{n=2}^{\infty} \langle 0| - \lambda \mathcal{W} [(\mathcal{H}_0 - z)^{-1}(-\lambda \mathcal{W})]^{n-1} |0\rangle' \qquad (31)$$

The "destruction term" is defined as the sum of all irreducible transitions starting from any initial state $|\kappa\mathbf{v}\rangle$ with $\kappa\mathbf{v} \neq 0$ and ending with the "vacuum" state $|0\rangle$:

$$\mathcal{D}_{0,\kappa\mathbf{v}}(z) = \sum_{\kappa\mathbf{v}} \sum_{n=1}^{\infty} \langle 0| [-\lambda \mathcal{W}(\mathcal{H}_0 - z)^{-1}]^n |\kappa\mathbf{v}\rangle' \qquad (32)$$

Finally in the calculation of the equilibrium correlation function $\rho_{\kappa\mathbf{v}}^{\mathrm{eq}}(P, \mathbf{N})$ we shall also need the so-called "creation fragment" which is defined by

$$\mathcal{C}_{\kappa\mathbf{v},0}(z) = \sum_{n=1}^{\infty} \langle \kappa\mathbf{v}| [(\mathcal{H}_0 - z)^{-1}(-\lambda \mathcal{W})]^n |0\rangle' \qquad (33)$$

If we now turn back to the perturbation expansion Eq. (30) for $\Delta\rho_0(P, \mathbf{N}; t)$, we may readily express the right-hand side as the sum of an arbitrary number of diagonal fragments preceded by a destruction term. We have identically:

$$\Delta\rho_0(P, \mathbf{N}; t) = -\frac{1}{2\pi i} \oint_C \frac{dz}{z} \exp(-izt) \sum_{\kappa'\mathbf{v}'} \sum_{n=0}^{\infty} \frac{1}{-z} \left[\Psi'_{00}(z) \frac{1}{-z} \right]^n$$
$$\times \{ [\delta_{\kappa'\mathbf{v}',0}^{\mathrm{Kr}} + \mathcal{D}_{0,\kappa'\mathbf{v}'}(z)] [\mathcal{H}^e \rho^{\mathrm{eq}}(P, \mathbf{N})]_{\kappa'\mathbf{v}'} \} \qquad (34)$$

$$\equiv \frac{1}{2\pi i} \oint_C \frac{dz}{z} \exp(-izt) [z + \Psi'_{00}(z)]^{-1}$$
$$\times \{ \mathcal{H}^e \rho_0^{\mathrm{eq}}(P, \mathbf{N}) + \sum_{\kappa'\mathbf{v}'} \mathcal{D}_{0,\kappa'\mathbf{v}'}(z) [\mathcal{H}^e \rho^{\mathrm{eq}}(P, \mathbf{N})]_{\kappa'\mathbf{v}'} \} \qquad (35)$$

where we have carried out the formal summation contained in Eq. (34).

To proceed with the discussion of Eq. (35) we need information concerning the analytic behaviour of the quantities $\Psi_{00}(z)$, $\mathscr{D}_{0,\kappa\nu}(z)$ or $\mathscr{C}_{\kappa\nu,0}(z)$ which will be all denoted formally by $F(z)$. We shall not discuss all the analytic properties of $F(z)$ in detail, but we shall simply give some important properties that may be deduced from complex analysis (for detailed justifications see references 11–13, 17).

$F(z)$ is an analytic function of z in the whole complex plane, except for a finite discontinuity along the real axis. As far as the destruction operator is concerned, this property is realized for a certain class of initial conditions, i.e. when the range of correlations in configuration space is finite. We have two functions $F^+(z)$ and $F^-(z)$ according to $\mathscr{I}m\,z > 0$ or $\mathscr{I}m\,z < 0$ which are analytic in their respective domains of definition, i.e. the first one in the upper half-plane S^+, the second one in the lower half-plane S^-. Moreover it is possible to show that $F^+(z)$ has an analytical continuation in the lower half-plane S^- and vice versa. This continuation has singularities which we shall always assume to be poles of finite order located at $z = z_i$ with the typical requirement:

$$\mathscr{I}m\,z_i = -\frac{1}{\tau_c} \tag{36}$$

where τ_c denotes the "collision time". This last property has been shown to be true for certain laws of interaction (for an explicit example see reference 13). In the following discussion Eq. (36) will be considered as a sufficient condition for the validity of the equations of evolution which we shall derive.

We first notice that because of the factor $\exp(-izt)$ the integral along the semi-circle at infinity is vanishing. We may thus formally replace $F(z)$ by $F^+(z)$ on that part of the contour. Moreover, along the real axis, we are in S^+, then we also need $F^+(z)$.

Second, as we are interested in the long time behaviour of the total momentum distribution function $\Delta\rho_0(P, \mathbf{N}; t)$, the integrations contained in Eq. (35) may be performed in the following asymptotic way:

$$I = \lim_{t\to\infty} \frac{1}{2\pi i} \oint_C \frac{\exp(-izt)}{z} F^+(z) \tag{37}$$

and then readily evaluated using the well-known residue theorem:

$$I = \lim_{t \to \infty} \left[F^+(0) + \sum_{z=z_i} \frac{\exp(-iz_i t)}{z_i} \operatorname{Res} F(z)\big|_{z=z_i} \right] \quad (38)$$

$$I = F^+(0) \quad (39)$$

Notice that as $F^+(z)$ is defined in S^+, the residue we denote simply $F^+(0)$ taken in the limit $z \to 0^+$ or $z \to i0$. Consequently the stationary momentum distribution function $\Delta\rho_0(P, \mathbf{N})$, which is realized after a long time when the system is submitted to a weak external electrostatic field, is given by the solution of the following time-independent transport equation:

$$i\Psi_{00}^+(0)\,\Delta\rho_0(P, \mathbf{N}) = i\mathscr{H}^e \rho_0^{eq}(P, \mathbf{N}) + i\sum_{\kappa'\nu'} \mathscr{D}_{0,\kappa'\nu'}^+(0)\,[\mathscr{H}^e \rho^{eq}(P, \mathbf{N})]_{\kappa'\nu'}$$

$$(40)$$

Before explaining the physical meaning of the different terms of this transport equation, let us quote a useful relation between the equilibrium correlations ($\kappa\nu \neq 0$) and the momentum distribution function ($\kappa\nu = 0$). One can show that

$$\rho_{\kappa\nu}^{eq}(P, \mathbf{N}) = \lim_{t \to \infty} \rho_{\kappa\nu}(P, \mathbf{N}; t) \equiv \mathscr{C}_{\kappa\nu,0}^+(0)\,\rho_0^{eq}(P, \mathbf{N}) \quad (41)$$

where the creation term $\mathscr{C}_{\kappa\nu,0}(z)$ is defined by Eq. (33). We do not wish to give a detailed proof of Eq. (41) here (see, for instance, reference 13); let us merely point out the fact that this dynamic formulation of the equilibrium correlations leads to results analogous to the usual methods based on the canonical distribution:

$$\rho_{\kappa\nu}^{eq}(P, \mathbf{N}) = \left[\frac{\exp(-\beta H)}{\operatorname{Tr} \exp(-\beta H)} \right]_{\kappa\nu} \quad (42)$$

but it has the advantage of being readily extended to non-equilibrium situations.

With Eq. (41) the basic transport equation (40) may be cast into the compact form:

$$i\Psi_{00}^+(0)\,\Delta\rho_0(P, \mathbf{N}) = i\mathscr{H}^e \rho_0^{eq}(P, \mathbf{N})$$

$$+ i\sum_{\kappa'\nu'} \mathscr{D}_{0,\kappa'\nu'}^+(0)\,\mathscr{H}^e \mathscr{C}_{\kappa'\nu',0}^+(0)\,\rho_0^{eq}(P, \mathbf{N}) \quad (43)$$

This equation describes exactly the linear response of the system to an external electrostatic field in the limit $t \to \infty$ and with equilibrium initial conditions. Bearing in mind that in this stationary situation one has

$$\partial_t \rho_0(P, \mathbf{N}; t) = \partial_t \rho_0(P, \mathbf{N}; t)|_{\text{coll}} + \partial_t \rho_0(P, \mathbf{N}; t)|_{\text{field}} \equiv 0$$

one may easily point out the physical meaning of the different terms of Eq. (43). The left-hand side contains the asymptotic collision operator which generalizes to a strongly coupled system the well-known quantum-mechanical collision Boltzmann operator for a dilute system. The first term of the right-hand side, where the external field acts on a "vacuum" state $|0\rangle$, is the usual flow term coming from the influence of the external perturbation on the momentum distribution function of the system. The second term corresponds to situations where the outer force acts on particles which are mutually interacting; in this sense, it represents the effect of the external field *during* the collision process. In the dilute gas limit, this effect disappears because the duration of the collision is very small. However, in a dense system, it can play an essential part.

It should be emphasized that because we are interested in a linear transport theory, the operators of the right-hand side of Eq. (43) are acting on the exact equilibrium distribution function which may "formally" be factorized:*

$$\rho_0^{\text{eq}}(P, \mathbf{N}) = \phi_0(\gamma; P) \rho_0^{\text{f}}(\mathbf{N}) \tag{44}$$

where $\phi_0(\gamma; P)$ is the complete quantum equilibrium distribution function of the Brownian particle which still depends on γ [because of the non-commutativity of H_0^{B} and $H^{\text{f}} = H_0^{\text{f}} + \lambda W$, see Eqs. (4) and (5)] and $\rho_0^{\text{f}}(\mathbf{N})$ denotes the equilibrium distribution function of the fluid particles.

Now we shall show that, although Eq. (43) is still purely formal—in the sense that it involves the complete N particle distribution function $\Delta\rho_0(P, \mathbf{N})$—it can be reduced to a transport equation for the single Brownian particle distribution function.

---

* By "formally" we mean the following: in the presence of interactions there is of course no strict factorization of the density matrix Eq. (44). However, if we compute an average value and denote $\langle\phi(P)\rangle$ the momentum distribution of the heavy particle while $\langle n_k\rangle = \text{Tr } \hat{n}_k \rho^{\text{f}}$, it may be shown that

$$\langle\phi(P) n_k\rangle = \langle\phi(P)\rangle \langle n_k\rangle + O(N^{-1})$$

(see reference 4 for a similar argument).

C. The Transport Equation for Quantum Brownian Motion

We assume that the departure from equilibrium may be written in the linear form:

$$\Delta\rho_0(P, \mathbf{N}) = \rho_0^f(\mathbf{N})\,\delta\phi(P) + \phi_0(\gamma;\, P)\,\delta\rho^f(\mathbf{N}) \tag{45}$$

The justification of this assumption of molecular chaos is now well established in the limit of an infinite system $(N \to \infty, \Omega \to \infty;$ $N/\Omega = \rho = $ finite) and we shall not discuss it here.

Let us now recall the main feature of our Brownian motion model, which is that when *a very few* heavy particles move through a fluid of N light particles they cannot cause it to depart from its equilibrium condition, the non-equilibrium situation of the Brownian particles being caused by an external perturbation. Thus, in our case, if we allow that the fluid particles are not affected by the presence of the Brownian particle we can set in Eq. (45):

$$\delta\rho^f(\mathbf{N}) = 0 \tag{46}$$

Let us also note that this rather obvious statement can be supported with a mathematical demonstration valid up to order $O(N^{-1})$ (see, for instance, reference 2).

Inserting Eqs. (44) and (45) into Eq. (43) and summing over the fluid variables $\mathbf{N}$, one can reduce the formal equation (43) to the following closed transport equation for the Brownian particle distribution function:

$$\langle \Psi_{00}^+(0) \rangle_f\,\delta\phi(P) = [\mathscr{H}^e + \langle \sum_{\kappa'\nu'} \mathscr{D}_{0,\kappa'\nu'}^+(0)\,\mathscr{H}^e\,\mathscr{C}_{\kappa'\nu',0}^+(0) \rangle_f]\,\phi_0(\gamma;\, P) \tag{47}$$

where we have averaged the various operator quantities in the variables P and $\mathbf{N}$ over the fluid equilibrium distribution function ρ^f (where Z_f denotes the fluid partition function):

$$\rho^f = Z_f^{-1}\exp(-\beta H^f); \quad H^f = H_0^f + V^f + V^{fB} \tag{48}$$

using the definition

$$\langle A \rangle_f = \mathrm{Tr}\,A\rho_f = \sum_{\mathbf{N}} A(\mathbf{N})\,\rho_0^f(\mathbf{N}) \tag{49}$$

The generalized transport equation for quantum Brownian motion, Eq. (47), will be the starting point of: (1) a microscopic derivation of the usual Fokker–Planck equation (Section 3) and (2) discussion of the validity of a Fokker–Planck type description for a Brownian motion situation in the very low temperature limit (Section 4).

D. Expansion in the Momentum Ratio

In the basic transport equation (47) we now take into account explicitly the Brownian motion condition (1). Let us first introduce a reduced variable

$$p = \gamma P \tag{50}$$

defined in such a way that if the "classical requirement" (1) is satisfied, the average value of p remains finite in the limit $\gamma \to 0$. This new variable enables us to take out the γ-dependence of the "von Neumann–Liouville" operators [see Eqs. (17)–(22)] and split them into

$$\mathscr{H} = \mathscr{H}^f + \gamma \mathscr{H}^I + \gamma^2 \mathscr{H}^{II} + \gamma^3 \mathscr{H}^{III} + O(\gamma^4) \tag{51}$$

$$\mathscr{H}_0 = \mathscr{H}_0^f + \gamma \mathscr{H}_0^I \tag{52}$$

$$\mathscr{W} = \mathscr{V}^f + \gamma \mathscr{U}^I + \gamma^2 \mathscr{U}^{II} + \gamma^3 \mathscr{U}^{III} + O(\gamma^4) \tag{53}$$

We have thus, parallel to the classical case:

$$\mathscr{H}_0^f = \boldsymbol{\epsilon} \cdot \boldsymbol{\nu} \tag{54}$$

(fluid unperturbed von Nuemann–Liouville operator)

$$\gamma \mathscr{H}_0^I = \mathscr{H}_0^B = \gamma \hat{\kappa} p \tag{55}$$

(Brownian particle unperturbed von Neumann–Liouville operator). Here $\hat{\kappa}$ is an operator defined by $\hat{\kappa} | \kappa \boldsymbol{\nu} \rangle = \kappa | \kappa \boldsymbol{\nu} \rangle$ [see Eq. (18)]. Moreover,

$$\mathscr{V}^f = \mathscr{V} + \mathscr{U}^{(0)} \tag{56}$$

(perturbation that the fluid would feel if the Brownian particle was fixed). Here $\mathscr{V}$ is given by Eq. (20) and $\mathscr{U}^{(0)}$ is zero-order in γ of the expression (21) where we simply replace the displacement operators $\zeta^{\pm\kappa}$ by the γ-expansion:

$$\zeta^{\pm\kappa} = \sum_{n=0}^{\infty} \frac{1}{n!} \left(\pm \tfrac{1}{2} \gamma \kappa \frac{\partial}{\partial p} \right)$$

Similarly, at order γ, the interaction of the Brownian particle with the fluid corresponds to

$$\langle \kappa \boldsymbol{\nu} | \gamma \mathscr{U}^I | \kappa' \boldsymbol{\nu}' \rangle$$

$$= \lambda \Omega^{-1} \sum_{k,l} u(k-l)\, \delta^{Kr}(k-l+\kappa-\kappa') \{[(N_k + \tfrac{1}{2}\nu_k)(N_l + \tfrac{1}{2}\nu_l + 1)]^{\frac{1}{2}}$$

$$\times \eta^{-1_k + 1_l} + [(N_k - \tfrac{1}{2}\nu_k + 1)(N_l - \tfrac{1}{2}\nu_l)]^{\frac{1}{2}}\, \eta^{+1_k - 1_l}\}$$

$$\times \tfrac{1}{2}\gamma(k-l)\frac{\partial}{\partial p}\, \delta^{Kr}_{\nu_k,\nu_k'+1}\, \delta^{Kr}_{\nu_l,\nu_l'-1}\, \delta^{Kr}_{\{\nu\},\{\nu'\}}. \tag{57}$$

Let us here explain the meaning of this expression. Let us express how this operator acts on any quantity $A(\mathbf{n})$ which we choose, for simplicity, diagonal in the occupation number $\mathbf{n}$:

$$\langle\kappa\nu|\gamma\mathscr{U}^{\mathrm{I}}|\kappa'0\rangle A_0(\mathbf{N})$$

$$= \lambda\Omega^{-1}\sum_{k,l} u(k-l)\,\delta(k-l+\kappa-\kappa')\,\tfrac{1}{2}\gamma(k-l)\frac{\partial}{\partial p}\delta^{\mathrm{Kr}}_{\nu_k,1}\,\delta^{\mathrm{Kr}}_{\nu_l,-1}\,\delta^{\mathrm{Kr}}_{\{\nu\}',0}$$

$$\times\left[(N_k+\tfrac{1}{2})(N_l+\tfrac{1}{2})\right]^{\frac{1}{2}}[A_0(N_k-\tfrac{1}{2},N_l+\tfrac{1}{2},\{N\}')$$

$$-A_0(N_k+\tfrac{1}{2},N_l-\tfrac{1}{2},\{N\}')]$$

If we now go back to the usual occupation number representation $(\mathbf{n}=\mathbf{N}+\tfrac{1}{2}\nu)$ we may write

$$\langle\kappa\nu|\gamma\mathscr{U}^{\mathrm{I}}|\kappa'0\rangle A_0(\mathbf{N})$$

$$= \lambda\Omega^{-1}\sum_{k,l} u(k-l)\,\delta(k-l+\kappa-\kappa')\,\tfrac{1}{2}\gamma(k-l)\frac{\partial}{\partial p}\delta^{\mathrm{Kr}}_{\nu_k,1}\,\delta^{\mathrm{Kr}}_{\nu_l,-1}\,\delta_{\{\nu\}',0}$$

$$\times[n_k(n_l+1)]^{\frac{1}{2}}[\langle n_k-1,n_l+1,\{n\}'|A|n_k-1,n_l+1,\{n\}'\rangle$$

$$-\langle\mathbf{n}|A|\mathbf{n}\rangle]$$

or, with the definition of the operators a_k^+, a_k:

$$\langle\kappa\nu|\gamma\mathscr{U}^{\mathrm{I}}|\kappa'0\rangle A_0(\mathbf{N})$$

$$= \lambda\Omega^{-1}\sum_{k,l} u(k-l)\,\delta(k-l+\kappa-\kappa')\,\tfrac{1}{2}\gamma(k-l)\frac{\partial}{\partial p}\delta^{\mathrm{Kr}}_{\nu_k,1}\,\delta^{\mathrm{Kr}}_{\nu_l,-1}\,\delta^{\mathrm{Kr}}_{\{\nu\}',0}$$

$$\times\langle\mathbf{n}|a_k a_l^+ A + A a_k^+ a_l|n_k-1,n_l+1,\{n\}'\rangle$$

This example shows that the operator $\gamma\mathscr{U}^{\mathrm{I}}$ may be represented as

$$\gamma\mathscr{U}^{\mathrm{I}} = \gamma\mathscr{F}^+\frac{\partial}{\partial p} \tag{58}$$

where $\mathscr{F}^+$ is an operator associated with the quantum-mechanical force F existing between the Brownian particle and the fluid, the subscript $(+)$ having been introduced as a reminder that it corresponds to an anticommutator, i.e.:

$$\mathscr{F}^+\mathscr{A}\to\frac{i}{2}[\hat{F}A+A\hat{F}] \tag{59}$$

with

$$\hat{F} = i\lambda\Omega^{-1}\sum_k qu(q)\,a_k^+ a_{k-q} = i[\hat{\kappa},V^{\mathrm{fB}}] \tag{60}$$

Let us add that the terms of $\mathscr{H}^{\mathrm{II}}$ (of order γ^2) can be obtained in a similar fashion but we shall not need their explicit form here.

If we note that $\mathscr{H}^{\mathrm{e}}$, as given by Eq. (22), is also of order γ we may easily obtain the following expansions:

$$\langle \Psi_{00}^{+}(\gamma^{-1}p, N; 0)\rangle_{\mathrm{f}} = \sum_{n=1}^{\infty} \gamma^{2n}\, \Omega^{(2n)} \tag{61}$$

$$\langle \sum_{\kappa\nu} \mathscr{D}_{0,\kappa\nu}^{+}(\gamma^{-1}p, N; 0)\, \gamma\mathscr{H}^{\mathrm{e}}\, \mathscr{C}_{\kappa\nu,0}^{+}(\gamma^{-1}p, N; 0)\rangle_{\mathrm{f}} = \sum_{n=3}^{\infty} \gamma^{n}\, G^{(n)} \tag{62}$$

From the symmetry properties of these expressions, one can show that

$$\Omega^{(0)} = \Omega^{(2n+1)} = 0 \tag{63a}$$

$$G^{(1)} = G^{(2)} = 0 \tag{63b}$$

For instance, by examining the explicit form of $\Omega^{(0)}$ one sees immediately that this contribution vanishes identically once the trace over the fluid variables is taken, indeed we have

$$\sum_{\mathbf{N}} (\eta^{\nu} - \eta^{-\nu})\, \mathrm{f}(\mathbf{N}) \equiv 0 \tag{64}$$

Moreover let us stress that in Eq. (47) the complete equilibrium distribution function of the Brownian particle is involved; it can itself be expanded in γ:

$$\phi_0(\gamma; P) = \sum_{n=0}^{\infty} \gamma^{n}\, \phi_0^{(n)}(p) \tag{65}$$

However, as is easily checked, the zero-order term of the Brownian particle equilibrium distribution function is simply the unperturbed Maxwellian distribution function

$$\phi_0^{(0)}(P) \equiv \left(\frac{M\beta}{2\pi}\right)^{\frac{3}{2}} \exp\left(-\frac{\beta P^2}{2M}\right) \tag{66}$$

and the higher order terms express quantum deviations (which may however become large in the low temperature limit, see Section 4.B).

When relations Eqs. (61), (62), and (65) are introduced in our basic transport equation Eq. (47), we obtain

$$\mathrm{i}[\gamma\mathscr{H}^{\mathrm{e}} + (\gamma^3\, G^{(3)} + \gamma^4\, G^{(4)} + \ldots)]\,[\phi_0^{(0)}(P) + \gamma^2\, \phi_0^{(2)}(P) + \ldots]$$
$$= \mathrm{i}(\gamma^2\, \Omega^{(2)} + \gamma^4\, \Omega^{(4)} + \ldots)\, \phi(P) \tag{67}$$

where we have turned back to the ordinary momentum variable P.

Let us stress again that the stationary Brownian particle momentum distribution function $\phi(P)$, which is realized in the presence of an external electrostatic field, obeys the transport equation (67) provided condition (1) is satisfied.

3. THE QUANTUM FOKKER–PLANCK APPROXIMATION

A. The Quantum Fokker–Planck Equation

Provided the γ-expansion appearing in Eq. (67) converges, it is easily shown that, to lowest order in γ, Eq. (67) reduces to:

$$i\gamma \mathscr{H}^e \phi_0^{(0)}(P) = i\gamma^2 \Omega^{(2)} \delta\phi(P) \tag{68}$$

[notice that $\delta\phi(P)$ is of $O(\gamma^{-1})$]. We shall discuss this equation here and postpone the analysis of the higher order corrections to the next Section. If we introduce Eqs. (52), (53) and (56) into the definition (32) of the collision operator and use the following expansion:*

$$
\begin{aligned}
(\mathscr{H}_0 - i\varepsilon)^{-1} &= \lim_{\varepsilon \to 0} (\mathscr{H}_0^f + \gamma \mathscr{H}_0^I - i\varepsilon)^{-1} \\
&= (\mathscr{H}_0^f - i\varepsilon)^{-1} \sum_{n=0}^{\infty} [-\gamma \mathscr{H}_0^I (\mathscr{H}_0^f - i\varepsilon)^{-1}]^n
\end{aligned} \tag{69}
$$

we get from Eq. (61):

$$
\begin{aligned}
i\gamma^2 \Omega^{(2)} = i\gamma^2 \sum_{N} \langle 0| \, \mathscr{U}^I \Big\{ (\mathscr{H}_0^f - i\varepsilon)^{-1} \sum_{n=0}^{\infty} [-\mathscr{V}^f (\mathscr{H}_0^f - i\varepsilon)^{-1}]^n \Big\} \\
\times (\mathscr{U}^I + \mathscr{H}_0^I) \Big\{ \sum_{m=0}^{\infty} [(\mathscr{H}_0^f - i\varepsilon)^{-1} (-\mathscr{V}^f)]^m \Big\} |0\rangle' \, \rho_0^f(\mathbf{N})
\end{aligned}
$$

To obtain this relation we have taken into account the fact that the collision operator $\Omega^{(2)}$ must begin with the "vertex" ($\mathscr{U}^I$) otherwise it would give zero by direct summation [see Eq. (64)]. We may also cast this expression into the compact form:

$$
\begin{aligned}
\gamma^2 \Omega^{(2)} = \gamma^2 \sum_{N} \sum_{\kappa'\nu'} \int_0^{\infty} dt \langle 0| \, \mathscr{U}^I \exp[-i(\mathscr{H}^f - i\varepsilon)t] \\
\times (\mathscr{U}^I + \mathscr{H}_0^I) |\kappa'\nu'\rangle \, \rho_{\kappa'\nu'}^f(\mathbf{N})
\end{aligned} \tag{70}
$$

* From now on, any quantity depending on $i\varepsilon$ will be understood in the limit $\varepsilon \to 0$.

where we have used the two following formal identities:

$$(\mathscr{H}_0^{\mathrm{f}} - \mathrm{i}\varepsilon)^{-1} \sum_{n=0}^{\infty} [-\lambda \mathscr{V}^{\mathrm{f}} (\mathscr{H}_0^{\mathrm{f}} - \mathrm{i}\varepsilon)^{-1}]^n \equiv (\mathscr{H}^{\mathrm{f}} - \mathrm{i}\varepsilon)^{-1}$$

$$= \mathrm{i} \int_0^{\infty} \mathrm{d}t \exp\left[-\mathrm{i}(\mathscr{H}^{\mathrm{f}} - \mathrm{i}\varepsilon)t\right]$$

where

$$\mathscr{H}^{\mathrm{f}} = \mathscr{H}_0^{\mathrm{f}} + \mathscr{V}^{\mathrm{f}}$$

$$\rho_{\kappa'\nu'}^{\mathrm{f}}(\mathbf{N}) \equiv \langle \kappa'\nu'| \sum_{n=0}^{\infty} [(\mathscr{H}_0^{\mathrm{f}} - \mathrm{i}\varepsilon)^{-1}(-\lambda \mathscr{V}^{\mathrm{f}})]^n |0\rangle \rho_0^{\mathrm{f}}(\mathbf{N})$$

$$= Z_{\mathrm{f}}^{-1} \int \mathrm{d}^3 R \exp(-\mathrm{i}\kappa' R)$$

$$\times \langle \mathbf{N} + \tfrac{1}{2}\nu' | \exp\left[-\beta(H_0^{\mathrm{f}} + V^{\mathrm{f}} + V^{\mathrm{fB}})\right] | \mathbf{N} - \tfrac{1}{2}\nu' \rangle \qquad (71)$$

This last expression is exactly the Fourier component of vector κ' of the complete equilibrium distribution function for the fluid particles including the interactions with the heavy particles at a fixed position; Z_{f} is the partition function of the fluid [see Eqs. (41) and (48)].

Let us now show that $\gamma^2 \Omega^{(2)}$ leads to the usual Fokker–Planck collision term. With definitions (55) and (58) we may easily write Eq. (70) as

$$\gamma^2 \Omega^{(2)} = \gamma^2 \left(\Omega_A^{(2)} \frac{\partial}{\partial p} \frac{\partial}{\partial p} + \Omega_B^{(2)} \frac{\partial}{\partial p} p \right) \qquad (72)$$

with

$$\Omega_A^{(2)} = \sum_{\mathbf{N}} \sum_{\kappa'\nu'} \int_0^{\infty} \mathrm{d}t \, \langle 0 | \mathscr{F}^+ \exp\left[-\mathrm{i}(\mathscr{H}^{\mathrm{f}} - \mathrm{i}\varepsilon)t\right] \mathscr{F}^+ | \kappa\nu \rangle \, \rho_{\kappa'\nu'}^{\mathrm{f}}(\mathbf{N})$$

$$(72a)$$

$$\Omega_B^{(2)} = \sum_{\mathbf{N}} \sum_{\kappa'\nu'} \int_0^{\infty} \mathrm{d}t \, \langle 0 | \mathscr{F}^+ \exp\left[-\mathrm{i}(\mathscr{H}^{\mathrm{f}} - \mathrm{i}\varepsilon)t\right] \hat{\kappa} | \kappa'\nu' \rangle \, \rho_{\kappa'\nu'}^{\mathrm{f}}(\mathbf{N})$$

$$(72b)$$

If we go back to the usual second quantization representation by using definition (59) and the following correspondence rule between the von Neumann–Liouville unitary operator and the Heisenberg unitary operator, i.e. for any operator A,

$$\exp(-\mathrm{i}\mathscr{H}t) A \equiv \exp(-\mathrm{i}Ht) A \exp(\mathrm{i}Ht) \qquad (73)$$

[this identity may be easily checked by direct time differentiation, see, for instance, the two forms of the von Neumann equation: Eqs. (8) and (15)] we may write

$$\Omega_A^{(2)} = \frac{1}{4} \int_0^\infty dt \, \mathrm{Tr} \, [\hat{F}, \exp{(-iH^t t)} \, [\hat{F}, \rho^t] \exp{(iH^t t)}]$$

$$= \frac{1}{2} \int_0^\infty dt \, \mathrm{Tr} \, [\hat{F}(t) \, \hat{F} \rho^t + \hat{F} \hat{F}(t) \, \rho^t]$$

$$\equiv \mathscr{R}e \int_0^\infty dt \, \langle \hat{F}(t) \hat{F}(0) \rangle_t \qquad (74a)$$

where
$$\hat{F}(t) = \exp{(iH^t t)} \, \hat{F} \exp{(-iH^t t)}$$

The last equality may be readily verified by evaluating $\Omega_A^{(2)}$, for instance in the representation that diagonalizes H^t. By the same way [see definition (55)] we get

$$\Omega_B^{(2)} = \int_0^\infty dt \, \mathrm{Tr} \, [\hat{F} \exp{(-iH^t t)} \, [\hat{\kappa}, \rho^t] \exp{(iH^t t)}]$$

Using now a formula due to Kubo[18] [see also Eq. (48)]:

$$[\hat{\kappa}, \rho^t] = \int_0^\beta d\lambda \, \rho^t \exp{(\lambda H^t)} \, [\hat{\kappa}, H^t] \exp{(-\lambda H^t)}$$

$$\equiv \int_0^\beta d\lambda \, \rho^t \exp{(\lambda H^t)} \, [\hat{\kappa}, V^{tB}] \exp{(-\lambda H^t)} \qquad (74b)$$

we may write [see Eq. (60)]:

$$\Omega_B^{(2)} = \mathscr{R}e \int_0^\infty dt \int_0^\beta d\lambda \, \mathrm{Tr} \, \{\hat{F} \exp{[-iH^t(t+i\lambda)]} \, \hat{F} \exp{[iH^t(t+i\lambda)]} \, \rho^t\}$$

Finally this result may be put in the following form [see, for instance, reference 19, Eq. (2.9) taken at $\omega = 0$]

$$\Omega_B^{(2)} = \mathscr{R}e \, \beta \int_0^\infty dt \, \langle \hat{F}(t) \hat{F}(0) \rangle_t \qquad (75)$$

Thus, going back to Eqs. (68) and (72) where we introduce Eqs. (74a) and (75) and evaluating the flow term [see Eq. (22)], we see that to the lowest order in γ, the Brownian particle distribution function obeys the Fokker–Planck equation:

$$eE \frac{\partial}{\partial \mathbf{P}} \phi_0^{(0)}(\mathbf{P}) = \zeta \frac{\partial}{\partial \mathbf{P}} \left[kT \frac{\partial}{\partial \mathbf{P}} + \frac{\mathbf{P}}{M} \right] \phi(\mathbf{P}) \qquad (76)$$

with the following definition for the friction coefficient:

$$\zeta = \tfrac{1}{3}\beta \, \mathscr{R}e \int_0^\infty dt \, \langle \hat{F}(t) \cdot \hat{F}(0) \rangle_f \tag{77}$$

where $\hat{F}(t)$ is the Heisenberg representation of the quantum-mechanical ion-fluid force operator and the bracket $\langle ... \rangle_f$ indicates the average over the equilibrium distribution function of the fluid at temperature $T = (k\beta)^{-1}$. Let us make some minor remarks:

(a) Equation (76) has been written explicitly in the usual momentum vector notation which should not be confused with the similar notation used in the definition of the script operators [see Eqs. (10) and (16)];

(b) We have reduced the friction tensor (which is diagonal) to its scalar form using the following identity valid for any function f of the scalar $|\kappa|$:

$$\int f(|\kappa|) \, \kappa\kappa \, d^3\kappa = \frac{1}{3}\left[\mathbf{1} \int f(|\kappa|) \, \kappa^2 \, d^3\kappa \right]$$

where $\mathbf{1}$ is the tensor unity.

We have thus shown, that in the limit of an infinitely heavy particle ($\gamma \to 0$) and provided condition (1) is satisfied, the generalized transport equation for quantum Brownian motion Eq. (67) reduces strictly to the classical Fokker–Planck equation but with a quantum friction coefficient ζ that contains the usual diffraction and statistical effects due to the quantum-mechanical nature of the fluid. Definition (77) may, for instance, be used for explicit calculation of the friction coefficient as an expansion in $\mathbf{h}$ for small quantum deviations.[7] In this case the quantum friction coefficient reads to the first order in $\mathbf{h}$:

$$\zeta = \zeta_0 + O(\mathbf{h}^2) \, \zeta_1 \tag{78}$$

where ζ_0 and ζ_1 depend on M in different fashions. Therefore quantum-mechanical effects on ζ might be observable experimentally and non-classical isotope effects would be expected from the measurement of the mobility of a heavy particle in a quantum fluid at very low temperatures, e.g. isotope diffusion of Ne in liquid He. This type of $\mathbf{h}$ correction, interpreted as "quantum fluctuations" in semi-classical cases, would allow investigation of the practical problem of the appearance of observable quantum effects when covering the intermediate range from classical to purely quantum fluids. Moreover, as was kindly pointed out to us by H. L. Frisch (private communication), there might be situations of physical interest involving a spatially *dependent* Wigner distribution function where **h** corrections would be meaningful. These effects are, however, not covered by the present theory, dealing only with spatially homogeneous systems, in which all corrections are of order (**h²**).

Another application of the quantum theory of Brownian motion concerns the theoretical interpretation of ionic mobilities in quantum liquids.[8, 9] Indeed, as we shall show in the next Sections, a calculation of the friction coefficient (in the Fokker–Planck approximation) of a heavy charged particle immersed in a Bose fluid or a Fermi system leads to numerical values which agree qualitatively with the experimental data obtained from measurements of the static mobilities of heavy ions in liquid ^{4}He or ^{3}He.[14, 15]

B. Calculation of the Friction Coefficient of a Heavy Particle

(a) *Fluid of Bosons*

We want to apply the general formula (77) to evaluate the friction coefficient of a heavy particle moving through a nearly perfect Bose gas. In order to make this calculation as simple as possible, we shall treat the interaction V^{fB} between the Brownian particle and the fluid as weak and thus we shall completely neglect it in H^f (which appear in the "propagators" and in the fluid equilibrium distribution function); this interaction will then merely appear in the operator $\hat{F}$ and we shall take

$$H^f = \sum_k \frac{k^2}{2} a_k^+ a_k + \frac{2\pi\lambda a}{\Omega} \sum_{klpr} a_k^+ a_l^+ a_p a_r \, \delta^{\mathrm{Kr}}(k+l-p-r) \quad (79)$$

which is the usual Hamiltonian of a Bose gas of hard spheres of radius a.

We then assume that our boson system is at a sufficiently low temperature to ensure Bose–Einstein condensation, i.e. the excited states of our system $|n_0, \{n\}'\rangle$ are such that

$$\lim_{N\to\infty} N_0 \simeq N$$

when N_0, the occupation number $(N_0 = \langle n_0 \rangle)$ of the fundamental state $(k = 0)$, is macroscopic. In this case the matrix elements of the force (60) and the fluid Hamiltonian (79) are such that for $k = 0$ we have

$$a_0 a_0^+ - a_0^+ a_0 = 1; \quad a_0^+ a_0 \simeq N$$

or, within an error of order (N^{-1}),

$$[a_0^+, a_0] = 0; \quad a_0 \simeq a_0^+ \simeq \sqrt{N} \tag{80}$$

The operators a_0^+ and a_0 are then considered as c-numbers.

Taking account of this assumption we may then write the ion-fluid force operator (60):

$$\hat{F}(0) = i\lambda\Omega^{-1}\{\sqrt{N}qu(q)\,[a_q^+ + a_{-q}] + \sum_{k \neq 0} qu(q)\,a_k^+ a_{k-q}\} \tag{81}$$

and show that the Hamiltonian (79) may be cast into the well-known form:

$$H_{\text{eff}}^t = 2\pi aN\rho + \sum_{k \neq 0} [(k^2/2 + 4\pi a\rho)\,a_k^+ a_k + 2\pi a\rho(a_k^+ a_{-k}^+ + a_k a_{-k})] \tag{82}$$

where ρ denotes the density of our system $(\rho = N/\Omega)$. Let us stress again that here only the ground state and states immediately above the ground state are significant. Thus the approximation leading to Eq. (82) becomes increasingly worse for higher excited states and the calculation we present here is only valid for a very low temperature state of our boson fluid.

The effective Hamiltonian (82) can be immediately diagonalized using the following linear transformation* first introduced by Bogoliubov:[21]

$$a_k^+ = g_k b_k^+ + f_k b_{-k}$$
$$a_k = g_k b_k + f_k b_{-k}^+ \tag{83}$$

with

$$g_k^2 - f_k^2 = 1$$

where g_k and f_k are the following functions of the absolute value of the momentum vector k:

$$g_k = (1 - \alpha_k^2)^{-\frac{1}{2}}; \quad f_k = \alpha_k(1 - \alpha_k^2)^{-\frac{1}{2}} \tag{84}$$

---

* It is clear that b_k^+ and b_k satisfy the same commutation rules as a_k^+ and a_k. Therefore b_k^+ and b_k can be interpreted respectively as creation and annihilation operators just as a_k^+ and a_k.

with

$$\alpha_k = 1 + x^2 - x(x^2 + 2); \quad x = k^2/8\pi a\rho \tag{85}$$

It is well known that this transformation applied to the Hamiltonian (82) leads to the familiar result:

$$H^t = E_0 + \sum_k \omega_k b_k^+ b_k \tag{86}$$

This result corresponds to the representation of our boson fluid by a system of non-interacting quasi-particles whose energies are given by

$$\omega_k = \frac{k}{2}(k^2 + 16\pi a\rho)^{\frac{1}{2}} \tag{87}$$

We shall only consider the very small wave number limit where these elementary excitations behave like phonons, i.e.:

$$\lim_{k \to 0} \omega_k = ck; \quad c = (4\pi a\rho)^{\frac{1}{2}} \tag{88}$$

Similarly when we introduce definition (83) into Eq. (81) we may show that the ion-fluid force operator becomes:

$$\hat{F}(0) = i\lambda\Omega^{-1} qu(q)[\sqrt{N}(f_q + g_q)(b_q^+ + b_{-q})$$
$$+ \sum_{k \neq 0}(f_k f_{k-q} b_{-k} b_{-k+q}^+ + f_k g_{k-q} b_{-k} b_{k-q}$$
$$+ f_{k-q} g_k b_k^+ b_{-k+q}^+ + g_k g_{k-q} b_k^+ b_{k-q})] \tag{89}$$

We then introduce the expressions (86) and (89) into formula (77) and perform the average in the phonon representation. After some simple algebra, we get

$$\zeta = \frac{1}{3MkT} \frac{\pi}{\Omega^2} \sum_{k,q} |u(q)|^2 [f_k^2 f_{k-q}^2 + g_k^2 g_{k-q}^2] m_k^0(m_k^0 + 1) \delta(\omega_k - \omega_{k-q}) \tag{90}$$

where

$$m_k^0 = [\exp(\beta ck) - 1]^{-1} \tag{91}$$

represents the phonon equilibrium distribution function. Note that in this latter formula we have neglected the conservation of particles condition, i.e. $N_0(T)$ has to be determined from

$$N = N_0 + \sum_k n_k$$

where n_k is the particle number, related to the phonon number through

$$\langle n_k \rangle = \langle (g_k b_k^+ + f_k b_{-k})(g_k b_k + f_k b_{-k}^+) \rangle$$

The summations in Eq. (90) are easily performed and as is shown in Appendix A, the following result is obtained:

$$\zeta = \frac{2\pi^3}{45} \frac{a^2(kT)^4}{M\mathbf{h}^3 c^4} \tag{92}$$

This result agrees with that of Abe and Aïzu[22] in the heavy ion limit.

It should be added that in the model of a heavy ion immersed in a sea of phonons one can immediately get a similar result from the usual definition of the friction coefficient:

$$\zeta \simeq \rho \sigma \bar{v} \tag{93}$$

where ρ denotes the number density of scatterers, σ the total scattering section and $\bar{v}$ represents here the mean relative velocity between the ion and the fluid particle.

Here we may use for ρ the well-known formula of the phonon density, $\rho \propto (kT/\mathbf{h}c)^3$, and set $\bar{v} \simeq c$ because the ion velocity is much smaller than the phonon velocity c. Moreover as the phonon momentum $p_t \propto kT/c$ is small compared to the thermal momentum of the heavy ion, the ion is scattered only through small angles and we may take for σ its small angles value, i.e. for hard spheres, $\sigma \propto a^2 \theta^2$, where θ is given by:

$$\theta \propto p_t/P \simeq (1/c)(kT/M)^{\frac{1}{2}}$$

(usually $\Delta P = 2P \sin \theta/2$). Thus when we rearrange all these parameters into definition (93) we arrive at a result similar to formula (92).

(b) *Fermi Systems*

We shall now extend to the case of a fluid of fermions the quantum Brownian motion theory we have presented previously. We shall consider the same simple model of a heavy ion moving in a sea of fermions under the influence of an electric field. Let us first remember that the "microscopic" derivation of the Fokker–Planck equation we gave in Section 3A was based on a systematic expansion in the ratio $[\langle p_{\mathrm{f}}^2 \rangle / \langle P^2 \rangle]^{\frac{1}{2}}$ of all terms contained in the generalized transport equation (47) [see Eq. (67)]. This development was assumed convergent, i.e.

$$[\langle p_{\mathrm{f}}^2 \rangle / \langle P^2 \rangle]^{\frac{1}{2}} \ll 1 \tag{94}$$

In the case of a fermion fluid one can immediately remark that the meaning of an expansion in the ratio $[\langle p_{\mathrm{f}}^2 \rangle / \langle P^2 \rangle]^{\frac{1}{2}}$ depends crucially on the range of temperature considered. Indeed there exist three different temperature regions characterized by the nature of the trajectory of the heavy particle:

(a) The high temperature region where the fluid behaves classically (whatever the statistics) and the heavy particle undergoes a Brownian motion, that is the Brownian particle moves through the fluid suffering only small random deflections from its linear trajectory. Condition (1) is satisfied and a Fokker–Planck description is valid. In this state the friction coefficient is given by the classical formula (recall Stoke's law, see, for instance, reference 20):

$$\zeta_{(\mathrm{a})} \propto \frac{\rho a^2 (mkT)^{\frac{1}{2}}}{M} \tag{95}$$

where ρ is the number fluid density.

(b) When the temperature decreases we attain a temperature region where the average fluid particle momentum is not of thermal range but rather proportional to the square root of its Fermi energy ε_{F}. Due to the exclusion principle the heavy ion only undergoes collisions with fermions whose energy is near the Fermi surface of energy ε_{F}. Thus the ratio becomes

$$[\langle p_{\mathrm{f}}^2 \rangle / \langle P^2 \rangle]^{\frac{1}{2}} \simeq [m\varepsilon_{\mathrm{F}}/MkT]^{\frac{1}{2}} \equiv \gamma\xi \tag{96}$$

where ξ denotes $(\varepsilon_{\mathrm{F}}/kT)^{\frac{1}{2}}$. Then we can conclude that the Brownian particle distribution function will obey a Fokker–Planck type equation only under the requirement:

$$\gamma\xi \ll 1 \tag{97}$$

Thus there exists a temperature region—when $T_{\mathrm{class}} \gg T \gg \gamma T_{\mathrm{F}}$ (T_{F} being Fermi temperature of the fluid)—where the statistics are important in determining the properties of the fluid but when the heavy particle nevertheless suffers only small deflections. We shall call this temperature range "quantum Brownian motion region" because the motion of the heavy ion is still described by a Fokker–Planck equation but with a friction coefficient ζ containing the quantum effects due to the nature of the fluid (as was the case for a Bose system).

(c) At very low temperature $(T \to 0)$ the average momentum of the fermion becomes larger than that of the Brownian particle and condition (97) must be replaced by the following requirement:

$$\xi^{-1} \ll \gamma \ll 1 \tag{98}$$

Consequently, when $T \leqslant \gamma T_F$, the heavy ion no longer experiences a Brownian motion through the fermion fluid and a Fokker–Planck description is no longer valid.* Let us point out that this extremely low temperature region has been studied independently by Abe and Aïzu[22] and Clark[27] in a weak coupling approximation. These authors start from the quantum-mechanical Boltzmann equation and evaluate approximatively the collision integral by using a variation principle and a simplified form of the fluid distribution function. We shall not give their calculation here but just quote their final expression for the friction coefficient of the heavy ion in this temperature range:

$$\zeta_{(c)} \simeq \frac{Ma^2(kT)^2}{h^3} \tag{99}$$

As $T \to 0$, the friction coefficient approaches zero as T^2. This phenomenon is due to the exclusion principle which allows collisions between the Brownian particle and only those fermions near the Fermi surface. Thus ζ, proportional to the number of available scatterers, is as T^2.

In order to put states (b) and (c) in perspective let us consider the case of ^{3}He and suppose the effective mass of the positive ion in liquid ^{3}He is about 30 times the mass of a helium atom.† Then at liquid densities $\rho \simeq 10^{22}$ molecules/cm^{-3} $(T \simeq 1.5^\circ$K; $\varepsilon_F/k = 5^\circ$K) one finds $\gamma\xi \simeq 0.5$. We may thus conclude that inequality (97) is

* Such a state does not exist in the case of perfect (or weakly coupled) boson fluid because the momenta of both kinds of particle are always of thermal range, thus condition (1) is satisfied and a Fokker–Planck description is valid even in the limit $T \to 0$.

† We shall see how this choice seems correct in the next Section where we consider the motion of an α particle in liquid helium.

no longer valid when $T \leqslant 1°K$ and the ion momentum distribution function no longer obeys the Fokker–Planck equation.

In order to reinforce the validity of such a temperature scaling in describing the behaviour of a Brownian particle moving through a fermion sea, we present in Appendix B the explicit derivation of the Fokker–Planck equation, starting from the Uehling–Uhlenbeck equation for a heavy ion immersed in a perfect Fermi gas. This calculation, made in a weak coupling approximation, suggests the existence of a well-defined "quantum Brownian motion region" also in the case of a Fermi fluid (remember Landau's quasi-particle picture for normal Fermi systems).

We now calculate the friction coefficient of a heavy ion in a perfect Fermi gas for temperature range (b), i.e. when condition (97) is realized. We shall use definition (191), Appendix B, valid for a weak coupling system and describe the ion-fluid particle interaction by a scattering length a which is treated within the Born approximation. Thus we start with

$$\zeta = -\frac{2\pi m^2 a^2}{3M} \left(\frac{m}{2\pi \mathbf{h}}\right)^3 \int_0^\pi (1 - \cos \theta) \sin \theta \, d\theta \int \frac{\partial n}{\partial \varepsilon} v^3 \, d^3 v \quad (100)$$

where v denotes the velocity of a fluid particle with energy ε. We may write the second integral as

$$\int \frac{\partial n}{\partial \varepsilon} v^3 \, d^3 v = \frac{16\pi}{m^3} \int_0^\infty \frac{\partial n}{\partial \varepsilon} \varepsilon^2 \, d\varepsilon = -\frac{32\pi}{m^3} \int_0^\infty n\varepsilon \, d\varepsilon \quad (101)$$

Bearing in mind the definition of the Fermi distribution function n (178) we can show that

$$\int_0^\infty n\varepsilon \, d\varepsilon = \tfrac{1}{2} \varepsilon_F^2 + \tfrac{1}{6} \pi^2 (kT)^2 \quad (102)$$

Thus finally, combining Eqs. (100) and (102), we get for the friction coefficient of the Brownian particle:

$$\zeta_{(b)} = \frac{8m^2 a^2}{3\pi M \mathbf{h}^3} \left[\varepsilon_F^2 + \frac{\pi^2}{3} (kT)^2 \right] \quad (103)$$

where the subscript (b) indicates that formula (103) is valid in the "quantum Brownian motion temperature region," i.e. when $T_{\text{class}} \gg T \gg \gamma T_F$.

It should be added that definition (103) can be transformed, following the ideas of Mazo and Kirkwood,[30] into the formula:

$$\zeta_{(b)} \propto \frac{\rho a^2 (mkT_q)^{\frac{1}{2}}}{M} \tag{104}$$

which has the classical form of the friction coefficient [see $\zeta_{(a)}$, Eq. (95)] but where the thermodynamic temperature T has been replaced by the "quantum temperature" T_q:

$$T_q = \frac{\varepsilon_F}{k}\left[1 + O\left(\frac{kT}{\varepsilon_F}\right)^2\right] \tag{105}$$

ρ being the number density of the Fermi gas, i.e.

$$\rho \propto \left(\frac{m\varepsilon_F}{\mathbf{h}^2}\right)^{\frac{3}{2}}$$

This way of writing the friction coefficient in this temperature range emphasizes the role of the zero point motion of the fluid particles. This concept of quantum temperature will be discussed in full detail in Section 4 where we shall study the very low temperature limit of quantum Brownian motion.

C. Discussion and Comparison with Experiment

In this Section we shall use the various formulae for the friction coefficient of a heavy particle immersed in a quantum fluid (either Bose or Fermi systems) to calculate the mobility of a heavy positive ion in both cases. We shall immediately get the mobility by the well-known definition:

$$\mu = e\tau/M = e/M\zeta \tag{106}$$

where τ denotes the relaxation time of the system.

The results so obtained will then be compared with the numerical values calculated from measurements of the ionic static mobilities in liquid ^{4}He and ^{3}He.

(a) Nearly Perfect Bose Gas

From Eqs. (92) and (106) we get immediately

$$\mu = \frac{45}{2\pi^3} \frac{e\mathbf{h}^3 c^4}{a^2 (kT)^4} \tag{107}$$

Although this formula is strictly valid for a weakly interacting Bose gas, it is worthwhile comparing our result tentatively with the experimental data obtained with ^{4}He. For instance, Meyer and Reif[14] have measured the static mobility of the ion $(He)_n^+$ in liquid ^{4}He at low temperatures. In the phonon regions, they found

$$\mu \propto T^k; \quad k = -3.3 \pm 0.3 \tag{108}$$

As a very recent experiment of Dahm and Sanders[23] has shown, the effective mass of the ion $(He)_n^+$ is from 20 to 40 ^{4}He atomic masses. The Brownian motion condition (1) is thus rather well satisfied and the temperature dependence of Eqs. (107) and (108) is in satisfactory agreement if one recalls the weakness of the model considered here.

Despite the approximations made in order to obtain formula (107), i.e.

$$\mu = \frac{4 \times 10^{-30}}{a^2} \left(\frac{c}{T}\right)^4 \text{ cm}^2\, \text{V}^{-1}\,\text{s}^{-1} \tag{109}$$

let us calculate the mobility of a positive ion in liquid ^{4}He at a temperature corresponding to the phonon region, for instance at $T = 0.55°$K. We substitute for c the real sound velocity at this temperature, i.e. $c = 237$ m s^{-1} [recall that formula (109) has been obtained in the long wavelength limit]. Moreover let us take the result of the calculation of Parks and Donnelly[24] for the radius of the positive ion $(He)_n^+$ in He II:

$$R_i^+ = (6.44 \pm 0.10)\ \text{Å}$$

a value in good agreement with that proposed by Kuper.[25] Then the scattering length appearing in Eq. (109) becomes

$$a = [R_i^+ + r_{^4He}] \simeq 7.7\ \text{Å} \tag{110}$$

With these numerical values we get

$$\mu \simeq 2.5 \times 10^3 \text{ cm}^2\, V^{-1}\,\text{s}^{-1} \tag{111}$$

while the experimental value is

$$\mu_{exp} = 5.9 \times 10^3 \text{ cm}^2\, \text{V}^{-1}\,\text{s}^{-1} \tag{112}$$

It should be emphasized, however, that the explicit form of Eq. (107) is very sensitive to the value of scattering length so a more refined value of the ionic radius would not change the *qualitative* aspect of our result.

(b) *Perfect Fermi Gas*

It is well known that the very low temperature properties of the normal Fermi systems may be described by the laws of the perfect Fermi gas provided the parameters (such as mass, for instance) are adjusted. Thus we can expect that the evaluation of the mobility of a heavy ion immersed in a perfect Fermi gas would give a result in qualitative agreement with the experimental value of the static mobility in liquid ^{3}He at very low temperature. In order to make this comparison let us first introduce definition (103) into Eq. (106). We find

$$\mu = \frac{3\pi e \mathbf{h}^3}{8m^2 a^2 [\varepsilon_{\mathrm{F}}^2 + \frac{1}{3}\pi^2 (kT)^2]} \tag{113}$$

or

$$\mu = \frac{7.5 \times 10^{-15}}{a^2 (T_{\mathrm{F}}^2 + 3.3\,T^2)} \ \mathrm{cm^2\,V^{-1}\,s^{-1}} \tag{114}$$

where T_{F} is the Fermi temperature of the fluid (for ^{3}He, $T_{\mathrm{F}} \simeq 5^\circ$K).

As no precise value of the radius of the positive ion $(\mathrm{He})_n^+$ in liquid ^{3}He is at present available* it is rather difficult to give a correct value for the scattering length a. Nevertheless let us suppose the radius R_{i}^+ is a little lower in the case of ^{3}He than for ^{4}He (see the previous calculation for ^{4}He) and set, for instance, the scattering length $a \simeq 7$ Å. Then for $T = 1.2^\circ$K we get

$$\mu_{(\mathrm{b})} \simeq 3.5 \times 10^{-2} \ \mathrm{cm^2\,V^{-1}\,s^{-1}} \tag{115}$$

This value agrees qualitatively with the experimental value:

$$\mu_{\mathrm{exp}} = 7.65 \times 10^{-2} \ \mathrm{cm^2\,V^{-1}\,s^{-1}} \tag{116}$$

obtained by Meyer and coworkers[26] for a positive ion in liquid ^{3}He at the same temperature.

It should be added that the best substituted value of the scattering length leads to an ionic radius value of about 4 Å. If one compares this value to that of $R_{\mathrm{i}}^+ \simeq 6.5$ Å proposed by Parks and Donnelly[24] in the case of ^{4}He one would be tempted to conclude that the effective mass of the ion $(\mathrm{He})_n^+$ in ^{3}He could be rather smaller than that in liquid ^{4}He. But before going so far let us

---

* The present literature does not furnish any value of the effective mass of the ion $(\mathrm{He})_n^+$ in liquid ^{3}He.

remember again the weakness of our model; the ion-fluid inter-action has been supposed weak and the corresponding scattering length treated within the Born approximation in the low energy limit.

Finally let us compare the mobility in state (b) Eq. (113) with that in the very low temperature region (c) [see Eqs. (99) and (106)]:

$$\mu_{(c)} \simeq \frac{e\mathbf{h}^3}{M^2\,a^2(kT)^2} \tag{117}$$

The difference between these two equations is remarkable. In the Brownian motion situation Eq. (113) shows that the mobility is independent of the ionic mass and quite insensitive to the temperature while, in the state $T \to 0$, the mobility varies inversely as the square of both these quantities. We may then suggest that measurements of the static mobility of an α particle in liquid ^{3}He made in both temperature regions (b) and (c) would give an order of magnitude of the effective mass M. Indeed one has

$$\frac{\mu_{(b)}}{\mu_{(c)}} \propto \frac{T^2_{(c)}}{T^2_{\mathrm{F}} + T^2_{(b)}} \left(\frac{M}{m}\right)^2 \tag{118}$$

if one supposes the scattering cross-section and the effective mass insensitive to the temperature [note that regions (b) and (c) cover the range $0 \leqslant T \leqslant 3^\circ\mathrm{K}$].

4. THE LOW TEMPERATURE LIMIT OF THE THEORY OF QUANTUM BROWNIAN MOTION

A. Introduction

In the previous Sections, we have extended to quantum situations the microscopic theory of classical Brownian motion. To summarize, let us recall that it has been shown that the stationary Brownian particle momentum distribution function $\phi(P)$, which is realized in the presence of an external electrostatic field, obeys the following transport equation [see Eq. (67)]:

$$\left\{eE \frac{\partial}{\partial P} [\phi_0^{(0)}(P) + \gamma^2 \phi_0^{(2)}(P) + \ldots]\right\} + \mathrm{i}[\gamma^3 G^{(3)} + \ldots]$$

$$= \mathrm{i}[\gamma^2 \Omega^{(2)} + \gamma^4 \Omega^{(4)} + \ldots]\phi(P) \tag{119}$$

where the various terms have been expanded according to their γ dependence. Moreover it has been also demonstrated that *provided the γ expansions appearing in Eq.* (119) *converge,* this transport equation reduces to the usual Fokker–Planck equation (see Section 3.A).

It must be emphasized that the Fokker–Planck equation (76) is only applicable when the higher order γ terms are small in Eq. (119). Until now, we have only considered this problem in detail for the very simple model of a Brownian particle moving through a fluid of non-interacting fermions (see Appendix B). It has been shown for this case that condition (1) should be replaced by the much more restrictive requirement (97), i.e.:

$$[\langle p_1^2\rangle/\langle P^2\rangle]^{\frac{1}{2}} \simeq \gamma\xi \ll 1 \qquad (120)$$

for the Fokker–Planck equation to be valid. Here $\xi = (\varepsilon_F/kT)^{\frac{1}{2}}$ is a parameter which can become very large at sufficiently low temperatures (recall that the Fermi energy ε_F is closely related to the zero-point motion of the fluid fermions).

It is clear that a rigorous mathematical analysis of the convergence of the series appearing in Eq. (119) is in general very difficult. As we shall report presently, it is, however, possible to get some information by looking at the first "corrections" to the Fokker–Planck equation (76) coming from the higher order γ terms in the general transport equation (119).

Before going into the explicit analysis of the various higher order γ contributions of Eq. (119) involving the operators $G^{(n)}$ and $\Omega^{(2n)}$ let us first discuss in detail the γ^2 correction to the equilibrium momentum distribution function $\phi_0(\gamma; P)$ of the Brownian particle.

B. Heavy Particle Equilibrium Distribution Function

At equilibrium, the Brownian particle momentum distribution function is given by

$$\phi_0(\gamma; P) = Z^{-1} \sum_{\mathbf{n}} \langle P\mathbf{n}| \exp{(-\beta H_0)}\, U(\beta)| P\mathbf{n}\rangle \qquad (121)$$

if the usual representation $|P\mathbf{n}\rangle$ is used [see Eq. (6)]. Z is the partition function of the total system and $U(\beta)$ is the well-known evolution operator in the temperature space:

$$U(\beta) = \exp(\beta H_0) \exp(-\beta H) \tag{122}$$

The distribution function $\phi_0(\gamma; P)$ has been explicitly formulated—up to order γ^2—in Appendix C. It is shown that the equilibrium momentum distribution function of the heavy particle may be cast into [see Eq. (208)]:

$$\phi_0(\gamma; P) = \phi_0^{(0)} + \gamma^2 \left[\frac{P^2}{2MkT} \left(\frac{kT_q}{kT} \right) - A \right] \phi_0^{(0)}(P) + O(\gamma^3) \tag{123}$$

where A denotes the contribution of all P independent terms which have only partly to ensure the normalization of $\phi_0(\gamma; P)$. Here $\phi_0^{(0)}(P)$ is the unperturbed Maxwellian distribution function and $T_q(\mathbf{h}, T)$ is *a characteristic quantum temperature** defined by

$$kT_q = 2\mathbf{h}^2 \beta^{-2} \left[\frac{1}{2} \int_0^\beta \langle \beta'^2 \, \overline{F}'(\beta') \rangle_f \, d\beta' \right.$$
$$\left. - \int_0^\beta \int_0^{\beta'} \langle \beta' \, \overline{F}(\beta') \beta'' \overline{F}(\beta'') \rangle_f \, d\beta' d\beta'' \right] \tag{124}$$

where the bracket $\langle ... \rangle_f$ as usual denotes the average over the fluid equilibrium distribution function [see Eq. (206) for the definition of $\overline{F}$].

It is of course very difficult to make general statements about the behaviour of $T_q(\mathbf{h}, T)$ as a function of $\mathbf{h}$ and T. Yet, it is immediately shown that it vanishes in the classical limit (see Appendix D):

$$\lim_{\substack{T \to \infty \\ (\text{or } \mathbf{h} \to 0)}} kT_q(\mathbf{h}, T) = 0 \tag{125}$$

Moreover, it is clear that T_q will tend to a finite limit when $T \to 0$, for instance it is proved in Appendix D that for a weak coupling between the Brownian particle and the fluid, one has

$$\lim_{T \to 0} kT_q = \frac{8\rho}{3\pi^2 \mathbf{h}^2} \int_0^\infty |U(q)|^2 \, dq \tag{126}$$

where ρ is the fluid density and $U(q)$ the Fourier transform of the

* $T_q(\mathbf{h}, T)$ is qualitatively similar to the "kinetic" temperature introduced by Mazo and Kirkwood;[30] it does not, however, depend on any approximation like the superposition approximation. Equation (123) is exact in the limit where $\gamma \to 0$.

Brownian particle–fluid interaction potential. In Appendix D we also obtain the formal generalization of this result. We show that

$$\lim_{T \to 0} kT_q = \text{constant} \tag{127}$$

for an arbitrary dense fluid.

The physical implications of Eq. (127) are important. Let us evaluate, for instance, the average kinetic energy of the Brownian particle. We obtain from Eq. (123)

$$\left\langle \frac{P^2}{2M} \right\rangle = \tfrac{3}{2}(kT + \gamma^2 kT_q) \tag{128}$$

Thus, in the high temperature region ($T \gg T_q$), the kinetic energy of the Brownian particle is essentially of thermal origin, and its average momentum is very large compared to that of a fluid particle:

$$\langle P^2 \rangle^{\frac{1}{2}} \simeq \gamma^{-1}(kT)^{\frac{1}{2}} \tag{129}$$

in agreement with Eq. (1). On the contrary, in the zero temperature limit:

$$\left\langle \frac{P^2}{2M} \right\rangle \simeq \gamma^2 kT_q \tag{130}$$

and thus

$$\langle P^2 \rangle^{\frac{1}{2}} \simeq (kT_q)^{\frac{1}{2}} \tag{131}$$

which is γ independent. This is not surprising because, in this limit, the momenta of both the Brownian particle and of the fluid molecules are essentially determined by their zero-point motion, which is a consequence of the uncertainty principle:

$$\Delta p \simeq \mathbf{h}(\Delta x)^{-1} \tag{132}$$

As the spatial localization Δx is entirely determined by the inter-molecular forces and is completely independent of the masses of the particles, we arrive at definition (131).

As we shall see in the next Section, similar conclusions are reached when the dynamics of the Brownian particle is considered and this imposes very severe restrictions on the validity of Brownian motion theory in the low temperature limit.

Before analysing this problem, let us make one more remark about the validity of Eq. (123). In the derivation of this formula, we have tacitly assumed that γ was the only characteristic parameter of the problem and that in the limit $\gamma \to 0$ the expansion (123) was converging. However, it appears clearly in this formula as well as in Eq. (131), that the so-called "quantum corrections" become dominant in the low temperature limit. In other words, the expansion (123) is valid only when

$$\gamma^2 T_q / T \ll 1 \tag{133}$$

Yet, the qualitative conclusion, as stated for instance in Eq. (131), does not depend on exact calculation and is an immediate consequence of the uncertainty principle, Eq. (132). In other words while the explicit expression for the Brownian particle momentum distribution function is given by Eq. (123) only when condition (133) is satisfied, the mass independence of the thermal momentum in the low temperature limit is of much wider significance.

C. Higher Order Corrections to the Fokker–Planck Equation

Let us now turn to the analysis of the transport equation (119), including the first corrections to both the Brownian particle unperturbed distribution function and the usual Fokker–Planck collision term. We know from the analysis given in Section 2.D that the collision operator of the general transport equation obeys the following equilibrium condition

$$\langle \psi_{00}^+(\gamma^{-1}p, N; 0) \rangle_{\mathrm{f}} \phi_0(\gamma; P) = \sum_{n=1}^{\infty} \gamma^{2n} \, \Omega^{(2n)} \sum_{m=1}^{\infty} \gamma^m \, \phi_0^{(m)}(p) = 0 \tag{134}$$

Indeed it is immediate to verify this equation up to order (see Section 3.A), i.e.

$$i\gamma^2 \, \Omega^{(2)}(0) \, \phi_0^{(0)}(p) = \sum_{N} \langle 0 | \gamma \mathscr{U}^{\mathrm{I}} \frac{1}{\mathscr{H}^{\mathrm{f}} - i0} \gamma \Gamma^{\mathrm{I}} \rho^{\mathrm{f}}(\mathbf{N}) | 0 \rangle' \phi_0^{(0)}(p)$$

$$\equiv \zeta \frac{\partial}{\partial P} \left[kT \frac{\partial}{\partial P} + \frac{P}{M} \right] \phi_0^{(0)}(P)$$

$$= 0 \tag{135}$$

where $\phi_0^{(0)}$ denotes the Brownian particle Maxwellian distribution function; ζ is defined by Eq. (77) and $\gamma \Gamma^{\mathrm{I}}$, the abbreviation:

$$\gamma \Gamma^{\mathrm{I}} = \gamma \mathscr{H}_0^{\mathrm{I}} + \gamma \mathscr{U}^{\mathrm{I}} \tag{136}$$

To the immediately higher order, Eq. (134) gives

$$\gamma^4 \, \Omega^{(4)}(0) \, \phi_0^{(0)}(p) = \gamma^2 \, \Omega^{(2)}(0) \, \gamma^2 \, \phi_0^{(2)}(p) \tag{137}$$

When one analyses this equation one may easily see that in Eq. (137) the correcting collision operator $\Omega^{(4)}$ whould split into two parts: one "classical" term which, when applied to $\phi_0^{(0)}(p)$, gives zero [recall the classical situation where $\phi_0(\gamma; P) \equiv \phi_0^{(0)}(P)$] and another contribution, of quantum origin, that factorizes into parts as the right-hand side of Eq. (137) [remember the quantum character of the first correction $\phi_0^{(2)}(p)$ to the Brownian particle equilibrium distribution function, see Eq. (123)]. We shall see that this factorizable part of $\Omega^{(4)}(0)$ will be used to "renormalize" the Fokker–Planck collision term. It should be added that very similar situations have been analysed in great detail following the same process; see, for instance, the treatment of the three-body collision operator in quantum mechanics[31, 33] where this operator is shown to contain a factorizable part which describes a renormalization of the two-body Boltzmann operator.

With this remark in mind let us now examine more explicitly the first correction to the usual Fokker–Planck term. From Eqs. (31) and (61) one readily obtains

$$\gamma^4 \, \Omega^{(4)}(0) = \lim_{z \to i0} \sum_N \left\{ \sum_{n=1}^{\infty} \langle 0| - \lambda \mathscr{W} [(\mathscr{H}_0 - z)^{-1}(-\lambda \mathscr{W})]^n |0\rangle' \right\}^{(4)} \rho_0^f(\mathbf{N}) \tag{138}$$

where the subscript (4) in the right-hand side indicates that, once the various matrix elements have been written in terms of the reduced variable p [see Eq. (50)], only contributions of order γ^4 are retained. In order to perform explicitly this calculation, we first use definitions (52), (53), (69) and (71) and then extract the γ^4 contributions, noticing that any contribution in Eq. (138) starting with γ^f or with $\gamma^2 \mathscr{U}^{II}$ vanishes once the trace over the fluid variables is taken. This property has already been used previously [see Eq. (64)]. The result is, after some straightforward manipulations:

$$\gamma^4 \, \Omega^{(4)} = \gamma^4 \, \Omega_A^{(4)}(0) + \gamma^4 \, \Omega_B^{(4)}(0) \tag{139}$$

with

$$\gamma^4 \, \Omega_A^{(4)}(0) = \lim_{z \to i0} \sum_N \langle 0| \gamma \mathscr{U}^I \frac{1}{\mathscr{H}^f - z} \gamma \Gamma^I \frac{1}{\mathscr{H}^f - z} \gamma \Gamma^I \frac{1}{\mathscr{H}^f - z}$$
$$\times \gamma \Gamma^I \rho^f(\mathbf{N}) |0\rangle' \tag{140}$$

and

$$\gamma^4 \Omega_B^{(4)}(0) = \lim_{z \to i0} \sum_N \left[\langle 0 | \gamma \mathcal{U}^I \frac{1}{\mathcal{H}^f - z} \gamma^2 \mathcal{U}^{II} \frac{1}{\mathcal{H}^f - z} \gamma \Gamma^I \rho^f(\mathbf{N}) | 0 \rangle' \right.$$

$$+ \langle 0 | \gamma \mathcal{U}^I \frac{1}{\mathcal{H}^f - z} \gamma \Gamma^I \frac{1}{\mathcal{H}^f - z} \gamma^2 \mathcal{U}^{II} \rho^f(\mathbf{N}) | 0 \rangle'$$

$$+ \langle 0 | \gamma^3 \mathcal{U}^{III} \frac{1}{\mathcal{H}^f - z} \gamma \Gamma^I \rho^f(\mathbf{N}) | 0 \rangle'$$

$$\left. + \langle 0 | \gamma \mathcal{U}^I \frac{1}{\mathcal{H}^f - z} \gamma^3 \mathcal{U}^{III} \rho^f(\mathbf{N}) | 0 \rangle' \right] \qquad (141)$$

The first term $\Omega_A^{(4)}$ is the exact analogue of the classical term analysed by Résibois and Davis.[3] In contrast, the second comes from the higher order terms in the expansion (53) [in the classical situation $\mathcal{W}$ is reduced to the two first terms of Eq. (53)] and is of purely quantum-mechanical origin.

We establish in Appendix E the following explicit form for $\Omega_A^{(4)}$:

$$\gamma^4 \Omega_A^{(4)}(0) = \gamma^4 \overline{\Omega}^{(2)}(0) \Omega^{(2)}(0) + \gamma^4 \Omega'^{(2)}(0) \Omega^{(2)}(0) + \gamma^4 \chi \Omega^{(2)}(0)$$

$$+ \gamma^4 \zeta \frac{\partial^2}{\partial p^2} [kT(\bar{\chi} - \chi/kT)] + \gamma^4 \sum_{(\alpha)} \overline{\Omega}_{(\alpha)}^{(4)}(0) \qquad (142)$$

This formal result is defined with the use of the following auxiliary quantities:

$$\overline{\Omega}^{(2)}(z) = \sum_N \langle 0 | \gamma \mathcal{U}^I \frac{1}{\mathcal{H}^f - z} \gamma \Gamma^I \frac{1}{\mathcal{H}^f - z} \rho^f(\mathbf{N}) | 0 \rangle' \qquad (143)$$

$$\chi = \sum_N \langle 0 | \mathscr{F}^+ \frac{1}{(\mathcal{H}^f - i0)^2} \mathscr{F}^+ \rho^f(\mathbf{N}) | 0 \rangle \qquad (144)$$

$$\bar{\chi} = \sum_N \langle 0 | \mathscr{F}^+ \frac{1}{(\mathcal{H}^f - i0)^2} \hat{\kappa} \rho^f(\mathbf{N}) | 0 \rangle \qquad (145)$$

with $\mathscr{F}^+$ and $\hat{\kappa}$ respectively given by Eqs. (59) and (55). In Eq. (142) we have also used the notation:

$$\Omega'^{(2)}(0) = \frac{d}{dz} \Omega^{(2)}(z)|_{z=0} \qquad (146)$$

where $\Omega^{(2)}(z)$ is defined by Eq. (135); the friction coefficient ζ (77) has the following form in the Liouville–von Neumann

3

formalism (see Section 3.A):

$$\zeta = \beta \sum_{\mathbf{N}} \langle 0 | \mathscr{F}^+ \frac{1}{\mathscr{H}^{\mathrm{f}} - \mathrm{i}0} \mathscr{F}^+ \rho^{\mathrm{f}}(\mathbf{N}) | 0 \rangle$$

$$\equiv \sum_{\mathbf{N}} \langle 0 | \mathscr{F}^+ \frac{1}{\mathscr{H}^{\mathrm{f}} - \mathrm{i}0} \hat{\kappa} \rho^{\mathrm{f}}(\mathbf{N}) | 0 \rangle \tag{147}$$

Finally let us note that the operator $\overline{\Omega}^{(4)}_{(\alpha)}(0)$ is defined in great detail in Appendix E [see Eq. (253)]. It has only a formal importance for the following.

Eq. (142) will allow us to simplify considerably the explicit form of the transport equation (119) and to give it a very physical interpretation. However, before considering this point we still need an explicit expression for the field correction $G^{(3)}$. From definition (62) one obtains readily, along the same line that is followed for the calculation of $\Omega^{(4)}_A$ (see Appendix E), that

$$\mathrm{i}\gamma^3 G^{(3)} = \mathrm{i} \Big\{ \sum_{\mathbf{N}} \langle 0 | \gamma \mathscr{U}^{\mathrm{I}} \frac{1}{\mathscr{H}^{\mathrm{f}} - \mathrm{i}0} \gamma \Gamma^{\mathrm{I}} \frac{1}{\mathscr{H}^{\mathrm{f}} - \mathrm{i}0} \mathscr{H}^{\mathrm{e}} \rho^{\mathrm{f}}(\mathbf{N}) | 0 \rangle'$$

$$+ \sum_{\mathbf{N}} \langle 0 | \gamma \mathscr{U}^{\mathrm{I}} \frac{1}{\mathscr{H}^{\mathrm{f}} - \mathrm{i}0} \mathscr{H}^{\mathrm{e}} \frac{1}{\mathscr{H}^{\mathrm{f}} - \mathrm{i}0} \gamma \Gamma^{\mathrm{I}} \rho^{\mathrm{f}}(\mathbf{N}) | 0 \rangle' \Big\} \phi_0^{(0)}(p) \tag{148}$$

remembering that $\mathscr{H}^{\mathrm{e}} \propto \gamma$, Eq. (22). Because $\mathscr{H}^{\mathrm{e}}$ is independent of the fluid variables, we may still write, using definitions (143) and (146):

$$\mathrm{i}\gamma^3 G^{(3)} = \mathrm{i}\gamma^2 [\overline{\Omega}^{(2)}(0) + \Omega'^{(2)}(0)] \mathscr{H}^{\mathrm{e}} \phi_0^{(0)}(p) \tag{149}$$

Now combining Eqs. (135), (139), (142) and (149) with the transport equation (119) we easily get:

$$\gamma e E \frac{\partial}{\partial p} [\phi_0^{(0)}(p) + \gamma^2 \phi_0^{(2)}(p)] + \gamma^2 [\overline{\Omega}^{(2)}(0) + \Omega'^{(2)}(0)] \gamma e E \frac{\partial}{\partial p} \phi_0^{(0)}(p)$$

$$= \mathrm{i} \Big\{ \gamma^2 \Omega^{(2)}(0) + \gamma^4 \overline{\Omega}^{(2)}(0) \Omega^{(2)}(0) + \gamma^4 \Omega'^{(2)}(0) \Omega^{(2)}(0)$$

$$+ \gamma^4 \chi \Omega^{(2)}(0) + \gamma^4 \zeta \frac{\partial^2}{\partial p^2} [kT(\bar{x} - \chi/kT)] + \gamma^4 \Omega_{\mathrm{g}}^{(4)}(0) \Big\} \phi(p) \tag{150}$$

where we have introduced an operator $\Omega_{\mathrm{g}}^{(4)}(0)$ which we shall call the "fourth-order genuine collision operator". This will become clear presently.

$$i\gamma^4 \, \Omega_g^{\prime(4)}(0) = i\gamma^4 \sum_{(\alpha)} \overline{\Omega}_{(\alpha)}^{(4)}(0) + i\gamma^4 \, \Omega_B^{(4)}(0) \tag{151}$$

The transport equation (150) can be much simplified. Let us use the notation:

$$\{A(\gamma)\}^{(n,m)} \equiv \gamma^n \, A^{(n)} + \gamma^m \, A^{(m)} \tag{152}$$

in order to indicate that in the γ expansion of an arbitrary function $A(\gamma)$, we only retain terms of order n and m. We may then cast Eq. (150) into the form:

$$\left\{ [1 + \gamma^2 \, \overline{\Omega}^{(2)}(0) + \gamma^2 \, \Omega^{\prime(2)}(0)] \, \gamma e E \frac{\partial}{\partial p} \phi_0(\gamma; p) \right\}^{(1,3)}$$

$$= \{ [1 + \gamma^2 \, \overline{\Omega}^{(2)}(0) + \gamma^2 \, \Omega^{\prime(2)}(0)] \, [i\gamma^2 \, \widetilde{\Omega}^{(2,4)}(0)$$

$$+ i\gamma^4 \, \Omega_g^{(4)}(0)] \}^{(2,4)} \, \phi(p) \tag{153}$$

where we have introduced the *renormalized Fokker–Planck operator*:

$$i\gamma^2 \, \widetilde{\Omega}^{(2,4)}(0) = \gamma^2 \, \zeta^{(0,2)} \frac{\partial}{\partial p} \left[kT(1 + (\bar{\chi} - \chi/kT)) \frac{\partial}{\partial p} + p \right] \tag{154}$$

where $\zeta^{(0,2)}$ is the renormalized friction coefficient:

$$\zeta^{(0,2)} = \zeta(1 + \gamma^2 \, \bar{\chi}) \tag{155}$$

Dividing Eq. (153) by the operator

$$[1 + \gamma^2 \, \overline{\Omega}^{(2)}(0) + \gamma^2 \, \Omega^{\prime(2)}(0)] \tag{156}$$

we get simply:

$$\left\{ \gamma e E \frac{\partial}{\partial p} \phi_0(\gamma; p) \right\}^{(1,3)} = \{ i\gamma^2 \, \widetilde{\Omega}^{(2,4)}(0) + i\gamma^4 \, \Omega_g^4(0) \}^{(2,4)} \, \phi(p) \tag{157}$$

Finally, we show in Appendix F that the coefficients χ and $\bar{\chi}$ are related to the characteristic quantum temperature $T_q(\mathbf{h}, T)$ introduced in Section 4.B. We then have

$$\bar{\chi} - \chi/kT = \gamma^2 \, kT_q/kT \tag{158}$$

The renormalized Fokker–Planck operator $\widetilde{\Omega}^{(2,4)}$ thus takes the very simple form:

$$i\gamma^2 \, \widetilde{\Omega}^{(2,4)} = \zeta^{(0,2)} \frac{\partial}{\partial P} \left[(kT + \gamma^2 \, kT_q) \frac{\partial}{\partial P} + \frac{P}{M} \right] \tag{159}$$

where we have turned back to the ordinary momentum variable P. In Eq. (159), it is very simple to analyse the part of the "corrections" to the Fokker–Planck equation in the limit of high and low temperatures.

In the high temperature limit, we have seen that the equilibrium distribution function $\phi_0(\gamma; P)$ tends toward the unperturbed Maxwellian distribution function $\phi_0^{(0)}(P)$, while the characteristic temperature tends toward zero; similarly, it is easily shown that $\bar{\chi}$ vanishes as well as the quantum fourth-order operator $\Omega_B^{(2)}$. We are thus left with

$$eE\frac{\partial}{\partial P}\phi_0^{(0)}(P)+O(\gamma^4)$$

$$= \zeta\frac{\partial}{\partial P}\left[kT\frac{\partial}{\partial P}+\frac{P}{M}\right]\phi(P)+\sum_{(\alpha)}\Omega_{(\alpha)}^{(4)}(0)\,\phi(P)+O(\gamma^5) \quad (160)$$

This equation is precisely the equation derived by Lebowitz and Rubin[2] in the classical case (up to the first non-trivial correction in γ); indeed it can be shown that the operator $\sum_{(\alpha)}\Omega_{(\alpha)}^{(4)}(0)$ tends, when $T\to\infty$ (or $\mathbf{h}\to 0$), toward the operator $\eta^3(t)$ defined in reference 2, formula (4.13). We shall, however, not give this proof here.

If we now turn to the low temperature limit, it is very difficult to use Eq. (157) for explicit calculations, because the structure of the genuine collision operator $\Omega_g^{(4}(0)$ is very complicated. It is clear that it describes interaction processes between the Brownian particle and one single group of molecules. This property has been explicitly used in the definition of $\Omega_{(\alpha)}^{(4)}(0)$ (see Appendix E) and it is not difficult to show that the same holds for $\Omega_B^{(4)}(0)$. In other words, this contribution $\Omega_g^{(4)}(0)$ cannot be factorized into a product of γ^2 contributions, as was done for $\Omega_{(\alpha,\beta)}^{(4)}(0)$ (see Appendix E). $\Omega_g^{(4)}(0)$ thus represents genuine fourth-order dissipative effects. This allows us to conclude that the convergence difficulties found in the renormalized Fokker–Planck operator cannot possibly be compensated by a similar term in $\Omega_g^{(4)}(0)$.

From Eq. (159) one sees immediately that in the low temperature region we may reach a state where

$$\gamma^2 kT_q/kT \geqslant 1 \quad (161)$$

in which case the so-called "γ^4 corrections" become dominant! As a matter of fact this signifies that the whole γ expansion becomes meaningless.

A priori, one might of course hope that under these circumstances the renormalized operator Eq. (159) still furnishes a good description of the dissipative behaviour. It is, however, very easy to convince oneself that the whole framework of the theory breaks down. Brownian motion theory is essentially based upon the idea that the relative momentum transfers suffered by the Brownian particle are very small. This assumption is of course valid in the high temperature limit but, at low temperature, we have seen that the average momenta of both the Brownian particle and the fluid molecules are essentially determined by the intermolecular interactions and are all of the same order of magnitude.

In this case, we indeed expect that the relative momentum transfers $\Delta P/P$ should become large, in which case a Fokker–Planck description is no longer valid.

D. Approach to Equilibrium

In the above calculation, we have considered the stationary state which is realized in the presence of an external field. We could of course have taken other situations of interest. In particular, there is no difficulty in analysing along the same lines the approach to equilibrium of the Brownian particle distribution function $\phi(P; t)$. One then obtains at long periods of time $(t \to \infty)$, the following transport equation:

$$\partial_t \phi(P; t) = \bar{\zeta}^{(0,2)} \frac{\partial}{\partial P} \left[(kT + \gamma^2 kT_q) \frac{\partial}{\partial P} + \frac{P}{M} \right] \phi(P; t) + \gamma^4 \Omega_g^{(4)} \phi(P; t)$$

(162)

valid up to terms of order $(\gamma^6 t)$.

This equation has again the remarkable feature that parts of the γ^4 corrections may be incorporated in a renormalization of the Fokker–Planck operator while the remaining part $\Omega_g^{(4)}(0)$ may be interpreted as genuine fourth-order dissipative processes.

Let us stress again that this situation is completely analogous to what has been found recently in the analysis of the approach to equilibrium of weakly coupled and dilute systems (fermions,[31] phonons,[32] quantum gas[33]). The only difference lies in the fact that here the renormalized equation applies directly to the momentum distribution function $\phi(P; t)$, while in the other cases one had first to introduce a "quasi-particle" distribution function. This difference is related to the great simplicity of the equilibrium distribution (123) with respect to the weakly coupled case. Otherwise, the arguments developed in this latter case may be reproduced here without difficulty. In particular, one can show that Eq. (162) drives the system toward the correct equilibrium distribution function including the γ^2 correction, as given by Eq. (123).

Let us simply show here that $\phi_0(\gamma; P)$, as defined by Eq. (123), is indeed a stationary solution of the collision operator appearing in the right-hand side of (159). We have to show that

$$\tilde{\zeta}^{(0,2)} \frac{\partial}{\partial P}\left[(kT + \gamma^2 kT_q)\frac{\partial}{\partial P} + \frac{P}{M}\right]\phi_0(\gamma; P) = 0 \qquad (163)$$

Expanding this expression in power of γ, we get at order γ^4 [at order γ^2 we simply have Eq. (132)]:

$$\gamma^2 \tilde{\zeta}^{(2)} \frac{\partial}{\partial P}\left(kT \frac{\partial}{\partial P} + \frac{P}{M}\right)\phi_0^{(0)}(P) + \tilde{\zeta}^{(0)} \frac{\partial}{\partial P}\left(kT \frac{\partial}{\partial P} + \frac{P}{M}\right)\gamma^2 \phi_0^{(2)}(P)$$

$$+ \tilde{\zeta}^{(0)} \frac{\partial}{\partial P}\gamma^2 kT_q \frac{\partial}{\partial P}\phi_0^{(0)}(P) + \gamma^2 \Omega_g^{(4)}(0)\phi_0^{(0)}(P) = 0 \quad (164)$$

Using Eq. (123), one shows readily that Eq. (164) is valid provided that

$$\gamma^4 \Omega_g^{(4)}(0)\phi_0^{(0)}(P) = 0 \qquad (165)$$

A proof of this latter formula is sketched in Appendix G.

Let us finally remark that all the difficulties found in the case of an external field are also present in the transport equation (162). In particular, this latter equation fails to describe the dynamics of the system in the very low temperature state characterized by Eq. (161).

5. APPENDICES

A. Explicit Calculation of ζ for a Heavy Ion Immersed in a Nearly Perfect Bose Gas

We have to evaluate the following expression [see Eq. (90)]:

$$\zeta = \frac{1}{3MkT} \frac{\pi}{\Omega^2} \sum_{k,q} |u(q)|^2 q^2 [f_k^2 f_{k-q}^2 + g_k^2 g_{k-q}^2]\, m_k^0(m_{k-q}^0 + 1)\, \delta(\omega_k - \omega_{k-q})$$

$$(90)$$

If we use the asymptotic relation for small wave number:

$$\lim_{|k|,|k-q|\to 0} [f_k^2 f_{k-q}^2 + g_k^2 g_{k-q}^2] = \left[\frac{c^2}{2|k||k-q|} - \frac{c}{2|k|} - \frac{c}{2|k-q|} \right]$$

$$\simeq \frac{c^2}{2|k||k-q|} \tag{166}$$

we easily obtain from (166), within the limit of a continuous spectrum of k, q,

$$\zeta = \frac{1}{MkT} \frac{a^2 c^2}{96\pi^3} \iint q^2 \frac{1}{|k||k-q|} m_k^0 (m_{k-q}^0 + 1)\, \delta(\omega_k - \omega_{k-q})\, \mathrm{d}^3 k\, \mathrm{d}^3 q \tag{167}$$

To get this expression, we have treated the ion-fluid scattering length within the Born approximation in the low energies limit, i.e.

$$u(q) \simeq u(0) = 2\pi a \quad (\gamma \ll 1; \mathbf{h} = m = 1) \tag{168}$$

As the δ function implies $|k| = |k-q|$, we have

$$\zeta = \frac{1}{MkT} \frac{a^2 c}{96\pi^3} \int m_k^0 (m_k^0 + 1) \frac{1}{k^2} I_k\, \mathrm{d}^3 k \tag{169}$$

with

$$I_k = \pi \int_0^\infty \int_0^\pi q^3 \sin\theta\, \delta(q/2k - \cos\theta)\, \mathrm{d}q\, \mathrm{d}\theta \tag{170}$$

which trivially gives

$$I_k = \pi \int_0^{2k} q^3\, \mathrm{d}q = 4\pi k^4 \tag{171}$$

We are thus left with

$$\zeta = \frac{1}{MkT} \frac{a^2 c}{6\pi} \int m_k^0 (m_k^0 + 1)\, k^4\, \mathrm{d}k \tag{172}$$

Setting $x = \beta ck$ and using the property of the Bose equilibrium distribution function,

$$m_x^0 (m_x^0 + 1) = -\frac{\partial}{\partial x} m_x^0 \tag{173}$$

it is easy to show that Eq. (172) becomes in the new variable:

$$\zeta = -\frac{a^2 (kT)^4}{6\pi Mc^4} \int_0^\infty \left(\frac{\partial}{\partial x} m_x^0 \right) x^4\, \mathrm{d}x \tag{174}$$

which may be integrated by parts to give the final result

$$\zeta = \frac{2\pi^3 a^2 (kT)^4}{45 M \mathbf{h}^3 c^4} \tag{175}$$

It should be added that formula (175), obtained with the approximation (166), is correct within an error of order $\zeta(kT/mc^2)$, i.e. about 10% in the "phonon temperature region".

B. Fokker–Planck Equation in the Case of a Perfect Fermi System*

We consider a system, composed of one charged heavy particle (mass M, charge e) immersed in a sea of N fermions, on which is applied a constant electric field $\mathbf{E}$. The quantum-mechanical Boltzmann equation for this system is[20, 28]

$$\partial_t f(\mathbf{R}, \mathbf{V}; t) + \mathbf{V} \partial_\mathbf{R} f + \frac{e\mathbf{E}}{M} \partial_\mathbf{V} f = \partial_t f|_{\text{coll}} \tag{176}$$

where the collision integral is given by

$$\partial_t f|_{\text{coll}} = \left(\frac{m}{2\pi\mathbf{h}}\right)^3 \iint [f'n'(1-n) - fn(1-n')] g\sigma(g, \theta) \, d\Omega_{g'} \, d^3 v \tag{177}$$

where f and n represent the singlet distribution function of the heavy ion and fluid particle, respectively. They depend on the velocities $\mathbf{V}$ and $\mathbf{v}$. Also, $f(\mathbf{V}')$ and $n' = n(\mathbf{v}')$, where $\mathbf{V}'$ and $\mathbf{v}'$ are the velocities which result from a binary collision between the two particles initially having velocities $\mathbf{V}$ and $\mathbf{v}$. The relative velocity of a fluid particle and the heavy ion is $\mathbf{g} = \mathbf{v} - \mathbf{V}$, and $\sigma(g, \theta)$ is the differential cross-section at relative speed $\mathbf{g}$ for scattering into the solid angle $d\Omega_{g'}$ at θ, the angle between the initial and final relative velocities (so that $d\Omega_{g'} = \sin\theta \, d\theta \, d\phi$).

The distribution function of the fluid particles may be taken as the equilibrium distribution function,

$$n = \{\exp[\beta(\varepsilon - \varepsilon_\mathrm{F}) + 1]\}^{-1} \tag{178}$$

where $\varepsilon = \frac{1}{2}m\mathbf{v}^2$ is the kinetic energy of the fermion of velocity $\mathbf{v}$ and ε_F its chemical potential. This statement is physically obvious: a single heavy particle cannot cause N fluid particles to depart significantly from equilibrium (see Section 2.C).

* In this Appendix we use the notation of Chapman and Cowling.[20]

In order to put (176) into a more convenient form we define

$$f = f_0(1 + \Phi) \tag{179}$$

where f_0 is the Maxwellian distribution function of the Brownian particle and Φ represents the deviation from equilibrium due to the electric field. Using this definition and the condition of detailed balance we can transform Eq. (177) into

$$\partial_t f|_{\text{coll}} = f_0 J\Phi \tag{180}$$

with the definition

$$J\Phi = \left(\frac{m}{2\pi\mathbf{h}}\right)^3 \iint n(1 - n')(\Phi' - \Phi)g\sigma(g, \theta)\,d\Omega_{g'}\,d^3v \tag{181}$$

Now we are ready to exploit the condition of Brownian motion as given by [see Eqs. (96) and (97)]

$$[\langle p_t^2 \rangle / \langle P^2 \rangle]^{\frac{1}{2}} \simeq \gamma\xi \ll 1 \tag{182}$$

Under this condition the relative momentum transfer of the Brownian particle is small and we can expand

$$\Phi' = \Phi(\mathbf{V}') = \Phi(\mathbf{V} + \Delta\mathbf{V})$$

in a Taylor's series about $\Delta\mathbf{V} = 0$. Then

$$\Phi' - \Phi = \Delta\mathbf{V}\,\partial_\mathbf{V}\Phi + \tfrac{1}{2}\Delta\mathbf{V}\,\Delta\mathbf{V}\,\partial_\mathbf{V}\partial_\mathbf{V}\Phi + \dots \tag{183}$$

Using the relative and centre of mass velocities we can easily show that

$$\frac{\Delta\mathbf{V}}{\mathbf{V}} \simeq \gamma^2 \frac{(\mathbf{g}' - \mathbf{g})}{\mathbf{V}} \propto \gamma^2 \frac{\mathbf{v}}{\mathbf{V}} = O(\gamma\xi) \tag{184}$$

Then if condition (182) is valid, we can limit our expansion (183) to order $(\gamma\xi)^2$. Thus up to the desired approximation (181) becomes

$$J\Phi = (\mathbf{A}\,\partial_\mathbf{V} + B\,\partial_\mathbf{V}\,\partial_\mathbf{V})\,\Phi \tag{185}$$

where

$$\mathbf{A} = \left(\frac{m}{2\pi\mathbf{h}}\right)^3 \iint n(1 - n')\,\Delta\mathbf{V}\,g\sigma(g, \theta)\,d\Omega_{g'}\,d^3v \tag{186}$$

$$B = \frac{1}{2}\left(\frac{m}{2\pi\mathbf{h}}\right)^3 \iint n(1 - n')\,\Delta\mathbf{V}\,\Delta\mathbf{V}\,g\sigma(g, \theta)\,d\Omega_{g'}\,d^3v \tag{187}$$

The collision integral operator $\partial_t f|_{\text{coll}}$ has been replaced by a differential operator, Eq. (185), expressing the fact that only small deflections are suffered by the heavy particle.

A and B can be simplified, and subsequently $J\Phi$, by further exploiting the condition $\gamma\xi \ll 1$. We need the following expansions:

$$g = (v^2 + V^2 - 2\mathbf{v}\cdot\mathbf{V})^{\frac{1}{2}} \simeq v - \frac{\mathbf{v}\cdot\mathbf{V}}{v} + O\left(\frac{\mathbf{V}}{\mathbf{v}}\right)^2 \quad (188a)$$

$$\sigma(g, \theta) = \sigma(v, \theta) + \frac{\mathbf{v}\cdot\mathbf{V}}{\mathbf{v}}\,\partial_v\,\sigma(v, \theta) + O\left(\frac{\mathbf{V}}{\mathbf{v}}\right)^2 \quad (188b)$$

$$(1 - n') = (1 - n) - (\varepsilon' - \varepsilon)\frac{\partial n}{\partial\varepsilon} + O(\gamma\xi)^2 \quad (188c)$$

To justify the last expansion note that

$$(\varepsilon - \varepsilon') = \left(\frac{m}{1+\gamma^2}\right)(\mathbf{g}' - \mathbf{g})\cdot\mathbf{G}$$

and

$$\frac{\partial n}{\partial\varepsilon} = -\beta n(1 - n)$$

where $n(1 - n)$ is never greater than unity so that

$$(\varepsilon' - \varepsilon)\frac{\partial n}{\partial\varepsilon} \simeq \gamma\xi(1 + \gamma\xi) \quad (189)$$

G being the centre of mass velocity. Equation (189) may be easily generalized to higher order, namely,

$$\left[\frac{(\varepsilon' - \varepsilon)^k}{k!}\right]\frac{\partial^k n}{\partial\varepsilon^k} \simeq [\gamma\xi(1 + \gamma\xi)]^k$$

where k is any positive integer. Thus, using condition (182) we can limit the expansion (188c) to the linear term in $\Delta\varepsilon$ ($\Delta\varepsilon = \varepsilon' - \varepsilon$).

The next step is to put the truncated expansions of Eqs. (188) into the expressions of **A** and B and then to simplify the results keeping only the lowest order in $\gamma\xi$. Straightforward but tedious manipulations lead to

$$\mathbf{A} = \zeta\mathbf{V}$$
$$B = (kT/M)\,\zeta \quad (190)$$

with the definition

$$\zeta = -\frac{2\pi}{3}\frac{m^2}{M}\left(\frac{m}{2\pi\mathbf{h}}\right)^3\iint\frac{\partial n}{\partial\varepsilon}v^3(1 - \cos\theta)\,\sigma(v, \theta)\sin\theta\,\mathrm{d}\theta\,\mathrm{d}^3 v \quad (191)$$

Substituting the results of Eq. (191) into the expression for $J\Phi$, Eq. (185), we get, in the case of a steady state (uniform system), the following basic equation:

$$\frac{e\mathbf{E}}{M}\partial_{\mathbf{V}}f = \zeta f_0\left(-\mathbf{V}\partial_{\mathbf{V}}+\frac{kT}{M}\partial_{\mathbf{V}}\partial_{\mathbf{V}}\right)\Phi$$

$$= \zeta\partial_{\mathbf{V}}\left(\mathbf{V}+\frac{kT}{M}\partial_{\mathbf{V}}\right)f \qquad (192)$$

which has the form of the Fokker–Planck equation [see Eq. (76) in Section 3.A], where ζ represents the coefficient of friction which the heavy ion experiences in being pulled through the fluid. Here again the quantum-mechanical properties of the fluid affect the behaviour of the heavy particle only through its friction coefficient.

C. Evaluation of the Brownian Particle Equilibrium Distribution Function

This Appendix will be devoted to the calculation up to order γ^2 of the Brownian particle momentum distribution function $\phi_0(\gamma; P)$ as given by Eq. (121).

Let us first recall that the evolution operator (120) obeys the Bloch equation, i.e. in our case we may write

$$\partial_\beta U(\beta) = -[\tilde{V}^{\mathrm{f}}(\beta)+V^{\mathrm{fB}}(\beta)]\,U(\beta) \qquad (193)$$

with

$$\tilde{V}^{\mathrm{f}}(\beta) = \exp(\beta H_0^{\mathrm{f}})\,V^{\mathrm{f}}\exp(-\beta H_0^{\mathrm{f}}) \qquad (194a)$$

$$V^{\mathrm{fB}}(\beta) = \exp(\beta H_0^{\mathrm{B}})\,\tilde{V}^{\mathrm{fB}}(\beta)\exp(-\beta H_0^{\mathrm{B}}) \qquad (194b)$$

$$\tilde{V}^{\mathrm{fB}}(\beta) = \exp(\beta H_0^{\mathrm{f}})\,V^{\mathrm{fB}}\exp(-\beta H_0^{\mathrm{f}}) \qquad (194c)$$

In these equations we have split in an obvious way terms which depend either on the fluid or on the Brownian particle only and terms which involve the coupling between two subsystems. Moreover we have used the properties:

$$[H_0, H_0^{\mathrm{B}}] = 0; \quad [V^{\mathrm{f}}, H_0^{\mathrm{B}}] = 0 \qquad (195)$$

Clearly the only γ dependent terms in Eq. (193) will come from the explicit dependence of $V^{\mathrm{fB}}(\beta)$ on H_0^{B}. We thus formally expand the exponentials of Eq. (194b) and write Eq. (193) as

$$\partial_\beta U(\beta) = -\{(\tilde{V}^{\mathrm{f}}(\beta) + \tilde{V}^{\mathrm{fB}}(\beta)) + \beta[H_0, \tilde{V}^{\mathrm{fB}}(\beta)]$$
$$+ \tfrac{1}{2}\beta^2[H_0^{\mathrm{B}}, [H_0^{\mathrm{B}}, \tilde{V}^{\mathrm{fB}}(\beta)]] + \ldots\} U(\beta) \qquad (196)$$

Using Eqs. (4) and (5) as well as definition (60) for the quantum-mechanical force exerted by the fluid on the Brownian particle, it is easy to show that in the free particle representation we have simply

$$\langle P\mathbf{n}|[H_0^{\mathrm{B}}, \tilde{V}^{\mathrm{fB}}(\beta)]|P'\mathbf{n}'\rangle = -i\gamma\mathbf{h}p\langle\mathbf{n}|\tilde{F}(\beta)|\mathbf{n}'\rangle\,\delta_{p',p}^{\mathrm{Kr}}$$
$$+ \sum_k \mathbf{h}k(n_k - n_k') + \gamma^2\langle\mathbf{n}|\alpha|\mathbf{n}'\rangle \quad (197)$$

and

$$\langle P\mathbf{n}|\{H_0^{\mathrm{B}}, [H_0^{\mathrm{B}}, \tilde{V}^{\mathrm{fB}}(\beta)]\}|P'\mathbf{n}'\rangle = -\gamma^2\mathbf{h}^2 p^2\langle\mathbf{n}|\tilde{F}'(\beta)|\mathbf{n}'\rangle\,\delta_{p',p}^{\mathrm{Kr}}$$
$$+ \sum_k \mathbf{h}k(n_k - n_k') + O(\gamma^3) \quad (198)$$

In the right-hand sides of these equations, we have used the reduced variable p defined by Eq. (50) and the $\mathbf{h}$ dependence has been explicitly indicated; moreover $\tilde{F}'(\beta)$ denotes the spatial derivative of the force $\tilde{F}(\beta)$ and α is an unimportant, p-independent, factor whose only part is to ensure the normalization of $\phi_0(\gamma; P)$.

Up to order γ^2, we may thus write Eq. (196) as

$$\partial_\beta\langle P\mathbf{n}|U(\beta)|P'\mathbf{n}'\rangle$$
$$= \sum_{\mathbf{n}''p''}\{\langle\mathbf{n}|\tilde{V}^{\mathrm{f}}(\beta) + \tilde{V}^{\mathrm{fB}}(\beta)|\mathbf{n}''\rangle - i\gamma\mathbf{h}\beta p\langle\mathbf{n}|\tilde{F}(\beta)|\mathbf{n}''\rangle$$
$$- \tfrac{1}{2}(\gamma\mathbf{h}\beta p)^2\langle\mathbf{n}|\tilde{F}'(\beta)|\mathbf{n}''\rangle + \gamma^2\beta\langle\mathbf{n}|\alpha|\mathbf{n}''\rangle\}$$
$$\times \delta_{p'',p+\gamma\sum_k \mathbf{h}k(n_k - n_{k'})}^{\mathrm{Kr}}\,\langle P''\mathbf{n}''|U(\beta)|P'\mathbf{n}'\rangle \qquad (199)$$

Because of the conservation of the total momentum in the system, the only non-vanishing elements of $\langle P\mathbf{n}|U(\beta)|P'\mathbf{n}'\rangle$ are such that

$$p + \gamma\sum_k \mathbf{h}kn_k = p' + \gamma\sum_k \mathbf{h}kn_k' \qquad (200)$$

We may thus use the following definition

$$\langle P\mathbf{n}|U(\beta)|P'\mathbf{n}'\rangle \equiv \langle\mathbf{n}|U_p(\beta)|\mathbf{n}'\rangle\,\delta_{p,p'+\sum_k \mathbf{h}k(n_{k'}-n_k)}^{\mathrm{Kr}} \qquad (201)$$

Taking into account the fact that in Eq. (199) the variable p'' only differs from p by a small quantity of order γ we get finally:

$$\partial_\beta \langle \mathbf{n} | U_p(\beta) | \mathbf{n}' \rangle$$

$$= \sum_{\mathbf{n}''} \{ \langle \mathbf{n} | \tilde{V}^t(\beta) + \tilde{V}^{tB}(\beta) | \mathbf{n}'' \rangle - i\gamma \mathbf{h} \beta p \langle \mathbf{n} | \tilde{F}(\beta) | \mathbf{n}'' \rangle$$

$$- \tfrac{1}{2} (\gamma \mathbf{h} \beta p)^2 \langle \mathbf{n} | \tilde{F}'(\beta) | \mathbf{n}'' \rangle + \gamma^2 \beta \langle \mathbf{n} | \alpha | \mathbf{n}'' \rangle \} \langle \mathbf{n}'' | U_p(\beta) | \mathbf{n}' \rangle$$

$$+ \sum_{\mathbf{n}''} \{ i\gamma \mathbf{h} \langle \mathbf{n} | \tilde{F}(\beta) | \mathbf{n}'' \rangle + (\gamma \mathbf{h})^2 \beta p \langle \mathbf{n} | \tilde{F}'(\beta) | \mathbf{n}'' \rangle \}$$

$$\times \langle \mathbf{n}'' | \frac{\partial}{\partial p} U_p(\beta) | \mathbf{n}' \rangle \qquad (202)$$

We now expand $U_p(\beta)$ as a formal series in the parameter γ:

$$U_p(\beta) = \sum_{n=0}^{\infty} \gamma^n U_p^{(m)}(\beta) \qquad (203)$$

Inserting this expression into Eq. (202) and identifying equal powers of γ we readily obtain the explicit from for $U_p^{(m)}(\beta)$. The zero order contribution is simply

$$U_p^{(0)}(\beta) = \exp(-\beta H^t) \qquad (204)$$

where H^t is defined by Eq. (48). To first order in γ, we have

$$\gamma U_p^{(1)}(\beta) = i\gamma \mathbf{h} p \int_0^\beta \exp(-\beta H^t) \beta' \overline{F}(\beta') \, d\beta' + \gamma B \qquad (205)$$

with

$$\overline{F}(\beta) = \exp(\beta H^t) \hat{F} \exp(-\beta H^t) \qquad (206)$$

Finally, the γ^2 term is given by

$$\gamma^2 U_p^{(2)}(\beta) = -(\gamma \mathbf{h} p)^2 \int_0^\beta \int_0^{\beta'} \beta' \overline{F}(\beta') \beta'' \overline{F}(\beta'') \, d\beta' \, d\beta''$$

$$+ \tfrac{1}{2} (\gamma \mathbf{h} p)^2 \int_0^\beta \beta'^2 \overline{F}'(\beta') \, d\beta' + \gamma^2 p C + \gamma^2 D \qquad (207)$$

In these formulae, B, C and D denote p-independent terms we shall not need here.

If we now insert Eqs. (203–207) into definition (121) we see immediately that all terms linear in the momentum p vanish for symmetry reasons, and that all p-independent terms only contribute to a factor A that normalizes the momentum distribution function to unity. We obtain after some trivial manipulations:

$$\phi_0(\gamma; P) = \phi_0^{(0)}(P) + \gamma^2 \left[\frac{p^2}{2kT} \left(\frac{kT_q}{kT} \right) - A \right] \phi_0^{(0)}(p) + O(\gamma^3) \quad (208)$$

where the characteristic quantum temperature T_q is defined by Eq. (124) in Section 4.B.

D. Calculation of the Characteristic Quantum Temperature
$$T_q(\mathbf{h}, T)$$

We have from Eq. (124)

$$kT_q = kT_q^{(a)} + kT_q^{(b)} \quad (209)$$

with

$$kT_q^{(a)} = \beta \mathbf{h}^2/3 \int_0^\beta \langle \overline{F}(0) \overline{F}(\beta') \rangle_f \, d\beta' \quad (210)$$

$$kT_q^{(b)} = -2\beta^{-2} \mathbf{h}^2 \int_0^\beta \int_0^{\beta'} \langle \beta' \overline{F}(\beta') \beta'' \overline{F}(\beta'') \rangle_f \, d\beta' \, d\beta'' \quad (211)$$

where all the symbols have already been defined.

In order to arrive at Eq. (210) we have used the following manipulations:

$$\int_0^\beta \langle \beta'^2 \overline{F}'(\beta') \rangle_f \, d\beta'$$

$$= Z_f^{-1} \mathrm{Tr} \int_0^\beta \beta'^2 \overline{F}'(\beta') \exp(-\beta H^f) \, d\beta'$$

$$= Z_f^{-1} \mathrm{Tr} \int_0^\beta \beta'^2 \exp[-(\beta - \beta') H^f] \hat{F}' \exp(-\beta' H^f) \, d\beta' \quad (212)$$

Because of the cyclic invariance of the trace, this may be written as

$$\int_0^\beta \langle \beta'^2 \overline{F}'(\beta') \rangle_f \, d\beta' = Z_f^{-1} (\beta^3/3) \mathrm{Tr} \, \hat{F}' \exp(-\beta H^f) \quad (213)$$

It is easily verified that the spatial derivative of the force may also be expressed as

$$\hat{F}' = i/\mathbf{h} [\hat{p}_f, \hat{F}] \quad (214)$$

where $\hat{p}_f$ is the momentum fluid operator:

$$\hat{p}_f = \sum_k hk\hat{n}_k \quad (215)$$

We then use the following identity due to Kubo[18] (see text, Section 3.A):

$$[\hat{p}_t, \rho^t] = \int_0^\beta \rho^t \exp(\beta' H^t)[\hat{p}_t, H^t] \exp(-\beta' H^t) \, d\beta'$$

$$= \int_0^\beta \rho^t \exp(\beta' H^t)[\hat{p}_t, V^{tB}] \exp(-\beta' H^t) \, d\beta'$$

$$= \mathbf{h}/i \int_0^\beta \rho^t \exp(\beta' H^t) \hat{F} \exp(-\beta' H^t) \, d\beta' \qquad (216)$$

Combining Eqs. (213), (214) and (216) we immediately obtain the identity between (210) and (212).

Let us now evaluate formally (210) and (211) in the representation that diagonalizes H^t. We introduce the eigenstates $|m_i\rangle$ and eigenvalues E_{m_i} by the equation:

$$H^t |m_i\rangle = E_{m_i} |m_i\rangle \qquad (217)$$

We thus get

$$kT_q^{(a)} = \frac{\beta \mathbf{h}^2}{3 Z_f} \sum_{m_i m_j} |\langle m_i|\hat{F}|m_j\rangle|^2 \exp(-\beta E_{m_i})$$

$$\times \int_0^\beta \exp[\beta'(E_{m_i} - E_{m_j})] \, d\beta'$$

$$= \frac{\beta \mathbf{h}^2}{3 Z_f} \sum_{m_i m_j} |\langle m_i|\hat{F}|m_j\rangle|^2 \frac{\exp(-\beta E_m) - \exp(-\beta E_{m_j})}{E_{m_i} - E_{m_j}} \qquad (218)$$

and

$$kT_q^{(b)} = \frac{-2\mathbf{h}^2}{\beta^2 Z_f} \sum_{m_i m_j} |\langle m_i|\hat{F}|m_j\rangle|^2 \exp(-\beta E_{m_i})$$

$$\times \int_0^\beta \int_0^{\beta'} \beta' \exp[\beta'(E_{m_i} - E_{m_j})] \beta'' \exp[-\beta''(E_{m_i} - E_{m_j})] \, d\beta' \, d\beta''$$

$$= -\frac{\mathbf{h}^2}{\beta^2 Z_f} \sum_{m_i m_j} |\langle m_i|\hat{F}|m_j\rangle|^2 \left[\frac{\beta^3}{3} \frac{\exp(-\beta E_{m_i}) - \exp(-\beta E_{m_j})}{E_{m_i} - E_{m_j}} \right.$$

$$+ \frac{\beta^2}{2} \frac{\exp(-\beta E_{m_i}) + \exp(-\beta E_{m_j})}{(E_{m_i} - E_{m_j})^2}$$

$$\left. - \beta \frac{\exp(-\beta E_{m_i}) - \exp(-\beta E_{m_j})}{(E_{m_i} - E_{m_j})^3} \right] \qquad (219)$$

Combining Eqs. (209), (218) and (219) we obtain

$$kT_{\mathrm{q}} = \frac{\mathbf{h}^2}{\beta Z_{\mathrm{f}}} \sum_{m_i} \sum_{m_j} |\langle m_i | \hat{F} | m_j \rangle|^2 \left[\frac{\beta}{2} \frac{\exp\left(-\beta E_{m_i}\right) + \exp\left(-\beta E_{m_j}\right)}{(E_{m_i} - E_{m_j})^2} \right.$$
$$\left. - \frac{\exp\left(-\beta E_{m_i}\right) - \exp\left(-\beta E_{m_j}\right)}{(E_{m_i} - E_{m_j})^3} \right] \tag{220}$$

From this result, one deduces immediately that in the low temperature limit, kT_{q} tends to a finite value:

$$\lim_{T \to 0} kT_{\mathrm{q}} = \mathbf{h}^2 \sum_{m_i \neq m_0} \frac{|\langle m_0 | \hat{F} | m_i \rangle|^2}{(E_{m_i} - E_{m_0})^2} \tag{221}$$

where m_0 denotes the ground state of the fluid. In the high temperature limit, application of the usual correspondence rules leads to

$$\lim_{T \to \infty} kT_{\mathrm{q}} = 0 \tag{222}$$

In the case of a weakly coupled system, where the fluid–fluid interactions are completely neglected, it is possible to get a compact explicit expression for $kT_{\mathrm{q}}(\mathbf{h}, T)$. We have indeed

$$H^{\mathrm{f}} = H_0^{\mathrm{f}}; \quad |m_i\rangle = |n\rangle \tag{223}$$

and Eq. (220) becomes

$$kT_{\mathrm{q}}^{(2)} = \frac{\mathbf{h}^2}{\Omega^2 \beta} \sum_{k,q} |u(q)|^2 \, \mathbf{q}\mathbf{q} n_k^0 (1 + n_{k+q}^0)$$
$$\times \left\{ \frac{\beta}{2} \frac{1 + \exp\left[\beta(\varepsilon_{k+q} - \varepsilon_k)\right]}{(\varepsilon_{k+q} - \varepsilon_k)^2} - \frac{1 - \exp\left[\beta(\varepsilon_{k+q} - \varepsilon_k)\right]}{(\varepsilon_{k+q} - \varepsilon_k)^3} \right\} \tag{224}$$

where we have introduced the Bose distribution:

$$n_k^0 = \{\exp\left[\beta(\varepsilon_k - \mu)\right] - 1\}^{-1} \tag{225}$$

In order to evaluate the sum involved in (224) we shall neglect the effect of statistics and consider instead the case of a Boltzmann gas. We get then

$$kT_{\mathrm{q}}^{(2)} = \mathscr{R}e \frac{\mathbf{h}^2 \rho}{8\pi^3 \beta} \frac{1}{3} \iint \mathrm{d}^3 k \, \mathrm{d}^3 q \, q^2 |u(q)|^2 f_k^0$$
$$\times \left\{ \frac{\beta}{2} \left[\frac{1}{(\varepsilon_{k+q} - \varepsilon_k - \mathrm{i}0)^2} + \frac{1}{(\varepsilon_{k+q} - \varepsilon_k + \mathrm{i}0)^2} \right] \right.$$
$$\left. - \left[\frac{1}{(\varepsilon_{k+q} - \varepsilon_k - \mathrm{i}0)^3} + \frac{1}{(\varepsilon_{k+q} - \varepsilon_k + \mathrm{i}0)^3} \right] \right\} \tag{226}$$

where f_k^0 is the Maxwellian distribution function with momentum $\mathbf{h}k$. Using polar coordinates for the q integration, Eq. (226) may be written as

$$kT_{\mathrm{q}}^{(2)} = \mathscr{R}e\,\frac{\mathbf{h}^5\,\rho}{3\pi^2\,\beta^2(2\pi kT)^{\frac{3}{2}}} \int \mathrm{d}^3\,k\,\exp\left(-\beta\varepsilon_k\right)\int_0^\infty \mathrm{d}q\,q^4\,|u(q)|^2$$

$$\times \int_{-1}^{+1}\mathrm{d}x\left[\beta^2\,\frac{1}{(\mathbf{h}^2\,qkx+\mathbf{h}^2\,q^2/2-\mathrm{i}0)^2}\right.$$

$$\left.-2\beta\,\frac{1}{(\mathbf{h}^2\,qkx+\mathbf{h}^2\,q^2/2-\mathrm{i}0)^3}\right] \tag{227}$$

After some elementary manipulations, we obtain

$$kT_{\mathrm{q}}^{(2)} = \frac{4\mathbf{h}^2\,\rho}{3\pi\beta^2}\int_0^\infty \mathrm{d}q\,|u(q)|^2\,q^4\,\mathscr{R}e\,\psi(q;\beta) \tag{228}$$

with

$$\psi(q;\beta) = \frac{\mathbf{h}}{q(2\pi kT)^{\frac{3}{2}}}\int_{-\infty}^\infty \mathrm{d}k\,\exp\left(-\beta\varepsilon_k\right)k$$

$$\times\left[\frac{\beta^2}{\mathbf{h}^2\,q^2/2-\mathbf{h}^2\,kq-\mathrm{i}0}-\frac{\beta}{(\mathbf{h}^2\,q^2/2-\mathbf{h}^2\,kq-\mathrm{i}0)^2}\right] \tag{229}$$

This function $\psi(q;\beta)$ may be written in a still more compact form

$$\psi(q;\beta) = \frac{\beta^2}{\mathbf{h}q^2(2\pi kT)^{\frac{3}{2}}}\int_{-\infty}^\infty \mathrm{d}k\,\frac{\exp\left(-\beta\varepsilon_k\right)k}{q/2-k-\mathrm{i}0}$$

$$-\frac{\beta}{\mathbf{h}^3\,q^3(2\pi kT)^{\frac{3}{2}}}\int_{-\infty}^\infty \mathrm{d}k\,\exp\left(-\beta\varepsilon_k\right)k\,\frac{\partial}{\partial k}\left(\frac{1}{q/2-k-\mathrm{i}0}\right)$$

$$\psi(q;\beta) = \frac{\beta^2}{\mathbf{h}^3\,q^3(2\pi kT)^{\frac{3}{2}}}\int_{-\infty}^\infty \mathrm{d}k\,\frac{(\beta^{-1}+\mathbf{h}^2\,k^2)\exp\left(-\beta\varepsilon_k\right)}{q/2-k} \tag{230}$$

If we notice that

$$\mathbf{h}^2\,k^2 f_k^0 = -2\,\frac{\partial}{\partial\beta}f_k^0 \tag{231}$$

we readily obtain

$$\psi(q;\beta) = (2\pi kT\mathbf{h}^2\,q^2)^{-\frac{3}{2}}\left[\beta-2\beta^2\,\frac{\partial}{\partial\beta}\right]\Phi(\alpha) \tag{232}$$

with

$$\Phi(\alpha) = 2\pi^{\frac{1}{2}}\exp\left(-\alpha^2\right)\mathrm{erf}\,\alpha$$

and

$$\alpha \equiv (\mathbf{h}^2\,q^2/8kT)^{\frac{1}{2}} \tag{233}$$

Thus the characteristic temperature $T_q(\mathbf{h}, T)$ becomes in this case:

$$kT_q^{(2)}(\mathbf{h}, T) = \frac{4\rho}{3\sqrt{2\pi^2}} \int_0^\infty |u(q)|^2 B(q, \mathbf{h}; \beta)\, q\, dq \qquad (234)$$

with

$$B(q, \mathbf{h}; \beta) = \frac{\beta^{\frac{1}{2}}}{\mathbf{h}} \left(1 - 2\beta \frac{\partial}{\partial \beta}\right) \exp\left(-\frac{\mathbf{h}^2 q^2}{8kT}\right) \mathrm{erf}\left(\frac{\mathbf{h}^2 q^2}{8kT}\right)^{\frac{1}{2}} \qquad (235)$$

Using known properties of the erf function, one shows easily that

$$\lim_{T\to\infty} B(q, \mathbf{h}; \beta) = \frac{q^3 \beta^2 \mathbf{h}^2}{12\sqrt{2}}; \quad \lim_{T\to 0} B(q, \mathbf{h}; \beta) = \frac{2\sqrt{2}}{\mathbf{h}^2 q} \qquad (236)$$

Thus, we obtain

$$\lim_{T\to 0} kT_q^{(2)} = \frac{\rho \mathbf{h}^2}{18\pi^2 (kT)^2} \int_0^\infty |u(q)|^2 q^4\, dq \qquad (237)$$

which tends to zero as $T\to\infty$.

While

$$\lim_{T\to 0} kT_q^{(2)} = \frac{8\rho}{3\pi^2 \mathbf{h}^2} \int_0^\infty |u(q)|^2\, dq \qquad (238)$$

which is a finite, temperature-independent quantity.

E. Explicit Formulation of the Correction $\Omega_A^{(4)}(0)$

We shall show here how the operator $\Omega_A^{(4)}(0)$ as given by definition (140) can be cast into the explicit form (142).

First we analyse the irreducibility condition in the operator $\Omega_A^{(4)}(0)$. Let us write down Eq. (140) more explicitly as

$$\Omega_A^{(4)}(0) = \lim_{\varepsilon\to 0} \sum_N \langle 0| \gamma \mathscr{U}^{\mathrm{I}} \left[\frac{1}{\mathscr{H}_0^{\mathrm{f}} - i\varepsilon} + \frac{1}{\mathscr{H}_0^{\mathrm{f}} - i\varepsilon}(-\mathscr{V}^{\mathrm{f}})\frac{1}{\mathscr{H}_0^{\mathrm{f}} - i\varepsilon} + \dots\right] \gamma \Gamma^{\mathrm{I}}$$

$$\times \left[\frac{1}{\mathscr{H}_0^{\mathrm{f}} - i\varepsilon} + \frac{1}{\mathscr{H}_0^{\mathrm{f}} - i\varepsilon}(-\mathscr{V}^{\mathrm{f}})\frac{1}{\mathscr{H}_0^{\mathrm{f}} - i\varepsilon} + \dots\right] \gamma \Gamma^{\mathrm{I}}$$

$$\times \left[\frac{1}{\mathscr{H}_0^{\mathrm{f}} - i\varepsilon} + \frac{1}{\mathscr{H}_0^{\mathrm{f}} - i\varepsilon}(-\mathscr{V}^{\mathrm{f}})\frac{1}{\mathscr{H}_0^{\mathrm{f}} - i\varepsilon} + \dots\right] \gamma \Gamma^{\mathrm{I}} \rho^{\mathrm{f}}(\mathbf{N}) |0\rangle' \qquad (239)$$

We now consider the *remaining* terms (with intermediate state $|0\rangle$).

If we remark that a state $|\kappa \mathbf{v}\rangle = 0$ in the first or in the last bracketed expression for the fluid propagator gives no contribution because

$$\langle \hat{F} \rangle = 0 \qquad (240)$$

we may conclude that the most general term in (239) involving an arbitrary number of state $|\kappa\nu\rangle = 0$ is

$$
\Omega_A^{(4)\,\text{red}}(0) = \lim_{\epsilon \to 0} \sum_N \langle 0 | \gamma \mathscr{U}^I \frac{1}{\mathscr{H}^f - i\epsilon} \gamma \Gamma^I \sum_{n=0}^{\infty} \left[\frac{1}{\mathscr{H}_0^f - i\epsilon}(-\mathscr{V}^f) \right]^n | 0 \rangle'
$$

$$
\times \frac{1}{-i\epsilon} \sum_{m=0}^{\infty} \left\{ \langle 0 | -\mathscr{V}^f \sum_{p=0}^{\infty} \left[\frac{1}{\mathscr{H}_0^f - i\epsilon}(-\mathscr{V}^f) \right]^p | 0 \rangle' \frac{1}{-i\epsilon} \right\}^m | 0 \rangle'
$$

$$
\times \langle 0 | \sum_{r=0}^{\infty} \left[-\mathscr{V}^f \frac{1}{\mathscr{H}_0^f - i\epsilon} \right]^n \gamma \Gamma^I \frac{1}{\mathscr{H}^f - i\epsilon} \gamma \Gamma^I \rho^f(\mathbf{N}) | 0 \rangle'
$$

$$
(241)
$$

We have thus shown that the only state in which the irreducibility condition has to be taken into account explicitly in the definition of $\Omega_0^{(4)}(0)$ is just before the third operator $\gamma \Gamma^I$ in (239): that is, $\Omega_A^{(4)\,\text{red}}(0)$ includes all *possible reducible transitions*.

This general expression simplifies considerably if we neglect contributions of order (N^{-1}). This simplification will be carried out by using the following property (which has been shown explicitly in reference 35, for instance): sequences of diagonal fragments $\psi_{00}^+(0)$ only give finite contributions if they are *semi-connected*, i.e. if they have only one particle in common with the *preceding* diagonal fragments.

This semi-connection condition obviously implies the following distinction:

(1) The class of terms where the Brownian particle interacts with one *single group* of fluid particles denoted by the subscript $(\alpha) = (1_\alpha, 2_\alpha, 3_\alpha, ...)$ for instance. A group of particles being defined as a set of particles which are mutually interacting, either in the dynamics (represented by the propagators) or in the equilibrium distribution [denoted by $\rho^f(\mathbf{N})$].

(2) The class of terms where the Brownian particle interacts with *two distinct groups* of fluid molecules denoted by $(\alpha) = (1_\alpha, 2_\alpha, 3_\alpha, ...)$, $(\beta) = (1_\beta, 2_\beta, 3_\beta, ...)$ which are thus mutually disconnected.* Moreover the following properties are evidently valid for any term of this class:

$$
\mathscr{H}^f = \mathscr{H}_{(\alpha)}^f + \mathscr{H}_{(\beta)}^f \tag{242}
$$

$$
\rho^f(\mathbf{N}) = \rho_{(\alpha)}^f \, \rho_{(\beta)}^f \tag{243}
$$

$$
\mathscr{H}_{(\alpha)}^f \, \rho_{(\alpha)}^f = 0 \tag{244}
$$

* It is obvious that this definition could be generalized to three, four, ... distinct groups of particles, if necessary.

With these remarks in mind one can write $\Omega_A^{(4)\,\text{red}}(0)$ in the following form (notice that the first and the third terms have already the Brownian particle in common!):

$\Omega_A^{(4)\,\text{red}}(0)$

$$= \lim_{\varepsilon \to 0} \sum_{N_\alpha} \sum_{N_\beta} \langle 0| \gamma \mathcal{U}_{(\alpha)}^{\text{I}} \frac{1}{\mathcal{H}_{(\alpha)}^{\text{f}} - i\varepsilon} \gamma \Gamma_{(\alpha)}^{\text{I}} \left\{ \sum_{n=0}^{\infty} \left[\frac{1}{\mathcal{H}_0^{\text{f}} - i\varepsilon}(-\mathcal{V}^{\text{f}}) \right]^n \right\}_{(\alpha)} |0\rangle_{(1)}$$

$$\times \frac{1}{-i\varepsilon} \sum_{m=0}^{\infty} \left[\langle 0| -\mathcal{V}_{(\alpha)}^{\text{f}} \left\{ \sum_{p=0}^{\infty} \left[\frac{1}{\mathcal{H}_0^{\text{f}} - i\varepsilon}(-\mathcal{V}^{\text{f}}) \right]^p \right\}_{(\alpha)} |0\rangle' \frac{1}{-i\varepsilon} \right]^m |0\rangle_{(1)}$$

$$\times \langle 0| \left\{ \sum_{r=0}^{\infty} \left[-\mathcal{V}^{\text{f}} \frac{1}{\mathcal{H}_0^{\text{f}} - i\varepsilon} \right]^r \right\}_{(\beta)} \gamma \Gamma_{(\beta)}^{\text{I}} \frac{1}{\mathcal{H}_{(\beta)}^{\text{f}} - i\varepsilon} \gamma \Gamma_{(\beta)}^{\text{I}} \rho_{(\alpha)}^{\text{f}} \rho_{(\beta)}^{\text{f}} |0\rangle_{(1)}$$

$$\tag{245}$$

Here the subscript (1) specifies that each of the two subcontributions involves one single group, as defined above, which we denote respectively by the subscript (α) and (β). Now if one sums on $\mathbf{N}_\alpha$ and $\mathbf{N}_\beta$ separately and uses Eq. (64) (recall also the definition of $\mathcal{V}^{\text{f}}$) one readily remarks that the only non-vanishing contribution to (245) corresponds to

$$r = 0 \quad \text{and} \quad m = 0 \tag{246}$$

We are thus led to

$\Omega_A^{(4)\,\text{red}}(0)$

$$= \lim_{\varepsilon 0 \leftarrow} \sum_{N_\alpha} \langle 0| \gamma \mathcal{U}_{(\alpha)}^{\text{I}} \frac{1}{\mathcal{H}_{(\alpha)}^{\text{f}} - i\varepsilon} \gamma \Gamma_{(\alpha)}^{\text{I}} \left\{ \sum_{n=0}^{\infty} \left[\frac{1}{\mathcal{H}_0^{\text{f}} - i\varepsilon}(-\mathcal{V}^{\text{f}}) \right]^n \right\}_{(\alpha)} \rho_{(\alpha)}^{\text{f}} |0\rangle'$$

$$\times \frac{1}{-i\varepsilon} \sum_{N_\beta} \langle 0| \gamma \mathcal{U}_{(\beta)}^{\text{I}} \frac{1}{\mathcal{H}_{(\beta)}^{\text{f}} - i\varepsilon} \gamma \Gamma_{(\beta)}^{\text{I}} \rho_{(\beta)}^{\text{f}} |0\rangle' \tag{247}$$

If we now use Eq. (244) in connection with the following property:

$$\lim_{\varepsilon \to 0} \sum_{n=0}^{\infty} \left[\frac{1}{\mathcal{H}_0^{\text{f}} - i\varepsilon}(-\mathcal{V}^{\text{f}}) \right]^n |0\rangle' \frac{1}{-i\varepsilon} \rho_0^{\text{f}}(\mathbf{N})$$

$$= \frac{1}{\mathcal{H}^{\text{f}} - i0} |0\rangle' \rho_0^{\text{f}}(\mathbf{N})$$

$$= \frac{1}{\mathcal{H}^{\text{f}} - i0} \rho^{\text{f}}(\mathbf{N}) |0\rangle - \frac{1}{\mathcal{H}^{\text{f}} - i0} \rho^{\text{f}}(\mathbf{N}) |0\rangle' \tag{248}$$

where the second (dashed, $'$,) term eliminates all Fourier components of $\rho^\dagger$ except the one with $\sum\limits_k \nu_k = 0$.

$$\gamma^4 \, \Omega_A^{(4)\,\mathrm{red}}(0) = \lim_{z \to \mathrm{i}0} \left[\frac{1}{z} \Omega^{(2)}(z)\, \Omega^{(2)}(z) - \overline{\Omega}^{(2)}(z)\, \Omega^{(2)}(z) \right] \quad (249)$$

where $\Omega^{(2)}(z)$ is defined by Eq. (135) and $\overline{\Omega}^{(2)}(z)$ by

$$\overline{\Omega}^{(2)}(z) = \sum_N \langle 0 | \gamma \mathcal{U}^{\mathrm{I}} \frac{1}{\mathscr{H}^\dagger - z} \gamma \Gamma^{\mathrm{I}} \frac{1}{\mathscr{H}^\dagger - z} \rho^\dagger(\mathbf{N}) | 0 \rangle' \quad (250)$$

Let us now turn back to $\Omega_A^{(4)}(0)$ as defined by Eq. (140) [or Eq. (239)]. The first step consists of eliminating the awkward irreducibility condition (the dash) in Eq. (140) and subtracting instead explicitly the contribution which has been added in this way: that is the term $\Omega_A^{(4)\,\mathrm{red}}(0)$ we have just analysed and cast into the form (249). Thus we write

$$\gamma^4 \, \Omega_A^{(4)}(0) = \gamma^4 \, \overline{\Omega}^{(4)}(0) + \gamma^4 \, \Omega_A^{(4)\,\mathrm{red}}(0)$$

$$= \gamma^4 \, \overline{\Omega}^{(4)}(0) - \lim_{\varepsilon \to 0} \Omega^{(2)}(0) \frac{1}{-\mathrm{i}\varepsilon} \Omega^{(2)}(0) + \overline{\Omega}^{(2)}(0)\, \Omega^{(2)}(0)$$

$$(251)$$

where

$$\overline{\Omega}^{(4)}(0) = \sum_N \langle 0 | \gamma \mathcal{U}^{\mathrm{I}} \frac{1}{\mathscr{H}^\dagger - \mathrm{i}0} \gamma \Gamma^{\mathrm{I}} \frac{1}{\mathscr{H}^\dagger - \mathrm{i}0} \gamma \Gamma^{\mathrm{I}} \frac{1}{\mathscr{H}^\dagger - \mathrm{i}0} \gamma \Gamma^{\mathrm{I}} \rho^\dagger(\mathbf{N}) | 0 \rangle \quad (252)$$

with no irreducibility condition on the intermediate states (no dash). It should be stressed here that however both first terms of Eq. (251) diverge in the limit $\varepsilon \to 0$, their sum remains finite. Moreover, because of the irreducibility condition (dash) appearing in definition (249), the correction $\overline{\Omega}^{(2)}(0)$ is finite.

We now give a more explicit expression for $\overline{\Omega}^{(4)}(0)$. Clearly this operator describes interactions between the fluid and the Brownian particle as well as interactions between the fluid molecules themselves. Using the notion of *group* of particles, we introduced above, we may classify the contributions to $\overline{\Omega}^{(4)}(z)$ in the following way:

$$\overline{\Omega}^{(4)}(z) = \sum_{(\alpha)} \overline{\Omega}_{(\alpha)}^{(4)}(z) + \sum_{(\alpha)} \sum_{(\beta)} \overline{\Omega}_{(\alpha)(\beta)}^{(4)} \quad (253)$$

that is: (1) we first consider the class of terms where the Brownian particle interacts with one single group. This class corresponds to

$$\overline{\Omega}_{(\alpha)}^{(4)}(z) = \sum_N \langle 0 | \gamma \mathcal{U}_{(\alpha)}^{\mathrm{I}} \frac{1}{\mathcal{H}_{(\alpha)}^{\mathrm{f}} - z} \gamma \Gamma_{(\alpha)}^{\mathrm{I}} \frac{1}{\mathcal{H}_{(\alpha)}^{\mathrm{f}} - z} \gamma \Gamma_{(\alpha)}^{\mathrm{I}}$$

$$\times \frac{1}{\mathcal{H}_{(\alpha)}^{\mathrm{f}} - z} \gamma \Gamma_{(\alpha)}^{\mathrm{I}} \rho_{(\alpha)}^{\mathrm{f}}(\mathbf{N}) | 0 \rangle_{(1)} \tag{254}$$

(2) We then take the class of terms when the Brownian particle interacts with two distinct groups denoted by (α) and (β) which are mutually disconnected. Let us stress about this class of terms that each group of fluid particles has to involve two γ dependent vertices of $\overline{\Omega}^{(4)}(z)$ [i.e. $\gamma \mathcal{U}^{\mathrm{I}}$ or $\gamma \Gamma^{\mathrm{I}}$ in Eq. (252)] otherwise the contribution vanishes for symmetry reasons [see (240)]. Thus all contributions to $\overline{\Omega}^{(4)}(z)$ involving more than two groups of fluid particles vanish:

$$\overline{\Omega}_{(\alpha)(\beta)(\gamma)}^{(4)} = 0 \tag{255}$$

For the same reason, we have simply

$$\Omega^{(2)} = \sum_{(\alpha)} \Omega_{(\alpha)}^{(2)} \tag{256}$$

for the γ^2 operator.

It is then a simple matter to enumerate the various contributions of this class, which we denoted by $\overline{\Omega}_{(\alpha)(\beta)}^{(4)}(z)$. We can write

$$\gamma^4 \overline{\Omega}_{(\alpha)(\beta)}^{(4)}(z) = \gamma^4 \overline{\overline{\Omega}}_{(\alpha,\beta)}^{(4)}(z) - 1/z \, \Omega_{(\alpha)}^{(2)}(z) \, \Omega_{(\beta)}^{(2)}(z) \tag{257}$$

with

$$\gamma^4 \overline{\overline{\Omega}}_{(\alpha,\beta)}^{(4)}(z) = \sum_{N_\alpha} \sum_{N_\beta} \langle 0 | \gamma \mathcal{U}_{(\alpha)}^{\mathrm{I}} \frac{1}{\mathcal{H}_{(\alpha)}^{\mathrm{f}} - z} \gamma \mathcal{U}_{(\beta)}^{\mathrm{I}} \frac{1}{\mathcal{H}_{(\alpha)}^{\mathrm{f}} + \mathcal{H}_{(\beta)}^{\mathrm{f}} - z}$$

$$\times \gamma \Gamma_{(\alpha)}^{\mathrm{I}} \frac{1}{\mathcal{H}_{(\beta)}^{\mathrm{f}} - z} \gamma \Gamma_{(\beta)}^{\mathrm{I}} \rho_{(\alpha)}^{\mathrm{f}} \rho_{(\beta)}^{\mathrm{f}} | 0 \rangle_{(2)}$$

$$+ \sum_{N_\alpha} \sum_{N_\beta} \langle 0 | \gamma \mathcal{U}_{(\alpha)}^{\mathrm{I}} \frac{1}{\mathcal{H}_{(\alpha)}^{\mathrm{f}} - z} \gamma \mathcal{U}_{(\beta)}^{\mathrm{I}} \frac{1}{\mathcal{H}_{(\alpha)}^{\mathrm{f}} + \mathcal{H}_{(\beta)}^{\mathrm{f}} - z}$$

$$\times \gamma \Gamma_{(\beta)}^{\mathrm{I}} \frac{1}{\mathcal{H}_{(\alpha)}^{\mathrm{f}} - z} \gamma \Gamma_{(\alpha)}^{\mathrm{I}} \rho_{(\alpha)}^{\mathrm{f}} \rho_{(\beta)}^{\mathrm{f}} | 0 \rangle_{(2)} \tag{258}$$

Here the subscript (2) specifies that the terms of the right-hand side only involve two distinct groups of fluid particles, as defined above, which we label (α) and (β). Combining Eqs. (251) to (257) we arrive at

$$\gamma^4 \, \Omega_A^{(4)}(0) = \gamma^4 \sum_{(\alpha)} \overline{\Omega}_{(\alpha)}^{(4)}(0) + \gamma^4 \sum_{(\alpha)} \sum_{(\beta)} \overline{\overline{\Omega}}_{(\alpha,\beta)}^{(4)}(0) + \gamma^4 \, \overline{\Omega}^{(2)}(0) \, \Omega^{(2)}(0) \quad (259)$$

In this expression, each term has a well-defined limit when $z \to i0$.

The argument leading to Eq. (259) is of course extremely formal yet this result is easily verified on simple examples using the diagram technique of Prigogine and Balescu.[11-13, 35] For instance, we have explicitly verified this formula by a perturbation calculus up to fourth order in the coupling parameter λ. The calculations are very long but offer no special difficulty and will not be reproduced here.

Our last step toward the complete analysis of the fourth order correction $\Omega_A^{(4)}(0)$ consists in showing that the term $\overline{\overline{\Omega}}_{(\alpha,\beta)}^{(4)}(0)$ may be factorized as a product of two independent terms, depending respectively on (α) and on (β).

Using definitions (55) and (58), it is a matter of some simple algebra to cast Eq. (258) into the following form:

$$\gamma^4 \, \overline{\overline{\Omega}}_{(\alpha,\beta)}^{(4)}(0) = \lim_{\epsilon \to 0} \Bigg\{ \sum_{N_\alpha} \sum_{N_\beta} \gamma^4 \langle 0 | \mathscr{F}_{(\alpha)}^+ \frac{\partial}{\partial p} \mathscr{F}_{(\beta)}^+ \frac{\partial}{\partial p} \frac{1}{(\mathscr{H}_{(\alpha)}^f - i\varepsilon)^2 (\mathscr{H}_{(\beta)}^f - i\varepsilon)}$$

$$\times \, \Gamma_{(\alpha)}^I \, \Gamma_{(\beta)}^I \, \rho_{(\alpha)}^f \, \rho_{(\beta)}^f | 0 \rangle + \sum_{N_\alpha} \sum_{N_\beta} \gamma^4 \langle 0 | \mathscr{F}_{(\alpha)}^+ \frac{\partial}{\partial p} \mathscr{F}_{(\beta)}^+ \frac{\partial}{\partial p}$$

$$\times \left[\frac{1}{(\mathscr{H}_{(\alpha)}^f - i\varepsilon)^2} - \frac{1}{(\mathscr{H}_{(\beta)}^f - i\varepsilon)^2} \right] \frac{1}{\mathscr{H}_{(\alpha)}^f + \mathscr{H}_{(\beta)}^f - i\varepsilon}$$

$$\times \, \mathscr{F}_{(\alpha)}^+ \, \hat{\kappa}_{(\beta)} \, \rho_{(\alpha)}^f \, \rho_{(\beta)}^f | 0 \rangle \Bigg\} \quad (260)$$

where we have used the operator identity

$$[\Gamma_{(\alpha)}^I, \Gamma_{(\beta)}^I] = \hat{\kappa}_{(\alpha)} \mathscr{F}_{(\alpha)}^+ - \mathscr{F}_{(\beta)}^+ \hat{\kappa}_{(\beta)} \quad (261)$$

in commuting the p dependent factors in Eq. (258). We then apply the following formula (valid in the limit $\varepsilon \to 0$):

$$\frac{1}{\mathscr{H}_{(\alpha)}^f + \mathscr{H}_{(\beta)}^f - i\varepsilon} \left(\frac{1}{\mathscr{H}_{(\alpha)}^f - i\varepsilon} - \frac{1}{\mathscr{H}_{(\beta)}^f - i\varepsilon} \right) = \frac{1}{(\mathscr{H}_{(\alpha)}^f - i\varepsilon)(\mathscr{H}_{(\beta)}^f - i\varepsilon)}$$

$$(262)$$

and take account of the fact that the (α) dependent and (β) dependent quantities may be calculated separately. Finally we remark that

$$\left(\mathscr{F}^+_{(\beta)}\frac{\partial}{\partial p}, \Gamma^I_{(\alpha)}\right) = \mathscr{F}^+_{(\beta)}\hat{\kappa}_{(\alpha)} \tag{263}$$

After simple but tedious manipulations, $\overline{\overline{\Omega}}^{(4)}_{(\alpha,\beta)}(0)$ can be considerably simplified. Taking the sum over all possible groups (α) and (β) we obtain simply

$$\sum_{(\alpha)}\sum_{(\beta)}\overline{\overline{\Omega}}^{(4)}_{(\alpha,\beta)}(0) = \Omega'^{(2)}(0)\,\Omega^{(2)}(0) + \bar{\chi}\Omega^{(2)}(0) + \zeta\frac{\partial^2}{\partial p^2}[kT(\bar{\chi}-\chi/kT)] \tag{264}$$

where $\Omega^{(2)}(0)$, $\Omega'^{(2)}(0)$, χ, $\bar{\chi}$ and ζ are respectively defined in the text by Eqs. (135) and (144) to (147).

Finally combining Eqs. (259) and (264) one immediately sees that $\Omega^{(4)}_A(0)$ can be cast into the form (142).

F. Evaluation of χ and $\bar{\chi}$

We want to demonstrate here Eq. (157). In order to calculate χ, as given by Eq. (144), we use Eqs. (60), (73) and the well-known formula:

$$\lim_{\epsilon\to 0}\frac{1}{(x-i\epsilon)^2} = \lim_{\epsilon\to 0}\int_0^\infty \tau\exp[-i(x-i\epsilon)\tau]\,d\tau \tag{265}$$

We get then

$$\chi = -\tfrac{1}{2}\lim_{\epsilon\to 0}\int_0^\infty \tau\,\mathrm{Tr}[\hat{F}\exp(-iH^t\tau)(\hat{F}\rho^t + \rho^t\hat{F})$$
$$\times \exp(iH^t\tau)]\exp(-\epsilon\tau)\,d\tau \tag{266}$$

Working in the representation (217) that diagonalizes H^t, we have

$$\chi = Z_t^{-1}\mathscr{R}e\lim_{\epsilon\to 0}\int_0^\infty \tau\sum_{m_i m_j}|\langle m_i|\hat{F}|m_j\rangle|^2\exp(-\beta E_{m_i})$$
$$\times \exp[-i(E_{m_i}-i\epsilon)\tau]\exp[i(E_{m_j}-i\epsilon)\tau]\,d\tau \tag{267}$$

$$\chi = -Z_t^{-1}\mathscr{R}e\lim_{\epsilon\to 0}\sum_{m_i m_j}|\langle m_i|\hat{F}|m_j\rangle|^2\frac{1}{(E_m - E_{m_j} - i\epsilon)^2}\exp(-\beta E_{m_j}) \tag{268}$$

Similarly one finds from Eqs. (145) and (74b)

$$\bar{\chi} = -Z_{\mathrm{f}}^{-1} \mathscr{R}e \lim_{\varepsilon \to 0} \sum_{m_i m_j} |\langle m_i | \hat{F} | m_j \rangle|^2 \frac{1}{(E_{m_i} - E_{m_j} - i\varepsilon)^2}$$

$$\times \frac{\exp(-\beta E_{m_i}) - \exp(-\beta E_{m_j})}{E_{m_i} - E_{m_j}} \tag{269}$$

Combining Eqs. (268) and (269) we get (setting now explicitly the **h** factors):

$$\bar{\chi} - \chi/kT = -\frac{\mathbf{h}^2}{Z_{\mathrm{f}}} \sum_{m_i m_j} |\langle m_i | \hat{F} | m_j \rangle|^2 \left[\frac{\exp(-\beta E_{m_i}) - \exp(-\beta E_{m_j})}{(E_{m_i} - E_{m_j})^3} \right.$$

$$\left. - \frac{\beta}{2} \frac{\exp(-\beta E_{m_i}) + \exp(-\beta E_{m_j})}{(E_{m_i} - E_{m_j})^2} \right] \tag{270}$$

where we have used the fact that expressions (268) and (269) remain regular for $\varepsilon \to 0$.

Comparing with Eq. (220), we thus establish Eq. (157).

G. Proof of Equation (165)

The direct proof of Eq. (164) seems very difficult and we have not been able to verify this result in a direct manner. The main reason for this lies in the fact that there is no simple way to work with compact expressions [as for instance the operator $(\mathscr{H}^{\mathrm{f}} - i0)^{-1}$] while keeping the condition that only one group of particles should appear.

We may, however, use a formal argument, based on the remark that the total equilibrium density matrix ρ^{eq} of both the Brownian particle and the fluid is the solution of the Von Neumann–Liouville equation:

$$\mathscr{H} \rho^{\mathrm{eq}} = 0 \tag{271}$$

which tends to the factorized form:

$$\rho^{\mathrm{eq}} \to \rho^{\mathrm{f}} \phi_0^{(0)}(P) \delta_{P,P'}^{\mathrm{Kr}} \tag{272}$$

in the limit where $\gamma \to 0$.

We write Eq. (271) as

$$(\mathscr{H}^{\mathrm{f}} + \gamma \mathscr{H}_0^{\mathrm{I}} + \gamma \mathscr{U}^{\mathrm{I}} + \gamma^2 \mathscr{U}^{\mathrm{II}} + \gamma^3 \mathscr{U}^{\mathrm{III}} + \dots) \rho^{\mathrm{eq}} = 0 \tag{273}$$

and obtain from it the following integral equation:

$$\rho^{eq} = \rho^f \phi_0^{(0)}(P)\delta_{P,P'}^{Kr} + \frac{1}{\mathscr{H}^f - i0}(\gamma \mathscr{H}_0^I + \gamma \mathscr{U}^I + \gamma^2 \mathscr{U}^{II} + \gamma^3 \mathscr{U}^{III} + \dots)\rho^{eq}$$

(274)

Equation (274) is the analogue in the present problem of similar integral equations derived in reference 34 for deriving an H theorem, and we may apply here the formalism developed in these latter cases:

(1) Iterate this equation and separate off explicitly irreducible contributions.

(2) Take the trace with respect to the fluid variables.

If one considers separately the contributions of order γ^4 involving only one group of particles, we are then led to (164). We shall, however, omit this calculation here.

Finally, let us mention that we have also checked Eq. (165) directly to lowest order in the perturbation calculus.

References

1. For details and references see Chandrasekhar, S., *Rev. Mod. Phys.* **15**, 1 (1943); Wang, M. C., and Uhlenbeck, G. E., *Rev. Mod. Phys.* **17**, 323 (1945); see also *Selected Papers and Noise and Stochastic Processes*, Wax, N., Ed., Dover, 1954.
2. Lebowitz, J. L., and Rubin, E., *Phys. Rev.* **131**, 2381 (1963).
3. Résibois, P., and Davis, H. T., *Physica* **30**, 1077 (1964).
4. Lebowitz, J. L., and Résibois, P., *Phys. Rev.* **139**, A1101 (1965).
5. Dagonnier, R., Ph.D. Thesis, Free University, Brussels, 1964.
6. Davis, H. T., Hirotte, K., and Rice, S., *J. Chem. Phys.* **43**, 2623 (1965).
7. Frisch, H. L., and McKenna, J., *Phys. Letters* **19**, 112 (1965); *Phys. Rev.* **145**, 93 (1966).
8. Dagonnier, R., and Résibois, P., *Bull. Acad. Sci. Belg.* **52**, 299 (1966).
9. Davis, H. T., and Dagonnier, R., *J. Chem. Phys.* **44**, 4030 (1966).
10. Résibois, P., and Dagonnier, R., *Phys. Letters* **22**, 252 (1966); *Bull. Acad. Sc. Belg.* **52**, 1475 (1966).
11. Prigogine, I., *Non-equilibrium Statistical Mechanics*, Interscience–John Wiley, New York, 1962.
12. Balescu, R., *Statistical Mechanics of Charged Particles*, Interscience–John Wiley, New York, 1964.
13. Résibois, P., *A Perturbative Approach to Irreversible Statistical Mechanics in Many Particle Physics*, Meeron, E., Ed., Gordon and Breach, New York, 1967.

14. Meyer, L., and Reif, F., *Phys. Letters* **5**, 1 (1960); *Phys. Rev.* **123**, 727 (1964).
15. Meyer, L., Rice, S. A., and Davis, H. T., *Proc. 18th Int. Conf. Low Temperature Physics*, Butterworth, London, 1963.
16. Résibois, P., *Physica* **27**, 541 (1961).
17. Prigogine, I., and Résibois, P., *Physica* **27**, 629 (1961).
18. Kubo, R., *J. Phys. Soc. Japan* **12**, 570.
19. Verboven, E., *Physica* **26**, 1091 (1960).
20. Chapman, S., and Cowling, T. G., *The Mathematical Theory of Non-uniform Gases*, Cambridge University Press, New York, 1939.
21. Bogoliubov, N. N., *J. Phys. U.S.S.R.* **II**, 23 (1947).
22. Abe, R., and Aïzu, K., *Phys. Rev.* **123**, 10 (1963).
23. Dahm, A. J., and Sanders, T. M., *Phys. Rev. Letters* **17**, 126 (1966).
24. Parks, P. E., and Donnelly, R. J., *Phys. Rev. Letters* **16**, 45 (1966).
25. Kuper, C. G., "Liquid Helium", in *Proceedings of the Enrico Fermi International School of Physics Course XXI*, Carin, G., Ed., Academic Press, New York, 1963.
26. Meyer, L., Davis, H. T., Rice, S. A., and Donnelly, R. J., *Phys. Rev.* **126** (1962).
27. Clark, R. C., *Proc. Phys. Soc. (London)* **82**, 785 (1963).
28. Uehling, E. A., and Uhlenbeck, G. E., *Phys. Rev.* **43**, 552 (1933).
29. Helfand, E., *Phys. Fluids* **4**, 1 (1961).
30. Mazo, R. M., and Kirkwood, J. G., *J. Chem. Phys.* **28**, 644 (1958).
31. Résibois, P., *Phys. Rev.* **138**, B281 (1965); *Bull. Acad. Sci. Belg.* **51**, 1288 (1964); Watabe, M., and Dagonnier, R., *Phys. Rev.* **143**, 40 (1965).
32. Prigogine, I., *Physica* **32**, 1828 (1966); *Physica* **32**, 1873 (1966).
33. Résibois, P., *Physica* **31**, 645 (1965); *Physica* **32**, 1473 (1966).
34. Résibois, P., *Physica* **27**, 241 (1961).
35. Prigogine, I., and Balescu, R., *Physica* **25**, 281 (1959).

ORIENTATION OF TARGETS BY BEAM EXCITATION

RICHARD BERSOHN, *Department of Chemistry, Columbia University, New York, N.Y. 10027*

and

SHENG H. LIN, *Department of Chemistry, Arizona State University, Tempe, Arizona 85281*

CONTENTS

1. INTRODUCTION

A. Processes of Orientation

There is a large class of phenomena in which a beam of incident particles is directed onto a target of particles which are oriented isotropically. As a result of collisions with the beam particles the excited target particles are anisotropically distributed. The net result is that the beam gains entropy at the expense of the excited target particles which become orientated ("pumped") to some degree. The beam particles can be photons, electrons, neutrons, atoms and ions and the target particles can be atoms, nuclei, molecules and ions. The orientation of the target particles is of no particular interest unless it can be observed in some way; often this cannot be done unless secondary particles (photons, atoms, electrons, ions) are emitted from the orientated excited target particles. The orientation caused by the collision is manifested in an anisotropy of direction of emission and sometimes an orientation of the intrinsic angular momentum of the particle which is emitted. This anisotropy and polarization can be diminished either by relaxation processes within the target medium or by imposing a magnetic field or a resonant radio-frequency. The overall conceptual scheme can be represented by four processes:

$$B + T \text{ (isotropic)} \longrightarrow T^* \text{ (anisotropic)} \qquad \text{Orientation} \qquad (1)$$

$$T^* \text{ (anisotropic)} \xrightarrow[\text{laboratory fields}]{\text{natural or}} T^* \text{ (isotropic)} \qquad \text{Relaxation or resonance in excited states} \qquad (2)$$

$$T^* \rightarrow T' + P \qquad \text{Dissociation} \qquad (3)$$

$$T' \text{ (anisotropic)} \xrightarrow[\text{laboratory fields}]{\text{natural or}} T' \text{ (isotropic)} \qquad \text{Relaxation or resonance in final state} \qquad (4)$$

T' may or may not be identical with T.

In Eqs. (1)–(4), B represents a beam particle, T, a target particle, T^* an excited target particle, P an emitted particle, and T' the final state of T. The distinction between these four processes and a one-step scattering process, $B + T \rightarrow T' + P$ is the existence of an intermediate T^* whose properties can be investigated.

In this paper we will review a number of familiar experiments characterized by different beams, different targets and different emitted particles. However, all the experiments will be seen to be versions of the same scheme. Once this is realized it is easy to write down examples of completely new experiments. The point of these experiments is to obtain: (1) selection rules for the orientation process, (2) the natural relaxation times for the orientation in the ground and excited states, and (3) the magnetic resonance spectra of the orientated ground and excited states.

B. Description of the Orientation

In orientation experiments we are primarily concerned with the relative occupation probability $W_{J'M'}$ of different sublevels M' of the excited state J' where

$$\sum_{M'=-J'}^{J'} W_{J'M'} = 1; \quad W_{J'M'} = \frac{1}{2J'+1}$$

represent isotropy. To obtain these relative probabilities we write down rate equations for the populations $n_{J'M'}$ of the excited states and n_{JM} of the ground states:

$$\frac{dn}{dt} n_{J'M'} = \sum_{J,M} \mathcal{O}_{JM \to J'M'} n_{JM} - \sum_{M''} \mathcal{R}_{J'M' \longleftrightarrow J'M''}(n_{J'M'} - n_{J'M''})$$
$$- \sum_{J,M} \mathcal{D}_{J'M' \to JM} n_{J'M'} \tag{5}$$

The three terms on the right-hand side of this equation correspond respectively to the orientation, relaxation and dissociation process of Eqs. (1)–(3). The matrices $\mathcal{O}_{JM \to J'M'}$, $\mathcal{R}_{J'M' \longleftrightarrow J'M'}$ and $\mathcal{D}_{J'M' \to JM}$ are the generalized rate constants.

One simplification can be made at once. The excitation process represented by $\mathcal{O}_{JM \to J'M'}$ is anisotropic. On the other hand, assuming that the medium is isotropic and that no external fields are present, the dissociation process is isotropic. The term $\sum_{J,M} \mathcal{D}_{J'M' \to JM}$ is the probability of dissociation/unit time of a system in the state M'; this dissociation probability cannot depend on the choice of axis of quantization. Therefore, in general, we can write

$$\sum_{J,M} \mathcal{D}_{J'M' \to JM} = w = \frac{1}{\tau} \tag{6}$$

where τ is the lifetime of the excited state (independent of M').

By using the relations

$$W_{J'M'} = \frac{n_{J'M'}}{\sum\limits_{M'=-J'}^{J'} n_{J'M'}}; \quad W_{JM} = \frac{n_{JM}}{\sum\limits_{M=-J}^{J} n_{JM}} \tag{7}$$

one can obtain the desired relative probabilities. The necessary condition that there be orientation of the excited target particles by the beam is that $\sum_{J,M} \mathcal{O}_{JM \to J'M'} n_{JM}$ depends on M'. A major part of this paper is the discussion of this dependence in individual cases. The physical origin of the M' dependence is the direction of the beam velocity and/or the beam particles' intrinsic angular momenta. Some M' dependence is expected in all cases except for an interaction independent of spatial direction and depending only on the radial distance between the target and beam particles; a beam of slow unpolarized neutrons is the only example which comes to mind.

The relaxation matrix $\mathcal{R}_{J'M' \to J'M'}$ is distinctive for each physical situation. There are many cases, however, such as a dilute gas or a rigid glass where relaxation processes can be neglected. This paper will concentrate on these cases although the treatment can be generalized to include effects of relaxation. In the steady state and neglecting relaxation, we can solve Eq. (5) and obtain

$$W_{J'M'} = \frac{\sum\limits_{J,M} \mathcal{O}_{JM \to J'M'} n_{JM}}{\sum\limits_{M'} \sum\limits_{J,M} \mathcal{O}_{JM \to J'M'} n_{JM}} \tag{8}$$

This or its classical analogue is the central quantity which we wish to evaluate. If relaxation processes which change the angular momentum of the excited target particles are infrequent as in a dilute gas, i.e. if an angular momentum J' can be defined, then the relative distribution over states of the excited target particles is given by $W_{J'M'}$.

When the target systems are in a condensed phase it is appropriate (except for nuclear and electronic spins) to describe their orientation by a classical distribution function $W(\theta, \phi)$ which gives the probability that an axis fixed in the target particle can be related to space fixed axes XYZ by the angles θ, ϕ. This function has the value $1/(4\pi)$ for isotropically distributed particles. The necessary condition that an anisotropic distribution of the T^*

particles is obtained is that the transition probability for the excitation process depends on the orientation of the target with respect to the beam direction or polarization if it is linearly polarized. For example, if T^* is formed with a probability proportional to the square of the cosine of an angle θ between a space-fixed axis and an axis fixed in the target particle, then

$$W(\theta) = \frac{3}{4\pi} \cos^2 \theta = \frac{1}{4\pi} [1 + 2P_2(\theta)] \qquad (9)$$

To obtain a distribution over angle from a distribution over angular momentum states one can write

$$W(\theta, \phi) = \sum_{J'} \sum_{M'} W_{J'M'} |\Psi_{J'M'}(\theta, \phi)|^2 \qquad (10)$$

which is independent of angle if $W_{J'M'}$ is independent of M' (Unsold's theorem).

2. OPTICAL PUMPING: PHOTON BEAM ON ATOMIC TARGET WITH PHOTONS EMITTED

A. Distribution in the Excited State

The interaction of a photon beam with a target of atoms which subsequently decay to the ground state is the process of optical pumping. Suppose that a weak magnetic field is applied to an atom with total angular momentum quantum number J in order to lift the $2J+1$ fold degeneracy. To simplify the description the nuclear spin is assumed to be zero. An unpolarized light beam directed along the magnetic field axis causes electric dipole transitions with selection rules $J' = J, J \pm 1$ and $M' = M \pm 1$ but for this direction of the light beam $M' = M$ transitions are forbidden. The general transition probability per unit time per unit incident energy flux is

$$B_\mu(JM; J'M + \mu) = \frac{2\pi e^2}{3\hbar^2} |\langle JM| T_\mu^{(1)} |J'M'\rangle|^2 \qquad (11)$$

where $\mu = +(-)1$ refers to right- (left-) hand circularly polarized light respectively. The dipole moment operator T_μ is defined as follows:

$$T_{\pm 1}^{(1)} = \mp (T_x \pm iT_y)/\sqrt{2}$$

and

$$T_0^{(1)} = T_z \qquad (12)$$

4

If we confine ourselves to those atoms where L, S coupling is a good approximation we have

$$\langle L'SJ'M_{J'}| T_{\mu}^{(1)}|LSJM_J\rangle$$

$$= \sum_{M_L, M'_L} \langle L'SJ'M_{J'}|L'SM'_L M_S\rangle \langle L'M_{L'}| T_{\mu}^{(1)}|LM_L\rangle$$

$$\times \langle LSM_L M_S|LSJM_J\rangle$$

$$= \sum_{M_L} \langle L'SJ'M_{J'}|L'SM_J - M_L, M \cdot + M_L - M_J\rangle$$

$$\times \langle L'M_{J'} + M_L - M_J| T_{\mu}^{(1)}|LM_L\rangle \langle LSM_L M_J - M_L|LSJM_J\rangle$$

$$\tag{13}$$

Now[1]

$$\langle L'SM_{L'} M_S| T_{\mu}^{(1)}|LSM_L M_S\rangle = (-1)^{L'+1+M_L} \delta_{\mu, M'_L - M_L}$$

$$\frac{1}{\sqrt{3}}\langle L'||T^{(1)}||L\rangle \langle L'L - M_{L'} M_L|L'L1 - \mu\rangle \tag{14}$$

so that by substituting this relation into the previous equation we find

$$\langle L'SJ'M_{J'}| T_{\mu}^{(1)}|LSJM_J\rangle$$

$$= (-1)^{L'+L+1} \delta_{\mu, M'_J - M_J}(2J+1)^{\frac{1}{2}}\langle L'||T^{(1)}||L\rangle$$

$$\times \langle 1JM_{J'} - M_J M_J|1JJ'M_{J'}\rangle W(1LJ'S; L'J) \tag{15}$$

where $W(1LJ'S; L'J)$ is Racah's W function.[2]
Hence

$$B_{M_{J'}-M_J}(JM_J; J'M_{J'})$$

$$= \frac{2\pi e^2}{3\hbar^2}(2J+1)|\langle L'||T^{(1)}||L\rangle|^2 W(1LJ'S;L'J)^2$$

$$\times |\langle 1JM_{J'} - M_J M_J|1JJ'M_{J'}\rangle|^2 \tag{16}$$

We are really only interested in the last factor of this equation. The quantities $\langle 1JM_{J'} - M_J M_J|1JJ'M_{J'}\rangle$ are tabulated in Condon and Shortley. A short calculation using these matrix elements and assuming no relaxation in the excited state shows that the relative probabilities in the excited state have the form

$$W_{J'M'} = \frac{3}{2}\frac{1}{2J'+1}\left[\frac{J'(J'-1)+\mu M'(2J'-1)+M'^2}{J'(2J'-1)}\right]; \quad J' = J+1$$

$$(17a)$$

$$W_{J'M'} = \frac{3}{2}\frac{1}{2J'+1}\left[\frac{J'(J'+1)+\mu M'-M'^2}{J'(J'+1)}\right]; \quad J' = J \qquad (17b)$$

$$W_{J'M'} = \frac{3}{2}\frac{1}{2J'+1}$$

$$\times \left[\frac{(J'+1)(J'+2)-\mu M'(2J'+3)+M'^2}{(J'+1)(2J'+3)}\right]; \quad J' = J-1$$

$$(17c)$$

All of these expressions are special cases of the general form

$$W_{J'M'} = \frac{1}{2J'+1}\left[1+\alpha M'+\beta\left(\frac{3}{2}\frac{M'^2}{J'(J'+1)}-\frac{1}{2}\right)\right]$$

$$= \sum_{L=0}^{2} W_{J'}^{(L)} P_L(M') \qquad (18)$$

(L in this equation does not, of course, refer to the atomic L.)

The quantities $W_{J'}^{(1)}$ and $W_{J'}^{(2)}$ are measures of the orientation and the alignment respectively of the atoms in the state J'. The excited state distribution is therefore uniquely characterized by these two quantities. Any arbitrary distribution over states is characterized by $2J'$ quantities

$$\sum_{L=0}^{2J'} W_{J'}^{(L)} P_L\left(\frac{M'}{J}\right) \quad \text{with} \quad W_{J'}^{(0)} = \frac{1}{2J'+1} \qquad (19)$$

and the simpler distribution obtained by optical pumping is a consequence of the electric-dipole selection rules.

A nice illustration of the quasi-independence of the orientation and alignment was given by Happer and Salomon[3] who showed that in a gas of excited $^3P_1^0$ lead atoms the orientation and alignment had different relaxation times; in other words $W_1^{(1)}$ and $W_1^{(2)}$ had different time dependence.

B. Polarization of Emission of the Optically Pumped Atoms

Suppose that we are given a distribution over excited states of the form

$$W_{J'M'} = \frac{1}{2J'+1}\left[1 + W_{J'}^{(1)}\frac{M'}{J'} + W_{J'}^{(2)}\left(\frac{3}{2}\frac{M'^2}{J'(J'+1)} - \frac{1}{2}\right)\right] \quad (20)$$

In a transition from the initial (excited) state J' to a final state J the light emitted at 90° to the exciting light has an intensity proportional to

$$\sum_{M'} W_{J'M'}|\langle J'M'|T_0^{(1)}|JM'\rangle|^2 \quad (21)$$

for light polarized parallel to the Z axis or

$$\tfrac{1}{2}\sum_{M'} W_{J'M'}(|\langle J'M'|T_1^{(1)}|JM'-1\rangle|^2 + |\langle J'M'|T_{-1}^{(1)}|JM'+1\rangle|^2) \quad (22)$$

for light polarized perpendicular to the Z axis.

If we define the polarization of the light by

$$R = \frac{I_\parallel - I_\perp}{I_\parallel + 2I_\perp}$$

$$= \frac{\sum\limits_{M'} W_{J'M'}(|\langle J'M'|T_0^{(1)}|JM'\rangle|^2 - \tfrac{1}{2}|\langle J'M'|T_2^{(1)}|J'M-1\rangle|^2 - \tfrac{1}{2}|\langle J'M'|T_{-1}^{(1)}|JM'+1\rangle|^2)}{\sum\limits_{M'} W_{J'M'}(|\langle J'M'|T_0^{(1)}|JM'\rangle|^2 + |\langle J'M'|T_1^{(1)}|J'M-1\rangle|^2 + |\langle J'M'|T_{-1}^{(1)}|JM'+1\rangle|^2)} \quad (23)$$

we can substitute Eq. (15) and find that

$$R = \frac{\sum\limits_{M'} W_{J'M'}(|\langle J1M'0|J1J'M'\rangle|^2 - \tfrac{1}{2}|\langle J1M'-11|J1J'M'\rangle|^2 - \tfrac{1}{2}|\langle J1M'+1-1|J1J'M'\rangle|^2)}{\sum\limits_{M'} W_{J'M'}(|\langle J1M'0|J1J'M'\rangle|^2 + |\langle J1M'-11|J1J'M'\rangle|^2 + |\langle J1M'+1-1|J1J'M'\rangle|^2)} \quad (24)$$

This expression can be evaluated using, for example, a table in Condon and Shortley.[1] The denominator is identically equal to one. We find in the three cases $J' = J+1$, J, and $J-1$ that

$$R = \sum_{M'} W_{J'M'} \left[\frac{J'(J'+1) - 3M'^2}{(2J'-1)(2J')} \right]; \quad J' = J+1 \quad (25a)$$

$$R = -\sum_{M'} W_{J'M'} \left[\frac{J'(J'+1) - 3M'^2}{(2J')(2J'+1)} \right]; \quad J' = J \quad (25b)$$

$$R = \sum_{M'} W_{J'M'} \left[\frac{J'(J'+1) - 3M'^2}{(2J'+1)(2J'+2)} \right]; \quad J' = J-1 \quad (25c)$$

and carrying out the sums over M'

$$R = W_{J'}^{(2)} \frac{(2J'+1)}{(2J'-1)(2J')} [\tfrac{3}{10} - \tfrac{2}{5}J'(J'+1)]; \quad J' = J+1 \quad (26a)$$

$$R = -W_{J'}^{(2)} \frac{(2J'+1)}{(2J')(2J'+1)} [\tfrac{3}{10} - \tfrac{2}{5}J'(J'+1)]; \quad J' = J \quad (26b)$$

$$R = W_{J'}^{(2)} \frac{1}{2J'+2} [\tfrac{3}{10} - \tfrac{2}{5}J'(J'+1)]; \quad J' = J-1 \quad (26c)$$

The key result is that the fluorescence polarization depends only on the alignment of the atomic state, i.e. only on $W_{J'}^{(2)}$. (This would also be true if we had used a definition of polarization,

$$P = (I_{\parallel} - I_{\perp})/(I_{\parallel} + I_{\perp}).$$

The latter definition gives rise to bulkier expressions.) Naturally the polarization of the emitted radiation depends only on the excited state distribution and not on how it was obtained. That is, whether we excite with photons, electrons or even protons[4] the polarization of the light depends only on the alignment.

3. PHOTON BEAMS AND MOLECULAR TARGETS

A. Angular Distribution of Excited Molecules

Optical pumping of a single $J \to J'$ transition in a gaseous molecule is in principle possible but in practice is likely to be difficult to observe because of the very large number of ground and excited angular momentum states available to the molecule. If we accept the fact that in most experiments a group of states is invariably oriented, then it is more convenient to use a classical description. This means that an angular distribution is to be used to describe the excited state rather than the discrete probabilities $W_{JM'}$. As the probability of excitation is proportional to the square of the component of the optical electric vector along the transition dipole axis in the molecule, the angular distribution of transition dipole axes of molecules excited by linearly polarized light is

$$W(\theta) = \frac{3}{4\pi} \cos^2 \theta = \frac{1}{4\pi}[1 + 2P_2(\theta)] \qquad (9)$$

where θ is the angle between the transition dipole axis and the optical electric field. There is one exception to this rule which occurs when the molecule has an n-fold axis ($n = 3, 4, 6, \infty$) and a degenerate transition is being excited so that all directions in the plane normal to the n-fold axis are equivalent. In this case

$$W(\theta) = \frac{3}{8\pi} \sin^2 \theta = \frac{1}{4\pi}[1 - P_2(\theta)] \qquad (27)$$

where θ is the angle between the n-fold axis and the optical electric field. The anisotropy of the distribution as measured by the coefficient of $P_2(\theta)$ is less when degenerate transitions are excited.

B. Depletion of the Ground State

The above distribution functions are valid for the usual case of weak pumping when the number of excited target particles is small compared to the number in the ground state so that depletion of the ground state is negligible. The general case can be studied by solving the rate equation[5] for the number of ground and excited species N_g and N_e. Let $N_g + N_e = N$ and define distribution functions $n_g(\theta)$ and $n_e(\theta)$ by the equations

$$N_g \equiv \int_0^\pi n_g(\theta) \sin \theta \, d\theta \qquad (28a)$$

$$N_e \equiv \int_0^\pi n_e(\theta) \sin \theta \, d\theta \qquad (28b)$$

The rate equations are:

$$\frac{dn_g}{dt} = -3I_0 \sigma \cos^2 \theta n_g + \frac{n_e}{\tau} = -\frac{dn_e}{dt} \qquad (29a)$$

$$n_e + n_g = N/2 \qquad (29b)$$

where I_0 is the flux of incident photons, $\sigma \cos^2 \theta$ is the absorption cross-section and τ is the excited state lifetime. The solution is

$$n_e(\theta, t) = \frac{\frac{3}{2}NI_0\,\sigma\tau\cos^2\theta}{1+3I_0\,\sigma\tau\cos^2\theta}\{1-\exp\left[-(1/\tau+3I_0\,\sigma\cos^2\theta)\,t\right]\} \quad (30)$$

If the decay time τ is infinite, i.e. a semi-permanent change has occurred such as photoionization, then

$$n_e(\theta, t) = 1-\exp\left(-3I_0\,\sigma\cos^2\theta t\right) \quad (31)$$

which reduces to the usual $\cos^2\theta$ law at small times and to an isotropic distribution at long times. This is reasonable because after the initial period of excitation the light beam samples a target whose ground-state particles are less favourably oriented than the original isotropic distribution.

For the steady state where $t\to\infty$, we have

$$n_e(\theta, \infty) = \frac{\frac{3}{2}NI_0\,\sigma\tau\cos^2\theta}{1+3I_0\,\sigma\tau\cos^2\theta} \quad (32)$$

It is instructive to derive a normalized excited state distribution function

$$\rho_e(\theta) = \frac{1}{2\pi}\frac{n_e(\theta, \infty)}{\displaystyle\int_0^\infty n_e(\theta, \infty)\sin\theta\,\mathrm{d}\theta}$$

$$= \frac{3/(4\pi)\cos^2\theta}{1+3I_0\,\sigma\tau\cos^2\theta}\times\frac{I_0\,\sigma\tau}{\left(1-\dfrac{\tan^{-1}\sqrt{3I_0\,\sigma\tau}}{\sqrt{3I_0\,\sigma\tau}}\right)} \quad (33)$$

The quantity $I_0\sigma\tau$, the product of the number of incident photons per unit area and the absorption cross-section and the lifetime of the excited atom, is the ratio of the lifetime to the average time between absorptions. In most cases this quantity is negligible and we have the usual distribution $3/(4\pi)\cos^2\theta$. Exceptions occur in the optical pumping of atoms where the resonant cross-sections are enormous or in molecular triplet state excitations where the lifetimes τ are exceptionally long while the cross-sections σ for singlet $\to$ singlet absorptions remain large. In the optical pumping of atoms the pumping light is so intense and the relaxation times of the atoms in the ground state are so long that appreciable orientation and alignment of the ground-state atoms is obtained. Moreover, there is a shift of the hyperfine structure of the ground state, jocularly called the "lamp shift",[6] because the atoms spend an appreciable fraction of the time in the excited state which has a much lower spin density at the nucleus.

C. Photoconverted Product

If the excited state T^* decays into a product T' which is different from T, one has then prepared an aligned array of molecules even though the parent molecules may have been randomly oriented as in a glass or liquid. This was first demonstrated by G. N. Lewis and coworkers.[7,8] Albrecht[9] has carefully reviewed many of the possibilities inherent in the photoselection method including the possibility of determining the angle between the polarization of different transitions by studying the polarization of fluorescence as a function of exciting wavelength. Polarizations of the triplet emissions and even of the various triplet–triplet absorptions relative to the polarization of the allowed singlet–singlet absorption could be studied. As Albrecht pointed out, the limitation is that only relative orientations of transition dipoles can be observed in this way. This limitation has been removed by the discovery of magnetophotoselection[10] which can be used to study the distribution of x, y, z axes of the triplet spin Hamiltonian

$$D[S_z^2 - \tfrac{1}{3}S(S+1)] + E(S_x^2 - S_y^2)$$

relative to the optical electric field. Fairly reliable calculations can now be made of the orientation in the molecule of the principal axes of the spin Hamiltonian.

As we know the distribution of transition dipole axes relative to the optical electric field, we can now determine the absolute orientation in the molecule of the transition dipoles. The power of the method consists in the fact that where triplet electron spin resonance and emission studies are possible, there is no longer any need to prepare extremely thin, single crystals. The principal concern of the present paper is, however, in the anisotropy of emission and polarization of the particles emitted be the excited target molecules, T^*. In the next two sections we discuss the emission of photons and atoms or radicals by the excited molecule.

D. Polarization of Emitted Photons: A Classical Treatment

The polarization of fluorescence and phosphorescence is an extensively investigated subject. A general treatment together with a large number of other references appears in references 11 and 12. Here we show how the phenomenon fits into the framework of angular distribution functions.

Suppose that we have a set of molecules whose transition dipoles are oriented with respect to a Z axis fixed in space. The angular distribution function of these dipoles can be represented by the general function of Eq. (19). An arbitrary transition dipole $\mathbf{P}$ can be written as

$$\mathbf{P} = P_0(\sin\theta\cos\phi\mathbf{i} + \sin\theta\sin\phi\mathbf{j} + \cos\theta\mathbf{k}) \tag{34}$$

and the electric vector of the radiated light wave can be put in the form

$$\mathscr{E} = \frac{\exp[i(\mathbf{k}\cdot\mathbf{r} - \omega t)]}{r} k^2(\mathbf{I} - \hat{r}_0\hat{r}_0)\cdot\mathbf{P} \tag{35}$$

For simplicity let us take the direction of observation $\hat{r}_0$ of the light beam to be the unit vector $\mathbf{i}$ perpendicular to the Z axis vector $\mathbf{k}$. (Had we observed the light emitted along the Z axis there would be no polarization at all.) If $\hat{r}_0 = \mathbf{i}$ then

$$\mathscr{E} = \frac{k^2}{r}\exp[i(kr - \omega t)](\sin\theta\sin\phi\mathbf{j} + \cos\theta\mathbf{k})P_0 \tag{36}$$

The intensities of the light polarized along the Z axis and the Y axis are proportional to $\cos^2\theta$ and $\sin^2\theta\sin^2\phi$ respectively. We must carry out an average of the intensity of emission over the angular distribution of the transition dipoles, i.e.

$$I_Z = A\iint d\Omega \frac{1}{4\pi}\left[1 + \sum_L W_L P_L(\theta)\right]\cos^2\theta = \frac{A}{3} + \frac{2}{15}AW_2 \tag{37a}$$

$$I_Y = A\iint d\Omega \frac{1}{4\pi}\left[1 + \sum_L W_L P_L(\theta)\right]\sin^2\theta\sin^2\phi = \frac{A}{3} - \frac{1}{15}AW_2 \tag{37b}$$

Using the previous definition of R, we find

$$R = \frac{I_\parallel - I_\perp}{I_\parallel + 2I_\perp} = \frac{I_Z - I_Y}{I_Z + 2I_Y} = \tfrac{1}{5}W_2 \tag{38}$$

The polarization of the emitted radiation is thus a function of only one parameter of the angular distribution, W_2. Indeed if the molecules had been initially excited by light this would be the only parameter occurring in the distribution.

E. Angular Distribution of Emitted Molecular Fragments

Let us consider a gas containing molecules being photolysed by a polarized light beam. The distribution of transition dipoles in the molecules which are photolysed is given by the expression $3/(4\pi)\cos^2\theta_d$ for a non-degenerate transition. For a degenerate transition the distribution of axes of symmetry is given by $3/(8\pi)\sin^2\theta_d$ as discussed earlier. Suppose that the excited molecule dissociates into two fragments (two atoms, an atom and a radical, or two radicals) whose velocities relative to the centre of mass velocity are antiparallel by the conservation of momentum. Furthermore, the directions of these velocities are assumed to make an angle ψ with the transition dipole axis. The distribution function of angles θ_v of the velocities is given by the distribution function of dipole angles, θ_d summed over all θ_d, ϕ_d consistent with a fixed ψ, i.e.

$$\frac{1}{2\pi}\int_\Omega \rho(\theta_d)\sin\theta_d\,d\theta_d\,d\phi = \frac{1}{2\pi}\int_0^{2\pi}\rho(\theta_d)\,d\phi \qquad (39)$$

where the symbol Ω indicates that the integrand is restricted in size. The various angles are shown in Fig. 1. As

$$\cos\theta_d = \cos\theta_v\cos\psi + \sin\theta_v\sin\psi\cos\phi \qquad (40)$$

$$\frac{1}{2\pi}\int_0^{2\pi}\tfrac{3}{2}\cos^2\theta_d\,d\phi = \tfrac{1}{2} + P_2(\psi)\,P_2(\theta_v) \qquad (41)$$

The distribution functions for the velocity directions are given by

$$\frac{1}{4\pi}[1 + 2P_2(\theta_v)\,P_2(\psi)] \quad \text{for non-degenerate transitions} \qquad (42a)$$

and

$$\frac{1}{4\pi}[1 - P_2(\theta_v)\,P_2(\psi)] \quad \text{for degenerate transitions.} \qquad (42b)$$

These are distribution functions for the velocity directions relative to the centre of mass. Equivalent formulae have been derived by Zare and Hershbach[13] in connection with a similar problem. If a molecule is dissociated by a photon which has precisely the threshold energy for dissociation, the fragments will separate with the Boltzmann velocity distribution due to the parent which is isotropic. There is a domain in which the added recoil

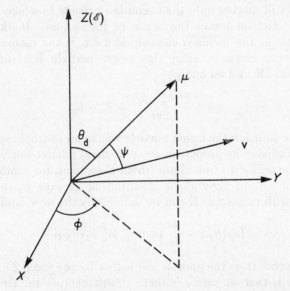

Fig. 1. The Z axis is the direction of the optical electric vector. μ is the electric dipole direction and v is the direction of dissociation.

kinetic energy acquired by the dissociating particles becomes comparable to and then greater than kT. In this domain the distribution function changes gradually from an isotropic function to the limiting function just derived.

The anisotropic fragment velocity distribution in a dilute gas caused by the photolysis causes an anisotropic flux of fragments to the walls of a container. In other words the molecular anisotropy is mapped onto the walls by the photolysis. This technique of photolysis mapping has been realized by J. Solomon[14] and C. Jonah[15] for a number of simple molecules.

4. ANISOTROPIC EXCITATION OF ATOMS AND MOLECULES BY ELECTRON IMPACT

A. General

Suppose that a target particle is struck by an electron travelling in a certain direction. Three inelastic processes can take place: an excitation, an electron capture, or an ionization. Each of these processes may or may not cause a subsequent dissociation. In this paper we will discuss only the excitation process in which one and only one electron leaves the scene of interaction. If $\hbar \mathbf{k}_a$ is the momentum of the incident electron and $\hbar \mathbf{k}_b$ is the momentum of the scattered electron, then the target particle has received a momentum $\hbar \mathbf{K}$ and an energy

$$\varepsilon_{ab} = \frac{\hbar^2(k_a^2 - k_b^2)}{2m} = E_a - E_b \tag{43}$$

Before the collision the target particle was in an isotropic space and after the collision its properties are all defined relative to a vector $\mathbf{K}$. This is the only direction along which alignment, for example, can take place. More generally a distribution of axes in the target particles with respect to $\mathbf{K}$ can be defined by the now familiar

$$W_K(\theta_K) = \frac{1}{4\pi}\left[1 + \sum_{L=1} W_L P_L(\theta_K)\right] \tag{44}$$

One difference from the photon case is that L can exceed 2. Another difference is that we rarely observe simultaneously the electron in the state $\mathbf{k}_b$ and the system excited to a state b from its initial state a. This means that in general we do not measure $W_K(\theta)$ but in effect average it over all directions of $\mathbf{K}$ with respect to $\mathbf{k}_a$. In other words, the total excitation cross-section is observed rather than the differential cross-section. What is generally measured then is an averaged distribution

$$W_{ka}(\theta_{ka}) = \frac{1}{4\pi}\left[1 + \sum_{L=2} W_L P_L(\theta_{ka})\right] \tag{45}$$

where L takes only even values. The reason is that once we do not measure the sign of the angular momentum of the scattered electron we cannot measure it for the target. Hence the W_L for odd L must vanish.

Granted that qualitatively anisotropic excitation can exist, to calculate the W_L is in general very difficult. We can distinguish between three regions of incident electron energy.

(1) Low energy: The electron energy E_a is at the theshold ε_{ab} for excitation or a small multiple of ε_{ab}. Electron exchange can occur and $\Delta S = \pm 1$ transitions are possible. Also electric-dipole allowed transitions are allowed but so are transitions forbidden by parity and higher multipole transitions. The Born approximation is not valid.

(2) Medium energy: The incident electron energy, E_a is fairly large compared to ε_{ab}. Electron exchange cross-sections are small. Both electric-dipole allowed and electric-dipole forbidden processes are allowed. The Born approximation is qualitatively valid at the lower limit of this energy region and quantitatively valid at the upper limit. Near the lower limit scattering of the electron takes place in all directions but near the upper limit forward scattering predominates.

(3) High energy: $E_0 \gg \varepsilon_{ab}$. For $\varepsilon_{ab} \sim 2\text{–}5$ ev, 100 volts might be large enough for E_0. In this region only electric-dipole allowed transitions have appreciable probability. The electron beam acts— aside from differences in cross-section—exactly like an unpolarized light beam.

The low-energy calculation is a very difficult one which we do not attempt here. The following sections deal with a general high-energy calculation which turns out to be very simple, a medium-energy calculation for atoms, and a discussion of selection rules for molecules.

B. The High-energy Limit

In the first Born approximation,[16, 17] the differential cross-section for the collision causing the transition $a \to b$ is given by

$$\sigma(a \to b) = \frac{m^2}{4\pi^2 \hbar^4} \frac{k_b}{k_a} |\langle \psi_b | V(\mathbf{r}, \mathbf{R}) | \psi_a \exp(i\mathbf{K} \cdot \mathbf{R}) \rangle|^2 \qquad (46)$$

where $\mathbf{K} = \mathbf{k}_b - \mathbf{k}_a$. Because the Coulomb interaction of the incident electron and the atomic nuclei does not excite the target, the interaction energy $V(\mathbf{r}, \mathbf{R})$ represents only the interaction between incident and target electrons, $\sum_j e^2 / |\mathbf{r}_j - \mathbf{R}|$. Integration of Eq. (46) over the coordinates of the colliding electron gives

$$\sigma(a,b) = X(k_a, k_b) |\langle \psi_b | \sum_j \exp{(i\mathbf{K} \cdot \mathbf{r}_j)} | \psi_a \rangle|^2 \qquad (47)$$

where

$$X(k_a, k_b) = \frac{4m^2 \, e^4}{\hbar^2 \, K^4} \frac{k_b}{k_a} \qquad (48)$$

In the high-energy limit a typical vector $\mathbf{K} = \mathbf{k}_a - \mathbf{k}_b$ has a small magnitude because the electron is mainly scattered in the forward direction. Under these conditions we can expand the operator

$$\sum_{j=1}^{n} \exp{(i\mathbf{K} \cdot \mathbf{r}_j)} = \sum_{j=1}^{n} (1 + i\mathbf{K} \cdot \mathbf{r}_j + \dots) = n - \frac{i}{e} \mathbf{K} \cdot \boldsymbol{\mu} + \dots \qquad (49)$$

where $\boldsymbol{\mu}$ is the electric-dipole moment operator of the target electrons. The constant term does not contribute to the transition probability.

Let us consider the validity of this expansion. If θ_K is the angle between the vectors $\mathbf{K}$ and $\mathbf{k}_a$, and θ is the angle between the vectors $\mathbf{k}_a$ and $\mathbf{k}_b$,

$$\sin \theta_K = \frac{k_b \sin \theta}{K} = \frac{k_b \sin \theta}{\sqrt{k_a^2 + k_b^2 - 2k_a k_b \cos \theta}} \qquad (50)$$

If $\delta = K/k_a \to 0$, i.e. in the high-energy limit

$$\lim_{\delta \to 0} \sin \theta_K = \cos \theta/2 \quad \text{and} \quad K \to k_a \theta$$

This means that when the scattering becomes peaked in the forward direction, i.e. $\theta \to 0$, $\theta_K \to \pi/2$ which means $\mathbf{K} \perp \mathbf{k}_a$. If this expansion is to converge $K a_0$ must be $\ll 1$ where a_0 is the Bohr radius. Now

$$K a_0 \to k_a a_0 \theta = \left(\frac{\hbar^2 k_a^2}{2m} \frac{2ma_0^2}{\hbar^2} \right)^{\frac{1}{2}} \theta = \sqrt{\frac{E_a}{E_H}} \theta \qquad (51)$$

This means that the width, θ of the forward scattering peak should decrease faster than $E_a^{-\frac{1}{2}}$ to make the series converge. We presume that K diminishes only slowly with increasing energy so that E_a must be quite large in order to make the series (49) converge.

In the high-energy limit the perturbation operator causing transitions is $\boldsymbol{\mu} \cdot \mathbf{K}$; this is similar to the operator $\boldsymbol{\mu} \cdot \mathbf{e}$ for photons where $\mathbf{e}$ is the polarization vector. For an initially unpumped target all values of ϕ_K, the azimuthal angle of $\mathbf{K}$ with respect to $\mathbf{k}_a$, are equally probable. While each target particle is aligned with respect to $\mathbf{K}$ the system of particles is aligned with respect to the incident direction $\mathbf{k}_a$. Thus in this limit the electron beam acts precisely like an unpolarized light beam.

C. Anisotropic Excitation of Atoms by Electron Impact

Free atoms have been aligned by electron beams by a number of workers.[18-22] In this section we will discuss the alignment of atoms and molecules by electron impact within the region of validity of the first Born approximation. The theory of the polarization of the radiation emitted by atoms excited by electron impact has been studied by Oppenheimer,[23] Lamb[24] and Percival and Seaton.[25] We repeat part of the Percival and Seaton calculations to show how they yield an anisotropic distribution function.

Assuming L–S coupling we can represent the initial and final states of the atom by the labels $\langle ALSJM_J |$ and $\langle BL'S'J'M_{J'} |$. A and B represent all quantum numbers other than those specified. The integral involved in the differential cross-section of Eq. (47) is

$$\langle BL'S'J'M_{J'} | \sum_j \exp(i\mathbf{K} \cdot \mathbf{r}_j) | ALSJM_J \rangle$$

$$= \sum_{M_L} \sum_{M_S} \sum_{M_{L'}} \sum_{M_{S'}} \langle L'S'J'M_J | L'S'M_{L'} M_{S'} \rangle \langle BL'S'M_{L'} M_{S'} |$$

$$\times \sum_j \exp(i\mathbf{K} \cdot \mathbf{r}_j) | ALSM_L M_S \rangle \langle LSM_L M_S | LSJM_J \rangle$$

$$(52)$$

$$(|L - S| \leqslant J \leqslant L + S; \ |L' - S'| \leqslant J' \leqslant L' + S'; \ M_J = M_L + M_S)$$

Because the operator V is independent of the spin, using the spherical harmonic addition theorem, $\exp(i\mathbf{K} \cdot \mathbf{r}_j)$ can be expanded as follows:

$$\exp(i\mathbf{K} \cdot \mathbf{r}_j) = \sum_{l_1=0}^{\infty} \sum_{m=-l_1}^{l_1} Y_{l_1 m}^*(\theta_K, \phi_K) Y_{l_1 m}(\theta_j \phi_j) f_{l_1}(Kr_j) \qquad (53)$$

where the $Y_{l_1 m}$ are the normalized surface harmonics and

$$f_{l_1} = 4\pi i^{l_1} j_{l_1}(Kr_j)$$

It is supposed that k_a is in the direction of the quantization axis.

Thus the angle θ_K is determined by the relation

$$\cos \theta_K = \frac{k_b \cos \theta - k_a}{(k_a^2 + k_b^2 - 2k_a k_b \cos \theta)^{\frac{1}{2}}} \tag{54}$$

where θ is the angle between $\mathbf{k}_a$ and $\mathbf{k}_b$. Substitution of Eq. (53) into Eq. (52) gives

$$\sigma(ALSJM_J, BL'SJ'M_J')$$

$$= X(k_a, k_b) \Big| \sum_{M_L} \sum_{M_L'} \langle L'SM_L' M_J' - M_L' | L'SJ'M_J \rangle$$

$$\times \langle BL'SM_L' M_J' - M_L' | \sum_j \sum_{l_1=0}^{\infty} \sum_{m=-l_1}^{l_1} Y_{l_1 m}^*(\theta_K, \phi_K) Y_{l_1 m}$$

$$\times (\theta_j, \phi_j) f_{l_1}(Kr_j) | ALSM_L M_J \rangle \langle LSM_L M_J - M_L | LSJM_J \rangle \big|^2 \tag{55}$$

To simplify this equation we make explicit the fact that the one electron operator V excites an electron in a state with angular momentum l' to a state with angular momentum l. The angular momentum quantum numbers of the remaining core are L_i, M_{L_i}. Thus

$$\langle BL'S'M_{L'} M_{J'} - M_{L'} | \sum_j \exp(i\mathbf{K} \cdot \mathbf{r}_j) | ALSM_L M_J - M_L \rangle$$

$$= \sum_{M_{L_i}} \sum_{l_1=0}^{\infty} \sum_{m_1=-l_1}^{l_1} \langle L_i l'LM_{L'} | L_i l'M_{L_i} M_{L'} - M_{L_i} \rangle \langle bl' | f_{l_1}(Kr) | al \rangle$$

$$\times \langle l'M_L - M_{L_i} | Y_{l_1 m_1}^* | lM_L - M_{L_i} \rangle \langle L_i lM_{L_i} M_L - M_{L_i} | L_i lLM_L \rangle$$

$$\times Y_{l_1 m_1}^*(\theta_K, \phi_K) \tag{56}$$

$$(|L' - l'| \leqslant L_i \leqslant L' + l'; \ |L - l| \leqslant L_i \leqslant L + l)$$

Using the tensor operator method,[2] the last integral in Eq. (56) can be expressed as

$$\langle l'M_L - M_{L_i} | Y_{l_1 m_1}^* | l_1 M_L - M_{L_i} \rangle$$

$$= (-1)^{M_L M_{L_i}} A_{ll_1 l'} \langle ll'M_L - M_L' M_L - M_{L_i} | ll'l_1 - m_1 \rangle \tag{57}$$

$$(|l - l'| \leqslant l_1 \leqslant l + l')$$

The constant $A_{ll_1 l'}$, defined by Racah,[2] does not depend on any magnetic quantum number. Substituting Eq. (57) into Eq. (56)

and using the definition of the Racah W coefficients we obtain

$$\langle BL'S'M_{L'}M_{S'}| \sum_j \exp{(i\mathbf{K}\cdot\mathbf{r}_j)}|ALSM_LM_S\rangle$$

$$= (-1)^{M_L+L+l}(2L+1)^{\frac{1}{2}}(2L'+1)^{\frac{1}{2}} \sum_{l_1} A_{ll_1l'} \, Y^*_{l_1M'_L-M_L}(\theta_K,\phi_K)$$

$$\times \langle bl'|f_{l_1}(Kr)|al\rangle \langle L'L-M'_L M_L|L'Ll_1 M_L-M_{L'}\rangle$$

$$\times W(L'L_i l_1 l; l'L) \tag{58}$$

$$(|L-L'| \leqslant l_1 \leqslant L+L')$$

Substituting this equation into Eq. (56) and using the relation between the Racah coefficients and the vector addition coefficients:

$$\sum_{\beta} \langle ab\alpha\beta|abe\alpha+\beta\rangle \langle ed\alpha+\beta\gamma-\alpha-\beta|edc\gamma\rangle \langle bd\beta\gamma-\alpha-\beta|bdf\gamma-\alpha\rangle$$

$$= (2e+1)^{\frac{1}{2}}(2f+1)^{\frac{1}{2}} \langle af\alpha\gamma-\alpha|afc\gamma\rangle W(abcd;ef)$$

we obtain

$$\langle BL'SJ'M'_J| \sum_j \exp{(i\mathbf{K}\cdot\mathbf{r}_j)}|ALSJM_J\rangle$$

$$= (-1)^{J+l+M_J} \sum_{l_1} (2J+1)^{\frac{1}{2}}(2J'+1)^{\frac{1}{2}}(2L+1)^{\frac{1}{2}}(2L'+1)^{\frac{1}{2}} A_{ll_1l}$$

$$\times Y^*_{lM'_J-M_J}(\theta_K,\phi_K)\langle bl'|f_{l_1}(Kr)|al\rangle W(LL_i l_1 l'; lL')$$

$$\times W(l_1 lJS; LJ')\langle JJ'-M_J M_{J'}|JJ'l_1 M'_J-M_J\rangle \tag{59}$$

$$(|J-J'| \leqslant l_1 \leqslant J+J'; |L-L'| \leqslant l_1 \leqslant L+L')$$

$$\sigma(ALSJM_J \to BL'SJ'M'_J)$$

$$= X^2(k_a,k_b)(2J+1)(2J'+1)(2L+1)(2L'+1)$$

$$\times |\sum_{l_1} A_{ll_1l}\langle bl'|f_{l_1}(Kr)|al\rangle W(LL_i l_1 l'; lL') W(l_1 lJS; LJ')$$

$$\times Y^*_{l_1,M'_J-M_J}(\theta_K,\phi_K)\langle JJ'-M_J M_{J'}|JJ'l_1 M'_J-M_J\rangle|^2 \tag{60}$$

This expression makes it clear that in this approximation electron collisions will populate the excited states $M_{J'}$ unequally. The transition probability involves a sum of amplitudes of monopole, dipole, quadrupole, ... character according as l_1 is $0, 1, 2, \dots$. The weighting factors for each transition depend on integrals involving the radial wave function as well as the function of the direction of the emerging electron. This expression can be simplified somewhat by averaging the cross-section over all directions of scattering of the electron. In the expression above this means

$$\bar{\sigma}(JM_J \to J'M_{J'})$$

$$= \iint \sigma \sin \theta_K \, d\theta_K \, d\phi_K = \sum_{l_1} \bar{\sigma}(l_1)$$

$$= X^2(k_a, k_b)(2J+1)(2J'+1)(2L+1)(2L'+1)$$

$$\times \sum_{l_1} |A_{l'l_1}\langle bl'|f_{l_1}(Kr)|al\rangle \, W(LL_i \, l_1 \, l'; \, lL') \, W(l_1 \, lJS; \, LJ')$$

$$\times \langle JJ' - M_J \, M_{J'}|JJ'l_1 \, M'_J - M_J\rangle|^2 \qquad (61)$$

In most experiments the direction of the scattered electron is not measured so that only the total cross-section $\bar{\sigma}$ is of interest. The latter is a set of partial cross-sections for each permissible multipole transition. All 2^{l_1} pole transitions are allowed for which

$$|L - L'| \leqslant l_1 \leqslant L + L'$$

An anisotropic distribution of the excited atom is defined by

$$W_{J'M'} = \frac{\sum_M \bar{\sigma}(JM_J \to J'M_{J'})}{\sum_M \sum_{M'} \bar{\sigma}(JM_J \to J'M_{J'})} \qquad (62)$$

assuming the atoms are isotropically distributed in their ground state. The polarization of the light as emphasized before will depend only on W.

D. Excitation of Molecules by Medium-energy Electrons

For medium-energy electrons the perturbing operator is effectively $\sum_j \exp(i\mathbf{K} \cdot \mathbf{r}_j)$ and not $\mathbf{K} \cdot \sum_j \mathbf{r}_j$ as it is for high-energy electrons. An immediate conclusion is that electrons can excite all the transitions that photons can excite and many more besides. An incisive discussion of the selection rules for diatomic molecules ($C_{\infty v}$ and $D_{\infty h}$) has been given by Dunn.[26] His analysis is immediately applicable to polyatomic molecules and as an example we will treat the point group C_{2v}.

The plane wave $\exp(i\mathbf{K} \cdot \mathbf{r})$ is symmetric with respect to all rotations about $\mathbf{K}$ and with respect to reflections in planes containing $\mathbf{K}$. The target molecule will have definite symmetries with respect to these operations depending on the state of the molecule and the orientation of the target with respect to the direction of the electron beams. In short there will be a new point group appropriate to the **K** vector *and* the molecule. We choose a coordinate system for the C_{2v} molecule as follows: the Z axis is the two fold (C_2) axis, the X axis is perpendicular to the Z axis and in the plane of the nuclei, and the Y axis is perpendicular to this plane. Thus if the target molecule has its C_2 axis aligned along **K**, then we see

TABLE I.   Selection Rules for Excitation by Electron (Photon) Impact for a C_{2v} Molecule

	A_1	A_2	B_1	B_2
A_1	exp (ikx) exp (iky) exp (ikz) (z)	0 (0)	exp (ikx) (x)	exp (iky) (y)
A_2		exp (ikx) exp (iky) exp (ikz) (z)	exp (iky) (y)	exp (ikx) (x)
B_1			exp (ikx) exp (iky) exp (ikz)	0
B_2				exp (ikx) exp (iky) exp (ikz) (z)

that there are symmetry operations, E, C_2, σ_v and σ_v'. If **K** is in the X direction, we have only two symmetry operations E, σ_v. Similarly if **K** is in the Y direction, we have the two symmetry operations E, σ_v'. (We recall that the medium-energy electron is moving orders of magnitude faster than the nuclei so that it is permissible to think of the molecule as frozen in space. The Franck–Condon principle is equally valid for medium-energy electrons.)

For the transition probability to be different from zero, the direct product of the initial and final state of the molecule and the plane wave $\exp(i\mathbf{K}\cdot\mathbf{r})$ should contain the totally symmetric representation of the group of the molecule plus the $\mathbf{K}$ vector. In this way we can prepare a table of the selection rules for excitation by electron impact, such as Table I for the C_{2v} molecule. Table I contains the operators which can induce specific transitions for the medium-energy electron (exponentials) and for the light wave

TABLE II. Selection Rules for Excitation by Electron Impact for Linear $C_{\infty v}$ Molecules

	Σ^+	Σ^-	π	Δ
Σ^+	$\exp(ikx)$ $\exp(iky)$ $\exp(ikz)$ (z)	0 (0)	$\exp(ikx)$ $\exp(iky)$ (x,y)	$\exp(ikx)$ $\exp(iky)$ (0)
Σ^-		$\exp(ikx)$ $\exp(iky)$ $\exp(ikz)$ (z)	$\exp(ikx)$ $\exp(iky)$ (x,y)	$\exp(ikx)$ $\exp(iky)$ (0)
π			$\exp(ikx)$ $\exp(iky)$ $\exp(ikz)$	$\exp(ikx)$ $\exp(iky)$ (x,y)
Δ				$\exp(ikx)$ $\exp(iky)$ $\exp(ikz)$ (z)

(components of $\mathbf{r}$.) From this table we see that in order to observe the transition by electron impact, those molecules which are coplanar with $\mathbf{K}$ and whose C_2 axes are perpendicular to $\mathbf{K}$ will be most favourable for the excitation although only molecules whose plane is perpendicular to $\mathbf{K}$ will have a forbidden transition. Thus there will be some degree of orientation of the excited molecules with respect to $\mathbf{K}$.

For the case of a linear $C_{\infty v}$ molecule, we chose the Z axis to be the internuclear axis. The selection rules for electron impact can be constructed just as for the molecule and are shown in

Table II. These selection rules are all based on the symmetry of the $\exp(i\mathbf{K}\cdot\mathbf{r})$ operator. They may be more generally valid than for just those scatterings which can be treated by the first Born approximation; as long as the wave functions of the incident and scattering electron have the same symmetry character as that of a plane wave, then the selection rules would remain valid.

5. POLARIZED NEUTRON BEAMS

A large variety of nuclear experiments fall within the framework of the scheme discussed in this paper. Often the excited target T^* may pass through a number of intermediate states of very short lifetime before reaching a state with an appreciable lifetime. This is somewhat analogous to a fluorescence which is excited by an excitation to a state above the level from which fluorescence occurs. In most cases the orientation of the intermediate state is with reference to the beam direction. In the case of slow neutrons the beam direction is of no importance and the orientation is with respect to the direction of spin polarization if any.

One particularly elegant experiment may suffice as an example in the nuclear domain. Tsang and Connor[27] irradiated nuclei of nuclear spin I with polarized neutrons. By conservation of angular momentum the product nucleus can have angular momentum $I' = I + \frac{1}{2}$ or $I - \frac{1}{2}$ and is formed in an oriented state. The relative probabilities $W_{I'M'}$ are given by

$$W_{I'M'} = \frac{|\langle IM' - 1\tfrac{1}{2}\tfrac{1}{2}|I\tfrac{1}{2}I \pm \tfrac{1}{2}M'\rangle|^2}{\sum\limits_{M'=-(I\pm\frac{1}{2})}^{I\pm\frac{1}{2}} |\langle IM' - 1\tfrac{1}{2}\tfrac{1}{2}|I\tfrac{1}{2}I \pm \tfrac{1}{2}M'\rangle|^2}$$

$$= \frac{1}{2I'+1}\left(1 \pm \frac{M'}{I' + \tfrac{1}{2} \mp \tfrac{1}{2}}\right) \tag{63}$$

From its definition $W_{I'}^{(1)} = 1$ for $I' = I + \frac{1}{2}$ and $-I'/I'+1$ for $I' = I - \frac{1}{2}$. These are the relative probabilities for the initial states formed on absorption of a neutron. Decay through intermediate states may diminish this orientation as discussed in Section 6.

The oriented nuclei can emit circularly polarized gamma rays whose intensity (for $\Delta I = 0, \pm 1$) follows the equations already derived for oriented atoms. More remarkably, as was first predicted by Lee and Yang,[28] the electrons (or positrons) emitted in β decay have an angular distribution given by

$$W(\theta) = \frac{1}{4\pi}\left(1 + \frac{v}{c}PA\cos\theta\right) \tag{64}$$

where v is the speed of the electron, P is an orientation defined by

$$P = \sum_{M'=-I'}^{I'} \frac{M'}{I'}W_{I'M'} = W_{I'}^{(1)}\tfrac{1}{3}(I'+1)(2I'+1) \tag{65}$$

and A is a number depending on the angular momentum change of the nucleus in the β decay. This means that the number of β particles emitted parallel to and antiparallel to the direction of polarization will be different. In this experiment it is convenient to set up an external field parallel to the direction of neutron spin polarization to minimize nuclear spin relaxation and to permit magnetic resonance measurements. In this way Connor and Tsang determined the gyromagnetic ratios of ^{8}Li and ^{20}F.

6. ORIENTATION OF INTERMEDIATES

Suppose we consider the following generalization of our basic processes

$$B + T \rightarrow T_1^*$$
$$T_1^* \rightarrow T_2^* \tag{66}$$

where T_1^* and T_2^* are different intermediate states each free to lose orientation by interactions with their environment. The intermediate process $T_1^* \rightarrow T_2^*$ is very general; it may involve an emission, a collision or a radiationless transition. Suppose further that the orientation of T_1^* is described by a known distribution function either quantum

$$W_{J_1M_1} = \sum_{L=0}^{2J_1} W_{J_1}^{(L)}P_L(M_1) \tag{67}$$

or classical

$$W(\theta) = \sum_{L=0}^{\infty} W^{(L)}P_L(\theta) \tag{68}$$

The object is to derive the distribution function for the state T_2^*. The state J_1 is converted (or decays) into the state J_2 with rate constants $\mathscr{D}_{J_1M_1 \to J_2M_2}$. First we write down the rate equations

$$\frac{dn_{J_1M_1}}{dt} = -\sum_{M_2} \mathscr{D}_{J_1M_1 \to J_2M_2} n_{J_1M_1} = -wn_{J_1M_1}$$

$$\frac{dn_{J_2M_2}}{dt} = \sum_{M_2} \mathscr{D}_{J_1M_1 \to J_2M_2} n_{J_1M_1} \qquad (69)$$

whose solutions are

$$n_{J_1M_1}(t) = n_{J_1M_1}(0) \exp(-wt) \qquad (70)$$

$$n_{J_2M_2}(t) = \sum_{M_1} \mathscr{D}_{J_1M_1 \to J_2M_2} n_{J_1M_1}(0)\,[1 - \exp(-wt)]/w \qquad (71)$$

These equations show that neither $W_{J_2M_2}$ nor $W_{J_1M_1}$ is a function of time during the decay. Suppose we let the state J_1 decay completely (i.e. $t \to \infty$) then

$$n_{J_2M_2}(\infty) = [\sum_{M_1} \mathscr{D}_{J_1M_1 \to J_2M_2} n_{J_1M_1}(0)]/w \qquad (72)$$

or

$$W_{J_2M_2}(\infty) = [\sum_{M_1} \mathscr{D}_{J_1M_1 \to J_2M_2} W_{J_1M_1}(0)]/w \qquad (73)$$

This equation will be applied to three phenomena:

(1) Transition of an atom (or a nucleus) from $J_1 M_1$ to $J_2 M_2$ by emission of a light ray.

(2) Energy transfer by the dipole–dipole mechanism from an array of donors excited by polarized light to an array of acceptors.

(3) Dissociation of molecules by atoms excited with polarized light.

A. Electric Dipole Transition from an Oriented State $J_1 M_1$ to a State $J_2 M_2$

If atoms are distributed among the sublevels of the state $J_1 M_1$ according to the probability law

$$W_{J_1M_1} = \frac{1}{2J_1+1}\left[1 + W_{J_1}^{(1)}\frac{M_1}{J_1} + W_{J_1}^{(2)}\left(\frac{3M_1^2}{J_1(J_1+1)} - 1\right)\right] \qquad (74)$$

and they decay by light emission to the state $J_2 M_2$, the distribution of final states is obtained directly by substituting the dipole moment matrix elements of Eq. (11) in Eq. (5). The results of the calculation are:

If $J_2 = J_1 - 1$

$$W_{J_2 M_2} = \frac{1}{2J_2 + 1} \left\{ 1 + \left(1 - \frac{1}{(J_2 + 1)^2} \right) W_{J_1}^{(1)} \left(\frac{M_2}{J_2} \right) \right.$$
$$\left. + \left[1 - \frac{3}{(J_2 + 1)(2J_2 + 3)} \right] W_{J_1}^{(2)} \left(\frac{3M_2^2}{J_2(J_2 + 1)} - 1 \right) \right\} \quad (75)$$

if $J_2 = J_1$

$$W_{J_2 M_2} = \frac{1}{2J_2 + 1} \left\{ 1 + \left(1 - \frac{1}{J_2(J_2 + 1)} \right) W_{J_1}^{(1)} \frac{M_2}{J_2} \right.$$
$$\left. + \left[1 - \frac{3}{J_2(J_2 + 1)} \right] W_{J_1}^{(2)} \left(\frac{3M_2^2}{J_2(J_2 + 1)} - 1 \right) \right\} \quad (76)$$

and if $J_2 = J_1 + 1$

$$W_{J_2 M_2} = \frac{1}{2J_2 + 1} \left\{ 1 + W_{J_1}^{(1)} \left(\frac{M_2}{J_2} \right) \right.$$
$$\left. + \left[1 - \frac{3}{J_2(2J_2 - 1)} \right] W_{J_1}^{(2)} \left(\frac{3M_2^2}{J_2(J_2 + 1)} - 1 \right) \right\} \quad (77)$$

The most remarkable aspect of these results is that the coefficient $W_{J_2}^{(1)}$ is proportional to $W_{J_1}^{(1)}$ and $W_{J_2}^{(2)}$ is proportional to $W_{J_1}^{(2)}$. In short the orientation and alignment of the daughter atoms are proportional to the orientation and alignment of the parent atoms respectively. For small quantum numbers ($J_2 \leqslant \frac{5}{2}$) the coefficients behave erratically, even changing sign, but for larger J_2 the coefficients $W_{J_2}^{(i)}$ steadily approach $W_{J_1}^{(i)}$ as they must.

B. Dipole–dipole Energy Transfer from Aligned Donors to Random Acceptors

Suppose that we have an array of molecules or atoms in a glass which have been excited by polarized light. The distribution of transition dipole axes relative to the electric vector of the light wave is

$$W(\theta_D) = \frac{1}{4\pi} [1 + W_D^{(2)} P_2(\theta_D)] \quad (78)$$

where θ_D is the angle between the transition dipole of the donor molecule and the electric vector of the light wave (as discussed in Section 3, $W_D^{(2)}$ can be as large as 2). In the same glass at random distances and orientations is a set of acceptor molecules and energy will be transferred to these acceptors by a dipole–dipole mechanism. We can then write

$$W(\theta_A) = \frac{\iint (\cos\theta_{AD} - 3\cos\theta_{AR}\cos\theta_{DR})^2 \, W(\theta_D) \, d\Omega_D \, d\Omega_R}{\iint (\cos\theta_{AD} - 3\cos\theta_{AR}\cos\theta_{DR})^2 \, d\Omega_D \, d\Omega_R} \qquad (79)$$

where θ_{AD} is the angle between a donor dipole and an acceptor dipole, θ_{AR} and θ_{DR} are the angles between the respective dipoles and the vector **R** joining the donor and acceptor and θ_A is the angle between the acceptor dipole and the electric field of the dipole. The integration is carried out by repeated use of the spherical harmonic addition theorem:

$$\cos\theta_{AD} = \cos\theta_A \cos\theta_D + \sin\theta_A \sin\theta_D \cos(\phi_A - \phi_D)$$
$$\cos\theta_{AR} = \cos\theta_A \cos\theta_R + \sin\theta_A \sin\theta_R \cos(\phi_A - \phi_R)$$
$$\cos\theta_{DR} = \cos\theta_D \cos\theta_R + \sin\theta_D \sin\theta_R \cos(\phi_D - \phi_R) \qquad (80)$$

with the result

$$W(\theta_A) = \frac{1}{4\pi}\left[1 + \tfrac{1}{25}W_D^{(2)}P_2(\theta_A)\right] \qquad (81)$$

Thus $W_A^{(2)} = \tfrac{1}{25}W_D^{(2)}$ and the excited acceptors are distributed almost completely isotropically. The angular dependence of the transfer process is sufficiently weak that very little of the original donor alignment is left.

C. Photosensitized Dissociation of Molecules

Many reactions have been studied in which an atom (Hg, Cd, Xe, ...) is excited with its own resonance radiation and the excited atom fragments a molecule on collision. A typical example is

$$Hg(^3P_1) + H_2 \rightarrow HgH + H \qquad (82)$$

Referring to Eq. (1) we see that the target is H_2 and the beam particle is the aligned mercury atom. If the atoms are excited with

linearly polarized light will the fragments be ejected anisotropically? The excited atom is described by the now familiar distribution:

$$W_{J'M'} = \frac{1}{2J'+1}\left\{1+W_{J'}^{(2)}\left[\frac{3M'^2}{J'(J'+1)}-1\right]\right\} \tag{83}$$

We are obliged to treat the atom quantum mechanically but the
rotating molecule which is being dissociated can be described
classically. The cross-section for its dissociation will depend on
the relative initial kinetic energy of the atom and molecule, on the
impact parameter of the collision and on the direction of approach
of the molecule relative to the angular momentum of the atom.
We describe the latter dependence by the expansion

$$\sigma(\chi,\omega) = \sigma_0(\chi)+\sigma_1(\chi)\frac{\mathbf{J'}\cdot\mathbf{v}}{J'v}+\sigma_2(\chi)\left[\frac{3(\mathbf{J'}\cdot\mathbf{v})^2}{J'(J'+1)\,v^2}-1\right]+\ldots \tag{84}$$

$\mathbf{v}$ is the initial relative velocity vector of the molecule and atom.
ω is the angle between $\mathbf{v}$ and the direction of the angular momentum
of the atom, i.e. $\cos\omega = \mathbf{J'}\cdot\mathbf{v}/J'v$. (The axis of quantization has
been chosen to be the direction of polarization of the light beam.)
χ is the angle between $\mathbf{v}$ and the direction of emission of the
fragment (the H atom in the example). The quantity $\sigma_1(\chi)$ is the
coefficient of a term which changes sign if the angular momentum
of the atom changes sign. The latter operation does not change
the charge distribution of the atom so that unless spin-dependent
forces are of some importance, $\sigma_1(\chi)$ would be expected to vanish
and is hereafter dropped. The way to investigate its possible
presence is to use circularly polarized light. The higher terms
$\sigma_3, \sigma_4, \ldots$ cannot contribute to the dissociation probability when the
excited state distribution is given by Eq. (83).

The distribution of velocities $\mathbf{v}$ which lead to dissociation is
given by

$$W(\omega,\chi) = \frac{1}{4\pi\sigma_0}\sum_{M'=-J'}^{J'}W_{J'M'}\langle J'M'|\sigma|J'M'\rangle$$

$$= \frac{1}{4\pi}\left\{1+\frac{\sigma_1(\chi)}{\sigma_0(\chi)}\,W_{J'}^{(1)}\frac{(J'+1)}{3J'}\cos\omega+\frac{\sigma_2(\chi)}{\sigma_0(\chi)}\,W_{J'}^{(2)}\right.$$

$$\left.\left[1-\frac{1}{3J'(J'+1)}\right]\left(\tfrac{3}{2}\cos^2\omega-\tfrac{1}{2}\right)\right\} \tag{85}$$

The quantities $\sigma_0(\chi), \sigma_2(\chi), \ldots$ are obtained from the dynamics of the collision. In our treatment we leave these undetermined and make an expansion

$$\frac{\sigma_2(\chi)}{\sigma_0(\chi)} = \sum_{l=0}^{\infty} \sigma_{20}^l P_l(\chi) \tag{86}$$

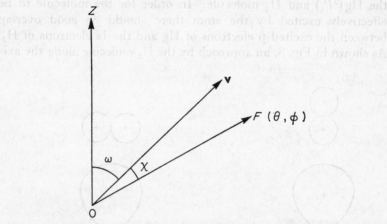

Fig. 2. The Z axis is the direction of polarization of the light beam and the axis of quantization. $\mathbf{v}$ is the initial relative velocity vector of the molecule and atom. $F(\theta, \phi)$ is the direction of emission of the fragment.

The final distribution which we require is the distribution of directions of fragment emission relative to the polarization of the light ray. This is obtained by integrating the distribution $W(\omega, \chi)$ over all possible directions $\mathbf{v}$ of the incoming molecule.

$$W(\theta) = \int d\Omega_v \, W(\omega, \chi) = \frac{1}{4\pi} \int d\Omega_v \bigg\{ 1 + \sum_{l=0}^{\infty} \sigma_{20}^l P_l(\chi)$$

$$\times W_{J'}^{(2)} \left[1 - \frac{1}{3J'(J'+1)} \right] P_2(\omega) \bigg\} \tag{87}$$

Using the spherical harmonic addition theorem (cf. Fig. 2) we obtain

$$W(\theta) = \frac{1}{4\pi} \bigg\{ 1 + \frac{\sigma_{20}^2}{5} W_{J'}^{(2)} \left[1 - \frac{1}{3J'(J'+1)} \right] P_2(\theta) \bigg\} \tag{88}$$

Thus without solving the mechanical problem of the trajectories of the collision one can show that the determination of the distribution of the fragment flux is characterized by a single dimensionless parameter

$$\sigma_{20}^2 = \frac{5}{2} \int_0^\pi P_2(\chi) \frac{\sigma_2(\chi)}{\sigma_0(\chi)} \sin\chi \, d\chi \qquad (89)$$

A valence theoretical model for the collision suggests that σ_{20}^2 should be a fairly large number. Consider again the specific case of the $Hg(^3P_1)$ and H_2 molecule. In order for the molecule to be effectively excited by the atom there should be good overlap between the excited p electrons of Hg and the 1s electrons of H_2. As shown in Fig. 3, an approach by the H_2 molecule along the axis

(a)

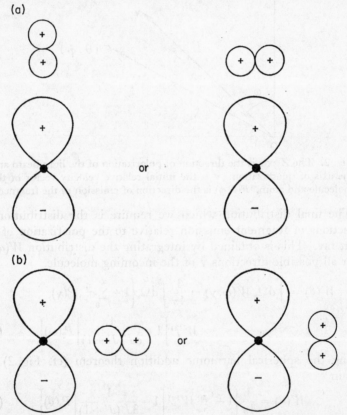

(b)

Fig. 3. (a) A hydrogen molecule approaches a mercury atom along the direction of polarization of the light in parallel and perpendicular orientation. (b) A hydrogen molecule approaches a mercury atom normal to the direction of polarization of the light in parallel and perpendicular orientation.

of the polarized p function will give a good overlap no matter how the H_2 molecule is oriented, but an approach by the H_2 molecule along the waist of the p function will give a poor overlap or even zero overlap. If we use this simple reasoning then the reaction is much more probable at values of ω near 0 or π than at $\pi/2$. This in turn means that σ_2/σ_0 will be $\simeq 1$ and presumably σ_{20}^2 will be of the same order. For linearly polarized light and for $J' = 1$, $\sigma_{20}^2 \simeq 1$ and

$$W(\theta) = \frac{1}{4\pi} [1 + \tfrac{1}{3}P_2(\theta)] \tag{90}$$

which is a rough prediction at best but illustrates what can be inferred from an experiment of this type.

References

1. Condon, E. U., and Shortley, G. H., *The Theory of Atomic Spectra*, Cambridge University Press, 1953, Chapters 3, 4, 5.
2. Racah, G., *Phys. Rev.* **62**, 438 (1942).
3. Happer, W., and Saloman, E., *Phys. Rev. Letters* **15**, 441 (1965).
4. Van Eck, J., de Heer, F. J., and Kistemaker, J., *Physica* **28**, 1184 (1962).
5. Stepanov, B. I., and Gribkovskii, V. P., *Optics and Spect.* **8**, 42, 114 (1962).
6. Kastler, A., *J. Opt. Soc. Am.* **53**, 902 (1963).
7. Lewis, G. N., and Lipkin, D., *J. Am. Chem. Soc.* **64**, 2801 (1942).
8. Lewis, G. N., and Bigeleisen, J., *J. Am. Chem. Soc.* **65**, 520 (1943).
9. Albrecht, A. C., *J. Mol. Spectroscopy* **6**, 84 (1961).
10. Lhoste, J. M., Haug, A., and Ptak, M., *J. Chem. Phys.* **44**, 648, 654 (1966).
11. Foerster, T., *Fluorescenz Organischer Verbindungen*, Gottingen, 1951.
12. Hercules, D. M., Ed., *Fluorescence and Phosphorescence*, Interscience, 1966.
13. Zare, R. N., and Hershbach, D. R., *Proc. I.R.E.* **51**, 173 (1965).
14. Solomon, J., *J. Chem. Phys.* **47**, 889 (1967).
15. Jonah, C., private communication.
16. Mott, N. F., and Massey, H. S. W., *The Theory of Atomic Collisions*, Oxford University Press, 1949, p. 224.
17. Massey, H. S. W., *Handbuch der Physik*, Vol. 36, Springer-Verlag, Berlin, 1956.
18. Skinner, H. W. B., *Proc. Roy. Soc.* (*London*) **A112**, 642 (1926).
19. Dehmelt, H. G., *Phys. Rev.* **103**, 1125 (1956).
20. Pebay-Peroula, J. C., *J. Phys. Radium* **20**, 669, 721 (1959).
21. Decomp. B., Pebay-Peroula, J. C., and Brossel, J., *C. R. Acad. Sci.* **251**, 941 (1960).
22. McFarland, R. H., *Phys. Rev.* **133A**, 986 (1964).

23. Oppenheimer, J. R., *Z. Phys.* **43**, 27 (1927); *Proc. Nat. Acad. Sci. U.S.* **13**, 800 (1927); *Phys. Rev.* **32**, 261 (1928).
24. Lamb, W. E., *Phys. Rev.* **105**, 559, 573 (1957).
25. Percival, I. C., and Seaton, M. J., *Phil. Trans. Roy. Soc.* **251A**, 113 (1958).
26. Dunn, G. H., *Phys. Rev. Letters* **8**, 62 (1962).
27. Tsang, T., and Connor, D., *Phys. Rev.* **132**, 1141 (1963).
28. Lee, T. D., and Yang, C. N., *Phys. Rev.* **104**, 254 (1956).

THERMAL DIFFUSION IN SYSTEMS WITH SOME TRANSFORMABLE COMPONENTS

B. BARANOWSKI, *Institute of Physical Chemistry, Polish Academy of Sciences, Warsaw, Poland*

A. E. DE VRIES and A. HARING, *FOM-Institute for Atomic and Molecular Physics, Amsterdam, The Netherlands*

R. PAUL, *Department of General Physics and X-rays, Indian Association for the Cultivation of Science, Calcutta 32, India*

CONTENTS

1. INTRODUCTION

In linear irreversible thermodynamics, thermal diffusion is treated as the cross-effect between two vectorial processes. No new cross-effect has to be expected if a molecular transformation, a scalar effect, like an excitation or a chemical process, is simultaneously taking place in the temperature gradient. This is due to the Curie symmetry principle. The presence of such a transformation, however, modifies considerably the stationary conditions for the system of reacting components. If the local equilibrium approximation can be applied the number of independent thermodynamic forces is reduced due to the equilibrium condition. Generally the presence of a transformation gives rise to two

101

additional changes which can be treated independently:

(a) The effective heat conductivity of a reacting system is always larger than in a non-reacting one. This is due to a transfer of the heat of reaction. This effect is well understood and we are not giving any references here.

(b) If, besides the reacting species, one or more non-reacting components are present an additional separation can appear as has been observed.[1] Contrary to the additional heat conductivity the additional separation is strongly dependent on the stoichiometric character of the reaction especially if one particle is transformed into another very similar in kinetic properties. For instance, in some excitation reactions almost no additional separation can be expected.[2]

Our present treatment presents a theory starting from rather general assumptions, distinguishing clearly the well-known "physical" part of the separation due to thermal diffusion from the new "chemical" contribution. Taking these two different effects into consideration we shall give quantitative experimental evidence of the additional "chemical" separation. Some parts of our treatment will be based on previously published papers.[1, 2] We shall show a very close similarity between the change of the thermal conductivity and the additional separation due to the chemical reaction with respect to the temperature dependence. In both cases the maximum effect observed occurs in the neighbourhood of the 1 : 1 molar ratio of the reacting components. Here a clear maximum of the "chemical" contribution appears. Thus both effects are observed only in the temperature range where the transformable components are present in appreciable amounts.

Before starting with our own results let us draw attention to some related papers. First of all the treatment of Prigogine and Buess must be mentioned.[3] In their paper, starting from thermodynamics of irreversible processes, a "chemical" contribution to the separation was clearly shown for the first time. A related idea had previously been published by Wagner in 1929.[4] The equations for the concentration gradients were derived for a three-component system with two reacting species and an inert component; the additional separation was discussed for an association reaction. However, no direct application of the equations derived has ever come to our knowledge. The theoretical treatment given by us has many related points with the paper mentioned[3] and we shall come back to this question in the Discussion.

The influence of reacting components on thermal diffusion was also discussed by Schäfer[5] but then the only influence he mentioned was the change of the temperature gradient due to the known effect on thermal conductivity. The treatment given, however, is based experimentally as well as theoretically on a rather qualitative level and therefore will not be further discussed here. A more extended treatment was presented by Whalley[6] applying a free-path theory. An additional separation is expected due to a continuous flow of the reacting components through the separating mixture, even in the stationary state. He treated a hydrogen–nitrogen mixture in the presence of nitrogen dioxide and dinitrogen tetroxide as reacting components. The magnitude and the sign of the separation depend on the diffusional cross-sections of the hydrogen and nitrogen in the mixture.

Finally, a paper by Taylor and Spindel must be mentioned. Here an additional separation due to a chemical kinetic isotope effect on thermal diffusion was investigated.[7] They made use of the $2NO_2 \rightleftharpoons 2NO + O_2$ reaction in which the isotope effect for ^{15}N against ^{14}N works in the same direction as the thermal diffusion: ^{15}N is concentrated in the nitrogen dioxide by the chemical reaction and the $^{15}NO_2$ moves preferentially to the colder part. Thus the two effects work together. The increase in the thermal diffusion factor due to the extra phenomenon is much smaller than in the present study.

2. THEORY

A. General Theory

Let us now derive the equation for the separation by thermal diffusion in the presence of a transformation. Because the experiments described subsequently were performed in gases we take here the most suitable formalism for this purpose. We could apply the method of irreversible thermodynamics as used by Prigogine and Buess[3] but a more straightforward way is the system of equations given by Waldmann.[8] While the two systems are equivalent the latter is preferable because of the clear meaning of the coefficients introduced.

5

We start with the simplest possible system, namely, two transformable and one inert, non-reacting component. The results obtained can be extended simply to a many-component system with several chemical reactions but we believe that no new result can be expected.

For component i the diffusion velocity $\mathbf{W}_i$ relative to the average velocity of the molecules $\mathbf{w}$ is given by

$$\mathbf{W}_i = \mathbf{v}_i - \mathbf{w} \tag{1}$$

where $\mathbf{v}_i$ denotes the average velocity of molecules i. The average molecular velocity $\mathbf{w}$ is given by

$$\mathbf{w} = \sum_i x_i \mathbf{v}_i \tag{2}$$

where x_i denotes molar fractions. From (1) and (2) follows:

$$\sum_i x_i \mathbf{W}_i = 0 \tag{3}$$

Let us suppose that a transformation between components **(1)** and **(2)** takes place, so that

$$\nu \underset{(1)}{A} \rightleftharpoons \underset{(2)}{A_\nu} \tag{4}$$

where ν denotes the number of monomers forming the molecule A_ν. For $\nu = 1$ Eq. (4) could represent an excitation reaction or an isomerism. For the diffusion velocities $\mathbf{W}_i$ we substitute the equations given by Waldmann:[8]

$$\mathbf{W}_i = -\sum_j D_{ij}\,\mathrm{grad}\,x_j - D_{T_i}\,\mathrm{grad}\ln T \tag{5}$$

in which D_{ij} is a polynary diffusion coefficient, different from a binary one, and D_{T_i} is a polynary thermal diffusion coefficient.

The Onsager reciprocity relations impose the condition

$$D_{ij} = D_{ji} \tag{6}$$

Also we have the additional conditions

$$\sum_i x_i D_{ij} = 0; \quad \sum_i x_i D_T = 0 \tag{7}$$

which are mathematically equivalent to the relation (3) due to the frame of reference chosen.

To specify more exactly the physical situation treated, let us formulate the balance equations for the masses of the three components and the overall momentum conservation equation. For the reacting and the inert component the continuity equations are

$$\frac{\partial \rho_1}{\partial t} = -\operatorname{div} \rho_1 \mathbf{v}_1 + v m_1 I \tag{8a}$$

$$\frac{\partial \rho_2}{\partial t} = -\operatorname{div} \rho_2 \mathbf{v}_2 - v m_1 I \tag{8b}$$

$$\frac{\partial \rho_3}{\partial t} = -\operatorname{div} \rho_3 \mathbf{v}_3 \tag{9}$$

where ρ_1, ρ_2, ρ_3 denote mass densities of the components, m_1 the molar mass of the monomer and I the chemical reaction rate for unit volume expressed in number of moles created in unit time.

The overall momentum conservation equation, neglecting external forces, reads

$$\rho \frac{d\mathbf{v}}{dt} = -\operatorname{div} \Pi \tag{10}$$

where ρ denotes the total mass density ($\rho = \sum \rho_i$), $\mathbf{v}$ is the barycentric (centre of mass) velocity

$$\mathbf{v} = \frac{\rho_1 \mathbf{v}_1 + \rho_2 \mathbf{v}_2 + \rho_3 \mathbf{v}_3}{\rho} \tag{11}$$

and Π is the total pressure tensor.

The mass conservation law for all components can be simply given by summing (8) and (9)

$$\frac{\partial \rho}{\partial t} = -\operatorname{div}(\rho_1 \mathbf{v}_1 + \rho_2 \mathbf{v}_2 + \rho_3 \mathbf{v}_3) \tag{12}$$

or in terms of the barycentric velocity and the substantial time derivative

$$\frac{d}{dt} = \frac{\partial}{\partial t} + \mathbf{v} \operatorname{grad} \tag{13}$$

we get

$$\frac{d\rho}{dt} = -\rho \operatorname{div} \mathbf{v} \tag{14}$$

If besides the component distribution we wanted to express the distribution of temperature, and thus the heat conductivity would be required, the energy conservation law would have to be considered too. As mentioned in the Introduction, however, the heat conductivity problem in a reacting system is a well-investigated subject and therefore we shall not give details here.

The system of Eqs. (5), (8), (9) and (10) should be suitable to express the mole fractions of all three components as a function of time and geometrical coordinates if the initial and boundary conditions of the problem should be known, knowing also the chemical reaction rate I as function of the concentrations and temperature (eventually also its value at the surface boundaries) and expressing the pressure tensor in a suitable way. Moreover, for this general programme a knowledge of the temperature field, due to its large influence on the chemical reaction rate and all phenomenological coefficients used, would be necessary. Because of its complexity, this solution would be of little interest for our purposes or for many other experimental situations. We intend to derive the maximum separations possible and this means the necessity of working with stationary conditions. On the other hand, the relaxation time in our experiments plays no role at all, being similar to the experiments on thermal diffusion without chemical reaction.

One can clearly see that the approach to a stationary state can be accompanied by a change of the hydrostatic pressure. This is due to the following reason. Starting from stationary conditions the barycentric velocity $\mathbf{v}$ [Eq. (11)] will be a function of time due to changes of the mean velocities of the components $\mathbf{v}_i$. This means, in terms of the momentum conservation equation (11), that the right-hand side has a non-vanishing contribution and this will be mainly the hydrostatic part of the overall pressure tensor due to negligible contribution of the viscous tensor. Thus one can expect that between the upper and lower volumes of the apparatus used a pressure difference will exist as long as we do not reach the stationary state. This state will be characterized by the following conditions

$$\frac{\partial \rho_i}{\partial t} = 0 \quad (i = 1, 2, 3) \tag{15}$$

$$\frac{d\mathbf{v}}{dt} = 0 \tag{16}$$

Condition (16) can be given even more strongly because the barycentric velocity has to vanish in each volume element for stationary states, in that we have no convection-current field as in the thermo-gravitational system

$$\mathbf{v} = 0 \qquad (17)$$

Condition (15) applied to the inert component [Eq. (9)] gives

$$\text{div}\,\rho_3\,\mathbf{v}_3 = 0 \qquad (18)$$

$\mathbf{v}_3$ vanishes at the boundaries of the system, however, so it also vanishes in any other volume element due to (18), thus

$$\mathbf{v}_3 = 0 \qquad (19)$$

Now, applying (19) in (17) and (11) one readily gets for both reacting components, as a condition for the stationary state, the relation

$$\rho_1\mathbf{v}_1 + \rho_2\mathbf{v}_2 = 0 \qquad (20)$$

or in terms of molar fractions

$$x_1\,m_1\,\mathbf{v}_1 + x_2\,m_2\,\mathbf{v}_2 = 0 \qquad (21)$$

For the reaction expressed in (4) we can write

$$m_2 = \nu m_1 \qquad (22)$$

thus condition (21) can be simplified. Further, we assume that locally the equilibrium condition for reaction (4) is fulfilled and that the mixture discussed can be treated as an ideal solution. Then we get

$$\nu x_2\,\text{grad}\,x_1 - x_1\,\text{grad}\,x_2 = -\frac{\Delta H}{RT}x_1\,x_2\,\text{grad}\ln T \qquad (23)$$

where ΔH denotes the reaction enthalpy for the transformation (4).

If there was no local chemical equilibrium one would have to start from the continuity equations and the rate equation and this would complicate the calculations considerably. This possibility will not be pursued.

Combining Eqs. (5), (6), (9), (21), (22) and (23) and the condition

$$\sum_i \text{grad}\,x_i = 0 \qquad (24)$$

we can express the gradients of transformable components as

$$\frac{\operatorname{grad} x_1}{\operatorname{grad} \ln T} = x_1 \begin{vmatrix} D^2 & D^P_T \\ +1 & \dfrac{\Delta H}{RT} x_2 \end{vmatrix} \Bigg/ \begin{vmatrix} D^1 & D^2 \\ \nu x_2 & -x_1 \end{vmatrix} \tag{25}$$

$$\frac{\operatorname{grad} x_2}{\operatorname{grad} \ln T} = x_2 \begin{vmatrix} D^1 & D^P_T \\ \nu & -\dfrac{\Delta H}{RT} x_1 \end{vmatrix} \Bigg/ \begin{vmatrix} D^1 & D^2 \\ \nu x_2 & -x_1 \end{vmatrix} \tag{26}$$

where the symbols D^1, D^2 and D^P_T stand for the following abbreviations

$$D^1 = (2x_1 + \nu x_2) D_{13} + \nu x_2 D_{23} - (x_1 + \nu x_2) D_{33} - x_1 D_{11} - \nu x_2 D_{12} \tag{27}$$

$$D^2 = x_1 D_{13} + (x_1 + 2\nu x_2) D_{23} - \nu x_2 D_{22} - (x_1 + \nu x_2) D_{33} - x_1 D_{12} \tag{28}$$

$$D^P_T = x_1 D_{T_1} + \nu x_2 D_{T_2} - (x_1 + \nu x_2) D_{T_3} \tag{29}$$

The molar fractions of the two transformable components are temperature-dependent, due to Eq. (23). In the experiment we are not interested in real concentrations but only in the overall ratio of the inert component to the reacting species, converted into monomer. The measured molar fractions are then

$$\bar{x}_1 = \frac{x_1 + \nu x_2}{1 + (\nu - 1) x_2}; \quad \bar{x}_3 = \frac{x_3}{1 + (\nu - 1) x_2} \tag{30}$$

The thermal diffusion factor α_T is defined by

$$\alpha_T = \frac{\operatorname{grad} \ln \bar{x}_1}{\bar{x}_3 \operatorname{grad} \ln T} \tag{31}$$

Making use of Eqs. (25), (26), (30) and (31) we get

$$\alpha_T = \frac{x_1 x_2 (\Delta H/RT)\{\nu - (\nu - 1) x_1 - [1 + (\nu - 1) x_2] b\} + a_1 [x_1 + x_1 x_2 (2\nu - \nu^2 - 1) + \nu^2 x_2]}{x_3 (x_1 + \nu x_2 b)(x_1 + \nu x_2)} \tag{32}$$

where the following abbreviations are introduced

$$b = \frac{D^2}{D^1}; \quad a_1 = \frac{D^P_T}{D^1}$$

These expressions can be simplified with the help of Reference 8, page 423. In terms of the binary diffusion coefficients D_{ij} we can express b as

$$b = b_0 \frac{1 + R_0/a_0 b_0}{1 + R_0/a_0} \tag{33a}$$

where b_0, a_0 and R_0 denote

$$b_0 = \frac{\nu D_{23}}{D_{13}}; \quad a_0 = \frac{D_{12}}{D_{23}}; \quad R_0 = \frac{x_1 + \nu x_2}{x_3} \tag{33b}$$

Equation (32) is the general formula for α_T valid for the reaction scheme (4). The term proportional to ΔH we simply call the "chemical" part $[\alpha_T \text{(chem)}]$ of α_T, the other term will be denoted by α_T (phys). α_T (chem) vanishes as soon as we are working with only one of the two transformable components (x_1 or $x_2 = 0$). In between, where both components are present in appreciable quantities, one has to expect a maximum contribution of the chemical part. This is just the same functional dependence as is known from the thermal conductivity in reacting systems. Let us discuss here some special forms of Eq. (32). If the stoichiometric coefficient ν equals 1, we are working with simple transformations like rotational or vibrational excitations or isomeric transformations. For these cases Eq. (32) takes the form:

$$\alpha_T = \frac{x_1 x_2 (\Delta H/RT)(1 - b) + a_1(x_1 + x_2)}{x_3(x_1 + x_2 b)(x_1 + x_2)} \tag{34}$$

If we further assume that the kinetic coefficients of both reacting components are the same, i.e.

$$D_{23} = D_{13} \quad \text{and} \quad D_{T_1} = D_{T_2} \tag{35}$$

we see from Eq. (33a) that $b = 1$, thus the term proportional to ΔH cancels. This is the same conclusion as discussed previously.[2] To this result we shall give below a clear physical interpretation.

If $\nu = 2$ the transformation discussed is a dimerization, we get from Eq. (32)

$$\alpha_T = \frac{x_1 x_2 (\Delta H/RT)[2 - x_1 - (1 + x_2) b] + a_1(x_1 - x_1 x_2 + 4x_2)}{x_3(x_1 + 2x_2 b)(x_1 + 2x_2)} \tag{36}$$

Unlike the case $v = 1$ there is now a chemical contribution. If we put $a_0 = 0$ or $b_0 = 1$, we get $b = 1$ and

$$\alpha_T = \frac{x_1 x_2 (\Delta H/RT)}{(x_1 + 2x_2)^2} + \frac{D_{T_1} - D_{T_3}}{2D_{13} - D_{33} - D_{11}} \cdot \frac{x_1 - x_1 x_2 + 4x_2}{x_3 (x_1 - 2x_2)^2} \qquad (37)$$

Equation (37) is identical with that derived previously[1] and applied to a discussion of the experimental data. The difference between the two equations (36) and (37) arises from an improper condition for the stationary state introduced in the previous treatment.

The chemical term as calculated by Eq. (37) gives too high a value in comparison with the experimental results.[1] We shall use here Eq. (36) for further discussion.

In view of experiments described in this paper we shall restrict ourselves to the following transformation

$$2NO_2 \rightleftharpoons N_2O_4$$

B. The Nitrogen Dioxide–Dinitrogen Tetroxide–Inert Gas System

For computational purposes we simplify Eq. (36) by introducing a new variable α_d, the degree of dissociation defined as

$$\alpha_d = \frac{x_1}{x_1 + 2x_2} \qquad (38)$$

Then, remembering (33b),

$$\alpha_T \text{ (chem)} = \left[\frac{(2-b) - (b-1) R_0}{b - (b-1) \alpha_d} \right] \frac{\alpha_d (1 - \alpha_d)}{2} \left(\frac{\Delta H}{RT} \right) \qquad (39)$$

The value of α_d can be calculated from known values of the equilibrium constant K_p,[9] which is related to α_d by

$$\frac{K_p}{P} = \frac{4\alpha_d^2}{(1 - \alpha_d)(1 + \alpha_d + 2/R_0)} \qquad (40)$$

P being the total pressure in the system.

In applying Eq. (40) it has been assumed that the gases are ideal. It is also seen from Eq. (33b) that b is a function of a_0, b_0 and R_0. a_0 and b_0 are assumed to be temperature-independent and are calculated at a constant temperature of $300°K$ using known values of potential parameters given in Table I.

The value of ΔH at 300°K is taken for all calculations. Figure 1 shows how α_T (chem) [Eq. (39)] changes with temperature and the ratio R_0, the total pressure of the system being always 1 atm and

TABLE I. Potential Parameters of Gases used in the Experiments

	He[a]	Ar[a]	NO$_2$[b]	N$_2$O$_4$[b]
ε/k (°K)	10.8	124.9	210	347
σ (Å)	2.57	3.42	3.77	4.62

[a] Reference 11, page 1212.
[b] Reference 10.

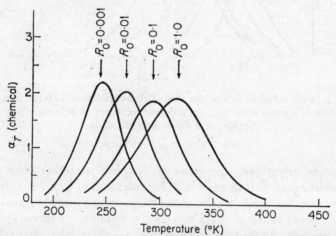

Fig. 1. Calculated values of the chemical contribution α_T (chem) as a function of temperature for four different mixtures of argon, nitrogen dioxide, and dinitrogen tetroxide,

$$R_0 = (x_{NO_2} + 2x_{N_2O_4})/x_{Ar}$$

the inert component being argon. The maximum in each curve occurs at the temperature at which $\alpha_d \simeq 0.5$. This is because the quantity in the square bracket in Eq. (39) is an insensitive function of α_d and the shape of α_T (chem) against temperature is controlled by the factor $\alpha_d(1 - \alpha_d)$. The maximum value of α_T (chem) decreases as R_0 increases because, although the term in the square bracket in

Eq. (39) increases with R_0, this increase is more than offset by the decrease of $\Delta H/RT$ through the increase in T. α_T (phys) is so small that it can be neglected. In Fig. 2 the calculated values of α_T (chem) are shown for the case of helium as the inert component.

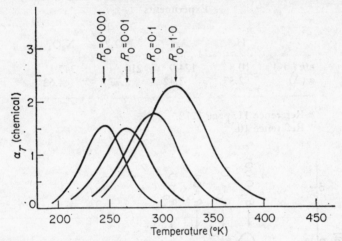

Fig. 2. Calculated values of the chemical contribution α_T (chem) as a function of temperature for four different mixtures of helium, nitrogen dioxide, and dinitrogen tetroxide

$$R_0 = (x_{NO_2} + 2x_{N_2O_4})/x_{He}$$

Unlike the argon case, the value of α_T (phys) is rather large and one has to calculate it to know α_T. An extension of Van der Valk's[12] method can be used for the calculations of α_T (phys), shown in Fig. 3. These values of α_T (phys) can be added to α_T (chem) given in the previous figure to obtain the total α_T. The calculated values of α_T are shown in Fig. 4. It is seen that the increase of α_T (chem) with change of R_0 is almost cancelled out by the decrease in α_T (phys).

3. EXPERIMENTS

A. Experimental Procedure

The experiments were done in conventional two-bulb apparatus, the higher temperature always on top to prevent convection. For variable temperatures below the constant temperature T_c the temperature of the upper bulb was kept constant, for variable temperatures above T_c the temperature of the upper bulb was changed. For constant T_c results of the experiments were plotted on the same graph. For small fluctuations in T_c a correction was made.

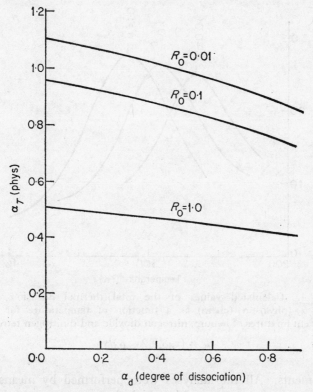

Fig. 3. Calculated values of α_T (phys) as a function of the degree of dissociation α_d for three different mixtures of helium, nitrogen dioxide, and dinitrogen tetroxide,

$$R_0 = (x_{NO_2} + 2x_{N_2O_4})/x_{He}$$

Nitrogen dioxide was made by two different methods. The first one by preparing nitric oxide from copper and nitric acid. The nitric oxide was left overnight with an excess of oxygen. The gas was then cooled to $-80°C$, and excess oxygen pumped off. This was repeated three times. The second method consisted in heating lead nitrate while passing oxygen over it. The nitrogen dioxide formed was led through a trap at liquid nitrogen temperature and the uncondensed part was pumped off. After repeating the heating, condensing and pumping off three times the gas was used for the

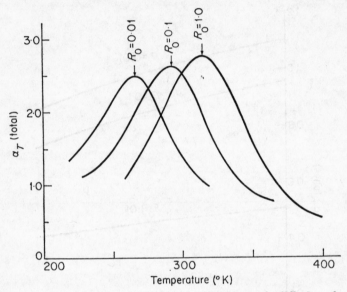

Fig. 4. Calculated values of the total thermal diffusion factor $\alpha_T = \alpha_T \text{(phys)} + \alpha_T \text{(chem)}$ as a function of temperature for three different mixtures of helium, nitrogen dioxide and dinitrogen tetroxide,

$$R_0 = (x_{NO_2} + 2x_{N_2O_4})/x_{He}$$

experiments. All the analyses were performed by means of an Atlas membrane manometer in which the pressure was so low that complete dissociation of the dinitrogen tetroxide into nitrogen dioxide took place.

First the pressure of the total mixture was measured, then the nitrogen dioxide was condensed and the inert gas pumped away. After warming the residue to room temperature the nitrogen dioxide pressure reading was taken. For the 1 : 1 mixture we measured the pressure of the inert gas instead of the nitrogen dioxide.

The analyses of a top and a bottom sample together give the separation factor:

$$q \equiv \frac{[\bar{x}_1/(1-\bar{x}_1)]_{\text{bottom}}}{[\bar{x}_1/(1-\bar{x}_1)]_{\text{top}}}$$

in which $\bar{x}_1$ is the molar fraction of the nitrogen dioxide.

From the definition of the thermal diffusion factor

$$\text{grad}\,\bar{x}_1 = \alpha_T\,\bar{x}_1\,\bar{x}_3\,\text{grad}\ln T$$

it can be easily derived that

$$\ln q = \int_{T_1}^{T_2} \alpha_T\,\text{d}(\ln T)$$

The experimental values of $\ln q$ are therefore plotted against $\ln T_1/T_2$. When either T_1 or T_2 has the constant value T_c in the experiments, it follows from the formula given above that the thermal diffusion factor α_T is equal to the slope of the line.

Although one can determine a value of the thermal diffusion factor α_T at any temperature in this way, the accuracy is rather poor.

For a comparison of theory and experiment we have therefore chosen a different method: from the theoretical values of α_T as a function of temperature the separation factor was determined graphically and plotted on a $\ln q$ against $\ln T/T_c$ graph.

B. Results

In Figs. 5–8 the logarithms of the separation factors $\ln q$ are given together with the calculated values.

These figures refer, of course, to the total separation given by both the physical thermal diffusion term and the chemical contribution. In the mixtures with argon the physical part α_T (phys) is smaller than 0.1 and therefore almost negligible. The situation is quite different in the mixtures with helium. Here, without the chemical contribution, there would still be a considerable separation.

As can be seen the agreement between theory and experiment is very good.

In Figs. 9 and 10 the thermal diffusion factor α_T is given as a function of temperature for two mixtures. As remarked earlier, the experimental values are rather inaccurate, the uncertainty in α_T being 0.2. The figures, however, show clearly the large increase in

α_T as compared to a mixture of the same gases in which no reaction takes place, so that α_T (chem) $= 0$. The maximum "chemical" contribution takes place at a temperature where the degree of dissociation $\alpha_d \simeq \frac{1}{2}$.

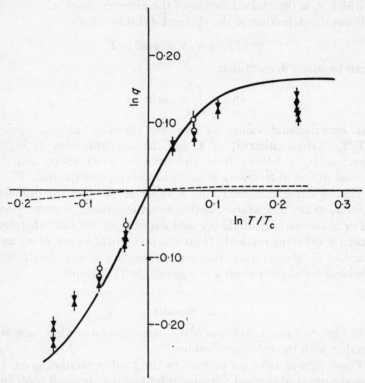

Fig. 5. The logarithm of the separation factor $\ln q$ for the nitrogen dioxide–dinitrogen tetroxide–argon system with $R_0 = \frac{1}{118}$ as a function of $\ln T/T_c$. $T_c = 273°K$, pressure $= 1$ atm,

$$R_0 \equiv (x_{NO_2} + 2x_{N_2O_4})/x_{Ar}$$

The continuous line is calculated from α_T (chem). The dashed line is $\ln q$ calculated from α_T (phys). $\bigcirc$, $\times$: results of two series of experimental measurements.

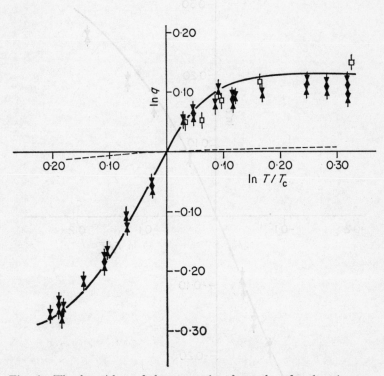

Fig. 6. The logarithm of the separation factor $\ln q$ for the nitrogen dioxide–dinitrogen tetroxide–argon system with $R_0 = \frac{1}{10}$ as a function of $\ln T/T_c$. $T_c = 304°K$, pressure = 1 atm,

$$R_0 \equiv (x_{NO_2} + 2x_{N_2O_4})/x_{Ar}$$

The continuous line is calculated from α_T (chem). The dashed line is $\ln q$ calculated from α_T (phys). $\square$, $\times$: two series of experimental measurements.

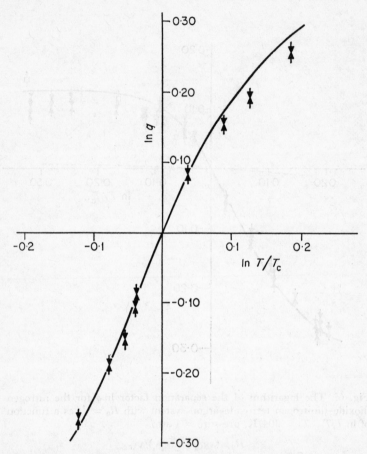

Fig. 7. The logarithm of the separation factor $\ln q$ for the nitrogen dioxide–dinitrogen tetroxide–helium system with $R_0 = \frac{1}{100}$ as a function of $\ln T/T_c$. $T_c = 273°K$, pressure 1 atm,

$$R_0 \equiv (x_{NO_2} + 2x_{N_2O_4})/x_{He}$$

The continuous line is calculated from α_T (total) $= \alpha_T$ (phys) $+ \alpha_T$ (chem).
×: experimental measurements.

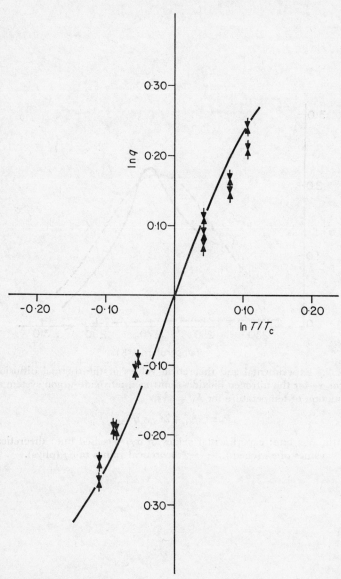

Fig. 8. The logarithm of the separation factor $\ln q$ for the nitrogen dioxide–dinitrogen tetroxide–helium system with $R_0 = 1$ as a function of $\ln T/T_c$. $T_c = 315°K$, pressure 1 atm,

$$R_0 \equiv (x_{NO_2} + 2x_{N_2O_4})/x_{He}$$

The continuous line is calculated from α_T (total) $= \alpha_T$ (phys) $+ \alpha_T$ (chem).
×: experimental measurements.

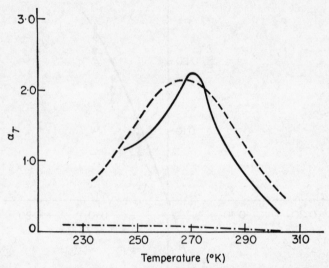

Fig. 9. Experimental and theoretical values of the thermal diffusion factor α_T for the nitrogen dioxide–dinitrogen tetroxide–argon system as a function of temperature for $R_0 = \frac{1}{115}$,

$$R_0 \equiv (x_{NO_2} + 2x_{N_2O_4})/x_{Ar}$$

Continuous line: experimental values of α_T. Dashed line: theoretical values of α_T (chem). —·—·—: theoretical values of α_T (phys).

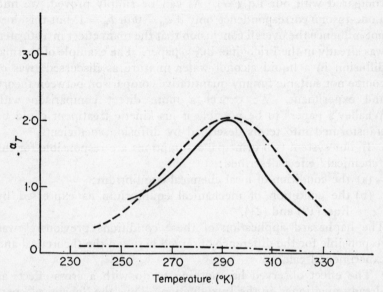

Fig. 10. Experimental and theoretical values of the thermal diffusion factor α_T for the nitrogen dioxide–dinitrogen tetroxide–argon system as a function of temperature for $R_0 = \frac{1}{10}$,

$$R_0 \equiv (x_{NO_2} + 2x_{N_2O_4})/x_{Ar}$$

Continuous line: experimental values of α_T. Dashed line: theoretical values of α_T (chem). —·—·—: theoretical values of α_T (phys).

4. DISCUSSION

The treatment that has been presented is related to the theory given by Prigogine and Buess.[3] Their general equation for the separation effect (Eq. 14,3 on page 859 of reference 3) can be directly compared with our Eq. (37). As can be simply proved, we find a one-to-one correspondence only if $a_0 = 0$ or $b_0 = 1$, but this does not influence the overall conclusion that the main effect investigated was already in the Prigogine–Buess paper. The example of thermal diffusion in a liquid alcohol–water mixture as discussed[3] was of course not suitable for any quantitative comparison between theory and experiment. We expect a more direct comparison with Whalley's paper[6] to be possible if his kinetic treatment could be transformed into terms described by diffusion coefficients.

In our system two principal assumptions are responsible for the "chemical" effect described:

(a) the condition of local chemical equilibrium;
(b) the condition of mechanical equilibrium as expressed by Eqs. (19) and (21).

The haphazard application of these conditions previously[1] was responsible for the difference observed between the theoretical and experimental values.

The effect observed has nothing to do with a cross-effect, as already mentioned in the Introduction. Thus the "chemical" part of the overall α_T is not a real thermal diffusion factor, as it is for the physical part. It would perhaps be better to speak about a "thermochemical separation".

Comparing Eqs. (37) and (34) we clearly see that the main part of the chemical contribution is due to the stoichiometric character of the chemical reaction. A simple physical interpretation corresponding to this behaviour can be given as follows.

If there is no thermal diffusion and the mixture is ideal, the partial pressure of the inert component, e.g. argon, in the steady state can be assumed to be equal in the two bulbs This means that the ratio $(N_2O_4 + NO_2)/Ar$ is the same for both bulbs. Due to the different temperatures in the upper and lower bulbs the ratio of $N_2O_4 : NO_2$, however, is not the same. As we analysed at low pressure the dinitrogen tetroxide was dissociated completely into nitrogen dioxide and we therefore measure the ratio $(2N_2O_4 + NO_2)/Ar$ which is not the same for the two bulbs. This physical picture accounts qualitatively for the chemical separation observed. The assumptions made here are of course very primitive and should not be taken too literally.

Let us now give some comparisons between the thermal diffusion separation discussed here and the well-known influence of a chemical reaction on the thermal conductivity. In both cases the temperature dependence of the "chemical" term has a similar shape, namely a negligible contribution in the temperature regions where one of the reacting species exists in minor concentration, and the maximum contribution in the neighbourhood of a degree of dissociation $\alpha_d \simeq \frac{1}{2}$, where the number of concentrations of dimer : monomer $= 1 : 2$. Let us also remember the observed difference between the experimental and theoretical results and that an unsuitable frame of reference was taken into account. It corresponds to a certain extent to the difference we observed between the inexact stationary condition applied previously and the exact condition for momentum conservation as applied in this paper. In both cases the theoretical values were larger than the experimental results.

In the treatment presented we did not consider the influence of the chemical reaction on the temperature field, because in the experimental method applied, only the temperatures of the two bulbs were important. This would not be the case if, for example, the thermogravitational method as in a thermal diffusion column were used. Then the temperature field would come directly into the Navier–Stokes equation and thus would influence the separation effect observed through the presence of a convection current.

From the results given here it is clear that the theory presented is confirmed by the experiments. The $N_2O_4 \rightleftharpoons 2NO_2$ equilibrium is of course a very special case. What happens for other equilibrium situations may be calculated along the same lines as has been done here. An interesting application which has been considered is the change in the thermal diffusion factor for moderately high pressures as compared to the low pressure values. The clusters or quasi-dimers which are formed have an energy of attraction in the order of 1–2 kcal/mole and an increase in the thermal diffusion factor of about 0.2 might be expected. Although the experiments show a larger increase in some cases, up to about 1 at pressures of 100 atm, a contribution by the effect studied here is certainly present.

In the case of the local equilibrium condition not being fulfilled, due to the slowness of the chemical reaction rates, one can expect that the separation measured will be smaller than the value calculated for the equilibrium case. If there is no equilibrium the geometry of the apparatus comes into the calculation. Dirac[14] treated the problem for a dissociating gas, without any additional component, slightly removed from equilibrium. His treatment may be extended to systems like ours.

ACKNOWLEDGEMENTS

We would like to thank Dr. W. A. Oost for pointing out an error in the calculation and also to thank him and Mr. F. Vitalis for help in programming. Many of the analyses were done by Mrs. I. Balster.

This work is part of the research programme of the Stichting voor Fundamenteel Onderzoek der Materie (Foundation for Fundamental Research on Matter) and was made possible by financial support from the Nederlandse Organisatie voor Zuiver-Wetenschappelijk Onderzoek (Netherlands Organisation for the Advancement of Pure Research).

APPENDIX

After the above review article was sent off for publication (June, 1967) there appeared three papers which we want to mention briefly as far as some new aspects in respect of the above treatment are concerned.

Based on the previously published results[1] systematic derivations and calculations were given by Brokaw.[15] In the introduction to his paper he briefly discusses the influence of a chemical reaction on the heat conductivity as well as Schäfer's and Whalley's papers, we mentioned above.[5, 6] He did not mention Prigogine's and Buess' results[3] which helped us considerably, treating the influence of internal excitations on the thermo-diffusional separation. Without this introductory step[2] we should probably never have started the investigation of a dissociating gas in the presence of an inert component. When we began our calculations and experimental investigations we did not even know of Schäfer's and Whalley's treatments, which came to our knowledge later. They had not initiated in the past any systematic, theoretical or experimental investigations. Brokaw's paper treats the following three types of reacting systems (the same classification will be applied for the two next papers to be considered):

1. A dissociating gas in the presence of one inert, non-reacting component.

2. A dissociating gas in the presence of two non-reacting components.

3. A reacting system, forming two different components from one $(A \rightleftharpoons B + C)$.

Type 1 is in detail just what we have treated in the above review and compared with experimental results. For this case Brokaw's formula for the additional separation, due to the chemical reaction, is given in terms of the additive heat conductivity coefficient (λ_v) due to the dissociating reaction taking place. In the notation used in the above review this equation reads [Eq. (9) in Brokaw's paper]

$$\alpha_T \,(\text{chem}) = \frac{RT}{\bar{x}_1 P} \left[\nu \left(\frac{1}{D_{23}} - \frac{1}{D_{13}} \right) + (\nu - 1) \left(\frac{1}{D_{21}} - \frac{1}{D_{23}} \right) \bar{x}_1 \right]$$

$$\times \left(\frac{\lambda_v}{R} \right) \left(\frac{RT}{\varDelta H} \right) \tag{41}$$

This equation corresponds to relation (39) in the above review article, illustrated by numerical calculations for the nitrogen dioxide–dinitrogen tetroxide–argon system (Figs. 9 and 10 of the review and Fig. 1 in Brokaw's paper). We see clearly, what was briefly discussed in the review, that both phenomena—the additive separation and the additive heat conductivity—are related in a quantitative way. This means also that the temperature-dependence of both effects has to be very similar as it is characterized by a well-shaped maximum, appearing in the neighbourhood of comparable concentrations of both reacting species, shown for the additional separation on Figs. 1, 2, 4, 9 and 10. Equation (41) confirms the principal importance of the stoichiometric coefficient of the reaction—since the second term on the right-hand side of (41) disappears for $\nu = 1$, i.e. for simple excitations. In deriving (41) the assumption of the local chemical equilibrium was made. The physical part of the separation, as in all cases treated by Brokaw, is not taken into account. Since Brokaw raised the question of too great a separation, put forward in the preliminary theoretical treatment in reference 1 (see, for example the footnote on page 3263), let us stress here that the only difference between the assumptions used in reference 1, on the one hand, and in Brokaw's paper and the above review article, on the other, is that in reference 1 we used $W_3 = 0$ as characterizing the stationary state instead of the more rigorous condition $v_3 = 0$ used later. This does

TABLE II

Reduced Thermal Diffusion Factors α_0 for Separation of Noble Gas Isotopes

Gas	Calculated for the presence of HF	Measured in the absence of HF
He	0.83	0.434
Ne	0.47	0.290–0.515
Ar	−0.22	0.065–0.438
Kr	−1.18	0.037–0.165
Xe	−1.59	0.038–0.13

not change the conclusion that the main results were already derived in reference 1 and that Brokaw's treatment as well as the above review article present mainly rederivations.

Brokaw gives new and interesting results for case 2, which is treated numerically in his paper for the separation of noble gas isotopes in the presence of the reaction:

$$(HF)_6 \rightleftharpoons 6HF$$

This was previously investigated by Franck and Spalthoff in respect of its influence on the effective heat conductivity.[13] We give here the table from Brokaw's paper, comparing the α_T coefficients for the isotopes of different noble gases, as calculated in the presence of the above reaction, and compared with known experimental results obtained in the absence of this reaction. In the table reduced α_0 coefficients are given related to the previous α_T quantity by

$$\alpha_0 \equiv \alpha_T \frac{M_3 + M_4}{M_3 - M_4} \qquad (42)$$

where M_3 and M_4 are the molar masses of both isotopes added to the reacting components as inert species.

The calculations were carried out for large additions of hydrogen fluoride at a temperature of $0°C$. In the absence of hydrogen fluoride the α_0 coefficients are always positive. This corresponds to the accumulation of the heavy isotope in the region of lower temperature. In the presence of the reaction mentioned, however, the heavy isotopes of argon, krypton and xenon are expected to be concentrated in the hot region of the separation device. On the other hand, let us remark that the α_0 coefficients are, in the presence of hydrogen fluoride, sometimes more than one order of magnitude higher than the known measured values in the absence of the chemical reaction. This may be of practical importance if using thermal diffusion for isotope separations.

The most interesting result of the above table is that the increase of the separation is considerable for heavy noble gases where the isotope separation by physical thermodiffusion becomes less effective.

Types 2 and 3 were also treated theoretically by Paul and de Vries.[16] Numerical calculations were carried out for the hydrogen–nitrogen system in the presence of the reacting nitrogen dioxide–dinitrogen tetroxide system and for the reacting set

$$2NOCl \rightleftharpoons 2NO + Cl_2 \quad \text{(type 3)}$$

The first system was studied by Schäfer in some experiments. The comparison of his results with the calculated[17] values shows that the separation factor measured by Schäfer is higher (1.68) than the theoretical separation (1.12). Also the accumulation of nitrogen in the hot region, as observed by Schäfer, could not be explained by the present theory. A rather high α_T coefficient was calculated for the nitrosyl chloride reacting component (9.6) due mainly to the large heat of reaction.

In the second paper[17] types 2 and 3 were investigated experimentally by Paul and de Vries. As an example of type 2 the mixture of helium and argon was taken in the absence of further components and in the presence of the reacting nitrogen dioxide–dinitrogen tetroxide mixture. A satisfactory agreement was found between the theoretical and experimental values. The reacting mixture caused, as usual, an increase of the separation, but its numerical values were not very large. For type 3 the reaction:

$$2NO_2 \rightleftharpoons 2NO + O_2$$

with an excess of oxygen was investigated. The theoretical and experimental results are shown in Fig. 11.

As can be seen, the theoretical values calculated for infinite reaction rate, which corresponds to the assumption of local chemical equilibrium used so far, are larger than the experimental results. It seems that for this system the local chemical equilibrium assumption is no longer true and a simple but effective theory was proposed to explain the indicated deviations of Fig. 11 quantitatively. The treatment was carried through along the following lines. The concentrations in both bulbs are assumed to be uniform but shifted from the equilibrium values by the diffusion flows through the connecting tube. The diffusion influx of nitrogen dioxide into the hot bulb of the separation device (J_{1h}) is approximated by

$$J_{1h} = \frac{nD_{12}A}{V_h}\left(\frac{\alpha_h - \alpha_c}{lR_0}\right) \tag{43}$$

where n is the total number density, D_{12} the binary diffusion coefficient of the nitrogen dioxide–oxygen combination, A the cross-section of the connecting tube, V_h the volume of the hot bulb, l the length of the connecting tube, α_h, α_c are the degrees of dissociation for the hot and cold bulb respectively, and R_0 denotes the initial total mole ratio x_{O_2}/x_{NO_2} (Fig. 11). Introducing further the conventional rate equations for the forward and reverse reactions and their relation to the equilibrum constant one derives finally the following equation for the degree of dissociation

$$\alpha_h^2\left(\frac{p}{K_p}-1\right) + \alpha_h(2+\phi_h) - (\phi_h\alpha_c+1) = 0 \tag{44}$$

where p denotes the total pressure, K_p the equilibrium constant as expressed in partial pressures and the abbreviation ϕ_h is given by

$$\phi_h = \frac{AD_{12}R_0}{nk_f V_h l} \tag{45}$$

where k_f is the rate constant of the forward reaction (dissociation rate of nitrogen dioxide in the hot bulb). A similar equation to (44) can be written for the cold bulb. The last equation was solved by a

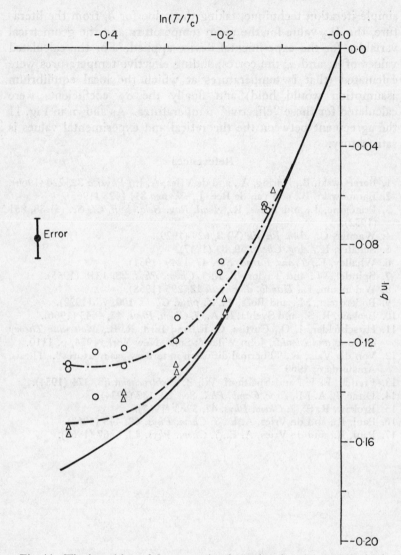

Fig. 11. The logarithm of the separation factor ln q for nitrogen dioxide/oxygen as a function of temperature with constant upper temperature $T_c = 799°K$. ○: experimental values for $R_0 = 95$. △: experimental values for $R_0 = 8.5$. ——: theoretical result for infinite reaction rate. – – –: theoretical result for $R_0 = 8.5$. –·–·–: theoretical result for $R_0 = 95$. $R_0 = $ mole ratio x_{O_2}/x_{NO_2}.

simple iteration technique, taking the value for k_t from the literature, the D_{12} value for the mean temperature and the geometrical variables for the experimental device applied. For the calculated values of α_h and α_c the corresponding effective temperatures were calculated (that is temperatures at which the local equilibrium assumption would hold) and finally the α_T coefficients were calculated for these "effective" temperatures. As shown in Fig. 11 the agreement between the theoretical and experimental values is satisfactory.

References

1. Baranowski, B., Haring, A., and de Vries, A. E., *Physica* **32**, 220 (1966).
2. Baranowski, B., and van de Ree, J., *Physica* **31**, 1428 (1965).
3. Prigogine, I., and Buess, R., *Acad. Roy. Belg. Bull. Cl. Sc.* (5) **38**, 851 (1952).
4. Wagner, C., *Ann. Physik* (V) **3**, 629 (1929).
5. Schäfer, K., *Ang. Chem.* **59**, 83 (1947).
6. Whalley, E., *Trans. Farad. Soc.* **47**, 1249 (1951).
7. Spindel, W., and Taylor, T. I., *J. Chem. Phys.* **23**, 1318 (1955).
8. Waldmann, L., *Handb. d. Physik* **12**, 295 (1958).
9. Bodenstein, M., and Boës, F., *Z. phys. Chem.* **100**, 75 (1922).
10. Brokaw, R. S., and Svehla, R. A., *J. Chem. Phys.* **44**, 4643 (1966).
11. Hirschfelder, J. O., Curtiss, C. F., and Bird, R. B., *Molecular Theory of Gases and Liquids*, John Wiley & Sons, New York, 1954, p. 1110.
12. Van der Valk, F., "Thermal diffusion in ternary gas mixtures". Thesis, Amsterdam, 1963.
13. Franck, E. U., and Spalthoff, W., *Z. Elektrochem.* **58**, 374 (1954).
14. Dirac, P. A. M., *Proc. Camb. Phil. Soc.* **22**, 132 (1924).
15. Brokaw, R. S., *J. Chem. Phys.* **47**, 3263 (1967).
16. Paul, R., and de Vries, A. E., *J. Chem. Phys.* **48**, 445 (1968).
17. Paul, R., and de Vries, A. E., *J. Chem. Phys.* **43**, 4867 (1968).

THE PHASE PROBLEM IN STRUCTURE ANALYSIS

JEROME KARLE, *Laboratory for the Structure of Matter, U.S. Naval Research Laboratory, Washington, D.C., 20390*

CONTENTS

1. INTRODUCTION

Most materials can occur in the crystalline state, forming regular three-dimensional lattices. Although it might be expected that simple materials would form regular arrays, it is somewhat surprising to find that extremely complex structures, such as proteins, can be made to crystallize. The readiness of materials to form crystals is of great importance, since the crystalline state is especially suited for analysis by means of X-ray diffraction which can reveal detailed atomic arrangements and vibrational motions. Correlations between fundamental structure and various physical and chemical properties are thus often facilitated by X-ray structure analyses.

Many problems in chemistry are concerned with the identification and the determination of the configuration of the intermediate and final products of chemical reactions. To a very large extent such problems have been handled in the past by methods of analysis and deduction which are largely chemical in nature. In recent years physical methods have played an increasing role in chemical studies. The many fields of spectroscopy are an outstanding example. A physical method becomes an important ally with classical chemical techniques when it attains a level of broad generality, simplicity of interpretation and relative ease of application. Such has been the recent history of the study of molecular and crystal structure by X-ray diffraction. The key to this development has been the considerable progress made in the phase problem and, in addition, the increasing availability of automatic diffractometers and high-speed computing facilities.

The phase problem is a central problem in crystal structure determination. Once phases can be associated with the structure factor magnitudes which are directly obtainable from experiment, the electron density distribution and therefore the atomic structure of the unit cell of a crystal may be readily computed.

Within the past twenty years, considerable changes have occurred both from the conceptual and technical points of view. Twenty years ago experiments were almost always carried out with photographic techniques, limited data were collected and computations were performed with a hand computer. Structure determinations were long-range projects in which use was made of intuitive insight, trial and error procedures and partial information obtained from the Patterson function, which is generally an incompletely resolved Fourier map of the collection of interatomic vectors in the crystal. Some special techniques such as the use of the heavy atom method and its further development in isomorphous replacement procedures were recognized to be quite useful, but required the availability of appropriate crystals.

Present-day instrumentation and the accelerating advances in the computer industry have extensively modified the experimental aspects of X-ray structure analysis. Complete sets of data are now collected as a matter of routine by photographic techniques or automated diffractometers, in which photon counting instruments measure the intensities of the diffracted X-ray beams. An extensive set of computer programmes is now in common use, including programmes for the direct determination of phase values from measurements of the diffracted intensities.

Owing to the fact that recent progress in phase determination has made it possible to solve almost all crystals of at least moderate complexity, i.e. up to several hundred atoms per unit cell, in many instances with routine simplicity, the question may be asked whether a simple and practical solution to the phase problem might have been anticipated twenty years ago. The answer to this question can provide a good introductory insight into the nature of the phase problem and the supporting theoretical framework on which the search for a solution may be based.

It will be seen that the structure problem in X-ray crystallography can be expressed in terms of a set of simultaneous equations in which the unknown quantities are the atomic positions and the known quantities are the structure factor magnitudes derivable from the measured X-ray intensities and the atomic scattering factors. It can be readily shown that this set of simultaneous equations is greatly overdetermined, by approximately a factor of fifty for a full sphere of reflection from copper radiation for centrosymmetric crystals, and a factor of about twenty-five for non-centrosymmetric ones. The unknown phases also occur in one manner of expressing the set of simultaneous equations, but they may be eliminated in order to demonstrate the overdeterminacy of the set. As a practical matter in structure analysis, it is generally quite expedient to determine first the values of the phases of the structure factors as an intermediate step and then use them in the computation of the atomic positions. Thus the overdeterminacy by a factor of twenty-five or fifty, which characterizes a structure determination, also applies to the intermediate step of determining phases. If this extremely favourable circumstance is coupled with the observation that mathematical formulations associated with physical problems often assume an aesthetic simplicity, it follows that the existence of a practical solution to the problem of phase determination could readily be expected. This expectation, based as it is on the strong theoretical support for the overdeterminacy, has played a significant background role in providing a motive in the search for and development of methods for direct phase determination from the measured X-ray intensities.

It is the intent of this article to describe the present status of phase determination both in terms of the theoretical background and a wide variety of illustrative examples which show the broad range of applicability of current methods. The most general method of phase determination, in which the phases are obtained directly from the measured X-ray intensities without taking advantage of any special structural features, is called the symbolic addition procedure. A particularly demanding problem to which this method of analysis can be applied is that of identifying and determining the structure of the end-products of rearrangement reactions. In such reactions materials of unknown configuration and composition are often obtained. Several examples of applications to this type of problem will be discussed.

When suitable crystals are available the methods of isomorphous replacement and anomalous dispersion are very useful and have found extensive application in attempts to probe the structures of protein molecules. Improvements in accuracy which have accompanied the development of automatic diffractometry have enhanced the opportunity to utilize the anomalous dispersion technique. The Patterson method plays an important auxiliary role in these procedures because of its use in the location of the heavy atoms in a crystal. In special circumstances there is an opportunity to combine various methods of structure determination readily. Increasing attention has recently been given to combining the isomorphous replacement and anomalous dispersion methods. Alternatively, initial information from such procedures may be combined with the direct phase determining formulae of the symbolic addition procedure to obtain new phase information.

The complete analysis of the interatomic vectors forming the Patterson function effectively avoids the phase problem by obtaining the atomic positions without the necessity for considering the phases. In actual practice a Patterson map for a moderately complex material has rarely been completely analysed, although there are some noteworthy exceptions as shown for example by the work of Hoppe[1-3] and Nordman [4,5] and their collaborators. Progress in this field has been recorded at a recent symposium on machine interpretations of Patterson functions.[6] The difficulties encountered in analysing vector maps are due to problems of accuracy and resolution which are readily comprehended when it is realized that for a crystal having N atoms per unit cell, there are $N(N-1)$ interatomic vectors. In the usual application of a vector map, partial structural information, such as the positions of heavy atoms, is readily obtained. This partial information may be used to compute approximate phases leading to a Fourier map of the electron density, from which additional structural information may be obtained. Reiteration of this procedure can lead to a complete structure determination. Methods of increasing sophistication have been developed for use with partial structural information in order to obtain the final result in fewer reiterative steps. One procedure involves the use of formulae for direct phase determination. It has proved to be very effective in developing a complete structure from partial structural information[7] consisting of the known positions of some light atoms which may be found by means of the symbolic addition procedure. In this way an effective method is obtained for studying equal atom non-centrosymmetric crystals. The latter type of structure has been the most difficult to analyse.

Several review articles and general discussions of special topics associated with the phase problem have appeared within recent years. A general review of the phase problem in X-ray crystal structure analysis covering the period up to 1962 has been published[8] and a description of the symbolic addition procedure[9] has appeared in 1966. Several reviews of the application of X-ray structure analysis to protein crystals have been written. Among the more recent ones are articles by Dickerson,[10] Holmes and Blow,[11] Phillips[12] and Vainshtein.[13] From the point of view of method, these concern the application and extension to protein crystals of the fundamental contributions of Patterson,[14] whose method leads to the location of the heavy atoms, and of Bijvoet and collaborators who initially developed the application of isomorphous replacement[15] and anomalous dispersion[16, 17] to non-centrosymmetric crystals.

The previously written general review article on phase determination[8] covers the period just previous to the introduction and the general adoption of the symbolic addition procedure for phase determination. It is therefore appropriate to write this article with the view toward reviewing the background ideas, content and diverse applications of this procedure. Recent developments in the other techniques of phase determination will also be reviewed.

6

Owing to the great proliferation of the literature on these subjects, the attempt will not be made to make an exhaustive survey. Rather the discussion will be confined to the main concepts involved and to illustrative examples which show the wide diversity of problems to which the current methods are applicable.

2. OVERDETERMINACY AND SOLVABILITY

The structure of the unit cell of a crystal may be described in terms of the electron density distribution in the cell. The maxima of the electron density function locate the positions of the atoms. Owing to the repetitive nature of the unit cell contents in a crystal lattice, the electron density distribution function $\rho(\mathbf{r})$ may be represented by a three-dimensional Fourier series

$$\rho(\mathbf{r}) = V^{-1} \sum_{\substack{\mathbf{h} \\ -\infty}}^{\infty} F_{\mathbf{h}} \exp(-2\pi i \mathbf{h} \cdot \mathbf{r}) \tag{1}$$

where the coefficients

$$F_{\mathbf{h}} = |F_{\mathbf{h}}| \exp(i\phi_{\mathbf{h}}) \tag{2}$$

are the crystal structure factors, the ratios of the amplitudes of coherent scattering from the plane segments in the unit cell, labelled with the vectors $\mathbf{h}$, to the amplitude of scattering from a free electron at the origin of the unit cell under the same conditions. The components of $\mathbf{h}$ are the integers h, k, l, called the Miller indices, which are inversely proportional to the intercepts of the corresponding planes on the chosen axes. The angle $\phi_{\mathbf{h}}$ is the phase associated with $F_{\mathbf{h}}$ and $\mathbf{r}$ is the position vector of any point in the unit cell.

If the coefficients $F_{\mathbf{h}}$ of the Fourier series were directly obtainable from experiment, the structure of any crystal could be immediately computed from Eq. (1). Unfortunately only the magnitudes $|F_{\mathbf{h}}|$ are ordinarily obtainable from experiment, whereas the phases $\phi_{\mathbf{h}}$ are not, giving rise to the so-called phase problem. Since in attempting to compute $\rho(\mathbf{r})$ from Eq. (1) only the magnitudes $|F_{\mathbf{h}}|$ are known, a more or less arbitrary function would be obtained from arbitrary specifications for the $\phi_{\mathbf{h}}$. This fact led to the belief that the phase problem was in principle unsolvable and that crystal structures could generally only be determined by the use of special devices, ranging from trial and error accompanied by experience to the various techniques which are applicable when heavy atoms are present. There is an important additional fact, however, which alters the latter conclusion, namely that $\rho(\mathbf{r})$ is not a completely unknown and arbitrary function, but may be represented as a sum of N discrete, known atomic electron distributions per unit cell. This may be expressed by performing the Fourier inversion of (1) and reducing the usual integral expression for the Fourier coefficient to the sum

$$|F_{\mathbf{h}}| \exp(i\phi_{\mathbf{h}}) = \sum_{j=1}^{N} f_{j\mathbf{h}} \exp(2\pi i \mathbf{h} \cdot \mathbf{r}_j) \tag{3}$$

where $f_{j\mathbf{h}}$ is the atomic scattering factor of the jth atom (the ratio of the amplitude of coherent scattering for the atom to that for a free electron at the atomic centre under the same conditions) in a unit cell containing N atoms and $\mathbf{r}_j$ is its position.

The solvability and overdeterminacy of the phase problem can be demonstrated by a study of the set of simultaneous equations comprising Eqs. (3). In Eqs. (3), the unknown quantities are the phases $\phi_{\mathbf{h}}$ and the atomic positions $\mathbf{r}_j$. The known quantities are the magnitudes of the structure factors $|F_{\mathbf{h}}|$, obtainable from experiment, and the atomic scattering factors $f_{j\mathbf{h}}$ which have been computed and tabulated.[18] Phases may be eliminated from Eqs. (3) by multiplying by the complex conjugate to obtain

$$|F_{\mathbf{h}}|^2 = \sum_{j=1}^{N} \sum_{k=1}^{N} f_{j\mathbf{h}} f_{k\mathbf{h}} \exp[2\pi i \mathbf{h} \cdot (\mathbf{r}_j - \mathbf{r}_k)] \tag{4}$$

The only unknown quantities in Eqs. (4) are the independent atomic coordinates in the asymmetric unit of the unit cell. Atomic positions in the unit cell may be obtained from a knowledge of the atomic positions in the asymmetric unit by application of the symmetry elements appropriate to the space group under consideration. The sphere of scattering obtained from X-ray radiation from a copper target generally provides the values of the intensities, $|F_{\mathbf{h}}|^2$, of more than 150 independent reflections for each atom in the asymmetric unit of a centrosymmetric crystal, and more than 75 independent reflections for each atom in the asymmetric unit of a non-centrosymmetric crystal. Since each atom has three positional coordinates, the simultaneous equations (4) are overdetermined by a factor of the order of 50 for centrosymmetric crystals and 25 for non-centro-symmetric ones. There is of course an experimental error in the values of the $|F_{\mathbf{h}}|^2$ which may range from a few percentage to 10 or 20%, depending upon the manner in which the experiment was carried out. There may also be an error of a few percentage in the values of the $f_{j\mathbf{h}}$ computed from theoretical considerations. However, the very great overdeterminacy of the simultaneous equations (4) makes these errors insignificant for phase determination.

Equations (3) which are an alternative form of Eqs. (4) contain the additional unknown phases as well as atomic positions, raising the question of the solvability and determinacy of this set. It is, in fact, seen to be the same as that of set (4) when it is realized that each equation in set (3) involving complex quantities is in fact two equations, one for the real part and one for the imaginary part. Thus, the great overdeterminacy in the determination of atomic positions applies equally well for the determination of phases. This is a matter of considerable practical significance since it is often more readily possible to determine phases and then obtain the structure by the computation of (1) than to find the atomic positions directly.

The general method for obtaining atomic positions directly is by means of the Patterson function,[14]

$$P(\mathbf{r}) = \sum_{\substack{\mathbf{h} \\ -\infty}}^{\infty} |F_{\mathbf{h}}|^2 \exp(-2\pi i \mathbf{h} \cdot \mathbf{r}) \tag{5}$$

the maxima of which represent the interatomic vectors. The coefficients of this Fourier series are proportional to the measured X-ray intensities and therefore the series can be readily calculated for any crystal. As mentioned previously, the analysis of this function is limited by a lack of resolution of the $N(N-1)$ interatomic vectors and uncertainties in the data. Nevertheless, many complex structures possessing heavy atom substituents have been readily solved with the aid of the Patterson function because the interatomic vectors associated with the heavy atoms can be relatively easily identified. The positions of the heavy atoms may be deduced from the interatomic vectors and the phases computed from only these positions employing Eqs. (3) can be used as an initial set of phases for carrying through the structure determination.

Although there may be limitations to the facility with which an arbitrary Patterson function may be analysed, it has been shown from algebraic considerations in this Section that the structure problem is greatly overdetermined. It is therefore not unexpected that an alternative procedure may facilitate the analysis and thereby extend the range of structural problems which can be readily solved from X-ray scattering data. In the later Sections of this chapter such a procedure, the symbolic addition procedure, and many of its applications will be described.

3. PROBABILITY DISTRIBUTION FOR A STRUCTURE FACTOR, CENTROSYMMETRIC CASE

Several aspects of direct procedures for structure determination depend upon the statistical properties of the structure factors or products of them. This is manifested in a structure analysis in the initial treatment of the measured intensities and also during the course of the phase determination. A general background familiarity with the probability concepts which are employed may be obtained from a discussion of the probability distributions of the structure factors for centrosymmetric and non-centrosymmetric crystals. In this Section we consider centrosymmetric ones. Probability distributions for structure factors were first derived by Wilson.[19, 20] They will be derived here in a manner which not only serves to yield the result of immediate interest but also forms the basis for a general procedure for obtaining the joint probability distributions, which have played an important role in suggesting phase determining relations and assessing the reliability of phase determining procedures.

We first consider a very simple problem containing the varied concepts which are generalized in the formalism for finding the joint distribution. Suppose we have the sum

$$F = \sum_{j=1}^{4} \alpha_j \qquad (6)$$

where α_j can be 0 or 1 with equal probability and, in considering all possible configurations, we ask what the probability is that the chance choice of a particular configuration will give $F < 2.1$. It can be readily verified that 11 of the 16 configurations satisfy $F < 2.1$

and since each of the configurations is equally probable, the desired probability is 11/16. In a more general problem, the variables α_j could take on a continuous range of values, not necessarily of equal probability. A facility would also be required for examining whether for a given configuration F is smaller than a preassigned number A or not and for taking proper account of this fact.

We now define the structure factor for a centrosymmetric crystal in an alternative fashion to (3)

$$F_{\mathbf{h}} = \sum_{j=1}^{N/n} f_{j\mathbf{h}}\, \xi(x_j, y_j, z_j;\, \mathbf{h}) \tag{7}$$

where the definition of the function ξ for the appropriate centro-symmetric space group may be found in the *International Tables for X-ray Crystallography*.[21] Equation (7) is obtained from (3) by combining the contributions to (3) from symmetry related atomic positions in the unit cell. These are also listed in the *International Tables*. If there are n symmetry related atomic positions in the unit cell, it is only necessary to sum over N/n terms as indicated in (7). The quantity n is called the symmetry number and there are N/n independent atomic positions, referred to as the asymmetric unit. The latter statement must be altered somewhat when atoms occupy special positions in the unit cell. We are considering only the case when a crystal contains atoms solely in general positions. The statistical effect of atoms in crystallographically special positions or in special arrangements has been the subject of much study, e.g. by Wilson,[20, 22] Lipson and Woolfson,[23] Hargreaves[24], Rogers and Wilson[25] and Parthasarathy.[26, 27] Since the results of these studies do not have a major impact on the current procedures for phase determination, the statistics of special positions will not be reviewed here.

There are two conceptually distinct structure factor distributions. The first arises by fixing the Miller indices $\mathbf{h} = (h, k, l)$ and allowing the atoms in the crystal to range uniformly throughout the unit cell, subject to the conditions imposed by symmetry. For finding the second distribution, the positions of the atoms are fixed and the Miller indices are allowed to range uniformly over the integers. The latter gives rise to the distribution observed experimentally for the structure factor magnitudes corrected for the vibrational motion of the atoms. It has been shown[28] that the two distributions are identical except for differences which occasionally arise when there are atoms in special positions in the unit cell. We shall obtain the general form of the probability distribution for all centrosymmetric crystals by the procedure of fixing the Miller indices and allowing the atoms in the crystal to range uniformly throughout the unit cell, the first procedure mentioned above. The distributions will be seen to be expressible in terms of the moments of the individual terms, $\xi(x_j, y_j, z_j; \mathbf{h})$, of the summation in (7) defining the structure factor.

We shall now find the probability $Q(A)$ that F as defined in (7) satisfies $F < A$. (The subscript $\mathbf{h}$ of $F_{\mathbf{h}}$ is omitted since it is cumbersome and disappears from the final result of this Section. Its presence in the intervening steps is understood.) The quantity $p(\alpha)\, d\alpha$ represents the probability that $\xi_{j,h} = \xi(x_j, y_j, z_j; \mathbf{h})$ lie in the interval $(\alpha, \alpha + d\alpha)$ if the x_j, y_j, z_j are uniformly and independently distributed in $(0, 1)$, the basis on which the calculation is to be made. We then obtain

$$Q(A) = \int_{-\infty}^{\infty} \cdots \int_{-\infty}^{\infty} p(\alpha_1) \cdots p(\alpha_{N/n})\, T(\alpha_1, \ldots, \alpha_{N/n})\, d\alpha_1 \cdots d\alpha_{N/n} \quad (8)$$

where T is the discontinuous integral,

$$T(\alpha_1, \ldots, \alpha_{N/n}) = \frac{1}{2} - \frac{1}{2\pi} \int_{-\infty}^{\infty} \frac{\exp\left[i(F-A)x\right]}{ix}\, dx \quad (9)$$

having the property that $T = 1$ if $F < A$ and $T = 0$ otherwise. The structure factor F is defined in terms of the integration variables α_j,

$$F = \sum_{j=1}^{N/n} f_j \alpha_j \quad (10)$$

for insertion into (9), and a comparison can now be made between the problem associated with (6) and the formulation in (8). In this more general case the α_j take on a continuous range of values weighted by the functions $p(\alpha_j)$. Every configuration of values of the α_j for which $F < A$ is appropriately counted by the function T which then multiplies the integrand by unity. It multiplies the integrand by zero otherwise. Note that if the function T is removed from the integrand of (8), the value of the integrals would be unity. Therefore $Q(A)$ represents the probability that $F < A$, since T counts only the weighted fraction of the total of configurations which satisfies this condition.

The probability distribution for the structure factor $P'(A)$ is readily obtainable from $Q(A)$ because $Q(A)$ is clearly the integral over the probability distribution function from $-\infty$ to A. Therefore

$$P'(A) = \mathrm{d}Q(A)/\mathrm{d}A \tag{11}$$

and $P'(A)\,\mathrm{d}A$ represents the probability that the structure factor might have a value in the interval $(A, A+\mathrm{d}A)$. The differentiation of (8) gives

$$P'(A) = \frac{1}{2\pi}\int_{-\infty}^{\infty} \exp(-\mathrm{i}Ax)\prod_{j=1}^{N/n} q(f_j x)\,\mathrm{d}x \tag{12}$$

where

$$q(f_j x) = \int_{-\infty}^{\infty} p(\alpha_j)\exp(\mathrm{i}f_j\,\alpha_j\,x)\,\mathrm{d}\alpha_j \tag{13}$$

Equation (12) can be readily evaluated by expanding the exponential in the integrand of (13). This can be justified by the fact that as x increases, the contribution of the product of the q-functions in the integrand of (12) decreases rapidly provided N/n is large enough. In the series expression to be obtained for $P'(A)$ this behaviour is manifested by successive terms containing increasing reciprocal powers of N, so that the rapidity of convergence of the series increases as N increases. Equation (13) is thus rewritten

$$q(f_j x) = \int_{-\infty}^{\infty} p(\alpha_j)\left[1 + \mathrm{i}f_j\,\alpha_j\,x - \frac{1}{2!}(f_j\,\alpha_j\,x)^2 - \ldots\right]\mathrm{d}\alpha_j \tag{14}$$

The individual terms in (14) are all of the form

$$m_k = \int_{-\infty}^{\infty} \alpha_j^k\,p(\alpha_j)\,\mathrm{d}\alpha_j \tag{15}$$

where m_k is the kth moment of $p(\alpha_j)$. Evidently the kth moment of $p(\alpha_j)$ is the expected (or average) value of $\xi_j^k \equiv \xi^k(x_j, y_j, z_j; \mathbf{h})$ and since the x_j, y_j, and z_j range uniformly throughout the unit cell, m_k can be rewritten

$$m_k = \int_0^1\int_0^1\int_0^1 \xi_j^k\,\mathrm{d}x_j\,\mathrm{d}y_j\,\mathrm{d}z_j \tag{16}$$

thus obviating the necessity for finding an explicit expression for the $p(\alpha_j)$. From (14), using the Taylor expansion of the logarithm, and the fact that $m_k = 0$ if k is odd for the functions involved

$$\log q(f_j x) = -\frac{f_j^2}{2!} m_2 x^2 - \frac{f_j^4}{4!}(3m_2^2 - m_4) x^4$$

$$-\frac{f_j^6}{6!}(30m_2^3 - 15m_2 m_4 + m_6) x^6 - \dots \tag{17}$$

and

$$\log \prod_{j=1}^{N/n} q(f_j x) = -\frac{s_2}{2! n} m_2 x^2 - \frac{s_4}{4! n}(3m_2^2 - m_4) x^4$$

$$-\frac{s_6}{6! n}(30m_2^3 - 15m_2 m_4 + m_6) x^6 - \dots \tag{18}$$

where

$$s_k = \sum_{j=1}^{N} f_j^k = n \sum_{j=1}^{N/n} f_j^k \tag{19}$$

Equation (18) may be rewritten

$$\prod_{j=1}^{N/n} q(f_j x) = \exp\left(-s_2 m_2 x^2/2n\right)\left[1 - \frac{s_4}{4! n}(3m_2^2 - m_4) x^4\right.$$

$$\left. -\frac{s_6}{6! n}(30m_2^3 - 15m_2 m_4 + m_6) x^6 - \dots\right] \tag{20}$$

where the square bracketed terms on the right-hand side of (20) are obtained from expanding a portion of the exponential term in a series. On substituting (20) into (12) it is found that the integrals are readily evaluated to give a series expression for the probability distribution for a structure factor in a centrosymmetric crystal

$$P'(F) = \left(\frac{n}{2\pi m_2 s_2}\right)^{\frac{1}{2}} \exp\left(-\frac{F^2 n}{2m_2 s_2}\right)\left[1 - \frac{ns_4(3m_2^2 - m_4)}{2\cdot 4 m_2^2 s_2^2}\right.$$

$$\times \left(1 - \frac{2nF^2}{m_2 s_2} + \frac{n^2 F^4}{3m_2^2 s_2^2}\right) - \frac{n^2 s_6(30m_2^3 - 15m_2 m_4 + m_6)}{2\cdot 4\cdot 6 m_2^3 s_2^3}$$

$$\left. \times \left(1 - \frac{3nF^2}{m_2 s_2} + \frac{n^2 F^4}{m_2^2 s_2^2} - \frac{n^3 F^6}{15 m_2^3 s_2^3}\right) - \dots\right] \tag{21}$$

where F has replaced A. The average value of F^2 is of interest and it can be immediately obtained from (21)

$$\langle F^2 \rangle = \int_{-\infty}^{\infty} F^2 P'(F)\, dF = m_2 s_2/n \tag{22}$$

Making use of the result in (22), we can define a form for the structure factor which is quite useful for phase determining procedures, the normalized structure factor E defined by

$$E_{\mathbf{h}} = F_{\mathbf{h}} \Big/ \left(\frac{m_2 s_2}{n}\right)^{\frac{1}{2}} \tag{23}$$

It is seen from (22) that this definition implies that $\langle E_{\mathbf{h}}^2 \rangle_{\mathbf{h}} = 1$. The usefulness of the normalized structure factor derives in part from the fact that the quantity $s_2^{\frac{1}{2}}$ in the denominator of (23) effectively eliminates the effect of the electron distributions around the atoms, so that the $E_{\mathbf{h}}$ act like structure factors from point atoms. In addition, the values of the normalized structure factors obey the same distribution function to a first approximation for all centro-symmetric space groups, and therefore the statistical properties of the normalized structure factors usually do not vary greatly from one centrosymmetric crystal to the other.

Examination of (21) shows that the definition in (23) would lead to a considerable simplification of the expression for the probability distribution. If $P(E)\,dE$ denotes the probability that E lie in the interval $(E, E + dE)$, then (21) becomes

$$P(E) = \left(\frac{1}{2\pi}\right)^{\frac{1}{2}} \exp\left(\frac{E^2}{2}\right)\left[1 - \frac{ns_4(3m_2^2 - m_4)}{2\cdot 4m_2^2 s_2^2}(1 - 2E^2 + \tfrac{1}{3}E^4)\right.$$
$$\left. - \frac{n^2 s_6(30m_2^3 - 15m_2 m_4 + m_6)}{2\cdot 4\cdot 6m_2^3 s_2^3}(1 - 3E^2 + E^4 - \tfrac{1}{15}E^6) - \cdots\right]$$
$$\tag{24}$$

It is seen that the first term in the series is the same for all centro-symmetric space groups. In principle they are distinguished by the terms of higher order. The values of the moments defined in (16) are evidently required in order to specialize (24) for the various space groups. The functions ξ_j appearing in (16) are trigonometric functions and the moments are easily evaluated. In actual practice the higher order terms have not been found generally useful for distinguishing the space groups, owing to the uncertainties that arise because of the limitations in accuracy and amount of experi-mental data and the effects of atoms specially placed in the unit cell. In connection with the above discussion concerning the convergence of successive terms in the series expressions (21) and (24), it may be pointed out that s_4/s_2^2 is of the order of N^{-1} and s_6/s_2^3 is of the order of N^{-2}. The main application of (24) is in the evaluation of expected values and other statistical properties which help to distinguish centrosymmetric from non-centrosymmetric crystals. Some are listed in Table I.

TABLE I. Selected Statistical Properties of Centrosymmetric Reflections obtained from Eq. (24)

$\langle |E_{\mathbf{h}}| \rangle_{\mathbf{h}}$	$\langle E_{\mathbf{h}}^2 \rangle_{\mathbf{h}}$	$\langle |E_{\mathbf{h}}^2 - 1| \rangle_{\mathbf{h}}$	Fraction of normalized structure factors		
			>1	>2	>3
0.798	1.000	0.968	0.32	0.05	0.003

4. PROBABILITY DISTRIBUTION FOR THE MAGNITUDE OF A STRUCTURE FACTOR, NON-CENTROSYMMETRIC CASE

In the preceding chapter, the probability distribution of the structure factor for centrosymmetric crystals was obtained. This chapter is concerned with the derivation of the probability distribution for the magnitude of the structure factor for non-centro-symmetric crystals. For centrosymmetric crystals the probability distribution for the magnitude of a structure factor can be readily obtained from (24) by multiplying by two and noting that $P(|E|)$ covers the range from 0 to ∞ rather than the range $-\infty$ to ∞ for $P(E)$. The method used to obtain (24) is now generalized to form the basis for deriving the joint probability distribution of jointly varying functions, i.e. functions of the same independent variables. We again average uniformly over all possible positions for the atoms in the unit cell. Except for certain circumstances when atoms occupy special positions, this gives the same result as averaging over all integral values for h, k and l with the atomic positions fixed.

The structure factor for non-centrosymmetric reflections can be written

$$F_{\mathbf{h}} = X_{\mathbf{h}} + iY_{\mathbf{h}} \tag{25}$$

where

$$X_{\mathbf{h}} = \sum_{j=1}^{N/n} f_j \xi(x_j, y_j, z_j; \mathbf{h})$$

$$Y_{\mathbf{h}} = \sum_{j=1}^{N/n} f_j \eta(x_j, y_j, z_j; \mathbf{h}) \qquad (26)$$

The functions ξ and η for the appropriate non-centrosymmetric space group may be found in the *International Tables for X-ray Crystallography*.[21] We are required to consider the joint probability distribution of the variates $X_{\mathbf{h}}$ and $Y_{\mathbf{h}}$ since they are functions of the same variables and therefore their values do not vary independently. If the values of the coordinates x_j, y_j, z_j are uniformly and independently distributed in $(0, 1)$, $p(\alpha, \beta)\,\mathrm{d}\alpha\,\mathrm{d}\beta$ is the joint probability that ξ lie in $(\alpha, \alpha + \mathrm{d}\alpha)$ and η lie in $(\beta, \beta + \mathrm{d}\beta)$. The probability $Q(A, B)$ that both $X_{\mathbf{h}} < A$ and $Y_{\mathbf{h}} < B$ may be written

$$Q(A, B) = \int_{-\infty}^{\infty} \cdots \int_{-\infty}^{\infty} p(\alpha_1, \beta_1) \ldots p(\alpha_{N/n}, \beta_{N/n})$$

$$\times\, T(\alpha_1, \beta_1, \ldots, \alpha_{N/n}, \beta_{N/n})\,\mathrm{d}\alpha_1\,\mathrm{d}\beta_1 \ldots \mathrm{d}\alpha_{N/n}\,\mathrm{d}\beta_{N/n} \qquad (27)$$

where T is the product of discontinuous integrals

$$T(\alpha_1, \beta_1, \ldots, \alpha_{N/n}, \beta_{N/n}) = \left\{\frac{1}{2} - \frac{1}{2\pi} \int_{-\infty}^{\infty} \frac{\exp[i(X_{\mathbf{h}} - A)\,x]}{ix}\,\mathrm{d}x\right\}$$

$$\times \left\{\frac{1}{2} - \frac{1}{2\pi} \int_{-\infty}^{\infty} \frac{\exp[i(Y_{\mathbf{h}} - B)\,y]}{iy}\,\mathrm{d}y\right\} \qquad (28)$$

having the property that $T = 1$ if $X < A$ and $Y < B$ and $T = 0$ otherwise. The structure factor F is defined in terms of the integration variables α_j and β_j (the subscript $\mathbf{h}$ is again omitted)

$$F = X + iY = \sum_{j=1}^{N/n} f_j(\alpha_j + i\beta_j) \qquad (29)$$

Equations (29) for the real and imaginary parts, X and Y are substituted into (28). The pairs of variables (α_j, β_j) take on joint values weighted by the joint probability distribution functions $p(\alpha_j, \beta_j)$. Every configuration of values of the α_j and β_j for which $X < A$ and $Y < B$ is appropriately counted by the function T which then multiplies the integrand by unity. It multiplies the integrand by zero otherwise. If the function T is removed from the integrand of (27), the value of the integrals would be unity. Therefore $Q(A, B)$ represents the probability that $X < A$ and $Y < B$ since T counts only the weighted fraction of the total of configurations which satisfies this condition.

The joint probability that X lie in the interval $(A, A + \mathrm{d}A)$ and Y lie in the interval $(B, B + \mathrm{d}B)$ is $P_{XY}(A, B)\, \mathrm{d}A\, \mathrm{d}B$, where in analogy to (11) we have

$$P_{XY}(A, B) = \partial^2 Q(A, B) / \partial A\, \partial B \tag{30}$$

The differentiation of (27) gives

$$P_{XY}(A, B) = \frac{1}{(2\pi)^2} \int_{-\infty}^{\infty} \int_{-\infty}^{\infty} \exp\left[-\mathrm{i}(Ax + By)\right] \prod_{j=1}^{N/n} q(f_j x, f_j y)\, \mathrm{d}x\, \mathrm{d}y \tag{31}$$

where

$$q(f_j x, f_j y) = \int_{-\infty}^{\infty} \int_{-\infty}^{\infty} p(\alpha_j, \beta_j) \exp\left[\mathrm{i}f_j(\alpha_j x + \beta_j y)\right] \mathrm{d}\alpha_j\, \mathrm{d}\beta_j \tag{32}$$

$$= \int_{-\infty}^{\infty} \int_{-\infty}^{\infty} p(\alpha_j, \beta_j)$$

$$\times \left[1 + \mathrm{i}f_j(\alpha_j x + \beta_j y) - \frac{f_j^2}{2!}(\alpha_j x + \beta_j y)^2 - \dots\right] \mathrm{d}\alpha_j\, \mathrm{d}\beta_j \tag{33}$$

The individual terms in (33) are all of the form

$$m_i^k = \int_{-\infty}^{\infty} \int_{-\infty}^{\infty} \alpha_j^i \beta_j^k p(\alpha_j, \beta_j)\, \mathrm{d}\alpha_j\, \mathrm{d}\beta_j \tag{34}$$

Evidently this is the ikth mixed moment of $p(\alpha_j, \beta_j)$ and is interpretable as the average value of the product

$$\xi_j^i \eta_j^k = \xi^i(x_j, y_j, z_j; \mathbf{h})\, \eta^k(x_j, y_j, z_j; \mathbf{h})$$

Since the x_j, y_j, and z_j range uniformly throughout the unit cell, m_i^k can be rewritten

$$m_i^k = \int_0^1 \int_0^1 \int_0^1 \xi_j^i \eta_j^k\, \mathrm{d}x\, \mathrm{d}y\, \mathrm{d}z \tag{35}$$

The first non-vanishing moments are m_0^0, m_2^0 and m_0^2, so that (33) becomes

$$q(f_j x, f_j y) = 1 - \frac{f_j^2}{2!}(m_2^0 x^2 + m_0^2 y^2) + \frac{f_j^4}{4!}$$

$$\times (m_4^0 x^4 + 6m_2^2 x^2 y^2 + m_0^4 y^4) - \cdots \qquad (36)$$

By an analysis similar to that carried out in Eqs. (17)–(20), we obtain from (31)

$$P_{XY}(A, B) = \exp \frac{\left(-\dfrac{A^2}{4\sum a_{j20}} - \dfrac{B^2}{4\sum a_{j02}}\right)}{4\pi(\sum a_{j20} \sum a_{j02})^{\frac{1}{2}}}$$

$$\times \Bigg(1 - \Bigg\{ \frac{[12(\sum a_{j20})^2 - 12A^2 \sum a_{j20} + A^4] \sum(\frac{1}{2}a_{j20}^2 - a_{j40})}{2^4(\sum a_{j20})^4}$$

$$+ \frac{(2\sum a_{j20} - A^2)(2\sum a_{j02} - B^2) \sum(a_{j20} a_{j02} - a_{j22})}{2^4(\sum a_{j20})^2(\sum a_{j02})^2}$$

$$+ \frac{[12(\sum a_{j02})^2 - 12B^2 \sum a_{j02} + B^4] \sum(\frac{1}{2}a_{j02}^2 - a_{j04})}{2^4(\sum a_{j02})^4} \Bigg\} - \cdots \Bigg)$$

$$(37)$$

where

$$\sum a_{jik} \equiv \sum_{j=1}^{N/n} a_{jik} = \sum_{j=1}^{N/n} m_i^k f_j^{i+k} \qquad (38)$$

We seek $P_{R\theta}(c, \phi)$ such that $P_{R\theta}(c, \phi) \, dc \, d\phi$ is the joint probability that R lie between c and $c + dc$ and that θ, the phase angle, lie between ϕ and $\phi + d\phi$ where

$$R = (X^2 + Y^2)^{\frac{1}{2}}; \quad \theta = \tan^{-1} Y/X \qquad (39)$$

and

$$X = R \cos \theta; \quad Y = R \sin \theta \qquad (40)$$

R will lie in $(c, c + dc)$ and θ will lie in $(\phi, \phi + d\phi)$ if and only if the point (X, Y) lies in the elementary area $c \, dc \, d\phi$ when $A = c \cos \phi$ and $B = c \sin \phi$. The probability of the latter event is

$$P_{XY}(c \cos \phi, c \sin \phi) c \, dc \, d\phi,$$

so that

$$P_{R\theta}(c, \phi) = cP_{XY}(c \cos \phi, c \sin \phi) \qquad (41)$$

which can be readily obtained from (37). Since in the absence of prior knowledge θ can take on any value between 0 and 2π, the probability that R lie between c and $c + \mathrm{d}c$ is obtained from (41) by means of

$$P_R(c) = c \int_0^{2\pi} P_{XY}(c \cos \phi, c \sin \phi) \, \mathrm{d}\phi \tag{42}$$

From (37), (38) and (42), including the first correction term in (37) generated by the non-zero moments, we obtain

$$P_R(c) = \frac{nc \exp\left(-nc^2 / 2m_2 s_2\right)}{m_2 s_2}$$

$$\times \left\{ 1 - \frac{ns_4}{8(m_2)^2 s_2^2} [8(m_2)^2 - M_4] \left[1 - \frac{nc^2}{m_2 s_2} + \frac{n^2 c^4}{8(m_2)^2 s_2^2} \right] - \ldots \right\} \tag{43}$$

where $M = m_4^0 + 2m_2^2 + m_0^4$ and $2m_2$ has replaced $m_2^0 + m_0^2$ since $m_2^0 = m_0^2$ for non-centrosymmetric reflections. If we set

$$E_{\mathbf{h}} = F_{\mathbf{h}} / [(m_2^0 + m_0^2) s_2 / n]^{\frac{1}{2}} \tag{44}$$

we have a single definition for the normalized structure factor which can serve for pure real reflections (centrosymmetric), pure imaginary reflections and the general complex non-centrosymmetric reflections. In the first case $m_0^2 = 0$, in the second case $m_2^0 = 0$ and in the last $m_2^0 = m_0^2 = m_2 \neq 0$.

Using (43) and (44), we obtain the probability distribution for the magnitude of the normalized structure factor for non-centro-symmetric reflections

$$P(|E|) = 2|E| \exp\left(-|E|^2\right) \left\{ 1 - \frac{ns_4}{8(m_2)^2 s_2^2} [8(m_2)^2 - M_4] \right.$$

$$\left. \times \left(1 - 2|E|^2 + \frac{|E|^4}{2} \right) - \ldots \right\} \tag{45}$$

The first term in the series is the same for the non-centrosymmetric reflections in all the non-centrosymmetric space groups. As in the centrosymmetric case, the space groups are in principle distinguished by the terms of higher order, but the same limitations prevail in practice. Several expected values and other statistical properties, which are readily computed from the first term of (45), are listed in Table II.

TABLE II.   Selected Statistical Properties of Non-centrosymmetric
Reflections obtained from Eq. (45)

$\langle |E_h| \rangle_h$	$\langle |E_h|^2 \rangle_h$	$\langle \|\, |E_h|^2 - 1\, | \rangle_h$	Fraction of normalized structure factors		
			>1	>2	>3
0.886	1.000	0.736	0.37	0.018	0.0001

A comparison of the leading terms in Eqs. (24) and (45) shows that a considerable difference exists between the shapes of the probability distributions for centrosymmetric and noncentrosymmetric crystals. In particular the distribution for centrosymmetric crystals has a maximum near the origin ($|E| = 0$) whereas that for the non-centrosymmetric reflections goes to zero at the origin and is very small in the vicinity. As a consequence it is often possible to distinguish X-ray photographs of centrosymmetric reflections from those of non-centrosymmetric ones by inspection, since in the centrosymmetric case there would be a much greater number of unobservable reflections on heavily exposed films. Unobservable non-centrosymmetric reflections should be relatively rare on such films. It is understood that a certain minimum intensity is required before a reflection becomes visible above the photographic background.

5. RELATIONS AND PROCEDURE FOR PHASE DETERMINATION

A. Phase Determining Relations

The main phase determining relations are relatively few in number. For simplicity it is our intention to derive them here mainly by algebraic means, although historically they have been obtained initially by means of inequality and probability methods. It is convenient to define the quasi-normalized structure factor

$$E_k = \sigma_2^{-\frac{1}{2}} \sum_{j=1}^{N} Z_j \exp(2\pi i \mathbf{k} \cdot \mathbf{r}_j) \tag{46}$$

and the quasi-normalized structure factor for the squared structure

$$E'_{\mathbf{h}} = \sigma_4^{-\frac{1}{2}} \sum_{j=1}^{N} Z_j^2 \exp\left(2\pi i \mathbf{h} \cdot \mathbf{r}_j\right) \tag{47}$$

where Z_j is the atomic number of the jth atom, having coordinates represented by the vector $\mathbf{r}_j$ in a unit cell containing N atoms, and

$$\sigma_n = \sum_{j=1}^{N} Z_j^n \tag{48}$$

The definition of the quasi-normalized structure factor (46) is a close approximation to the normalized structure factor, E, defined by (44) when $(m_2^0 + m_0^2)/n = 1$. The value of unity holds for all space groups when the Miller indices h, k, l are all different from zero. It may differ from unity otherwise.

By examining the substitution of the definition of the structure factor (3) into (44), the relationship between the normalized structure factor and the quasi-normalized structure factor can be clarified. We obtain

$$E_{\mathbf{h}} = [(m_2^0 + m_0^2) s_2/n]^{-\frac{1}{2}} \sum_{j=1}^{N} f_{j\mathbf{h}} \exp\left(2\pi i \mathbf{h} \cdot \mathbf{r}_j\right) \tag{49}$$

If it is assumed that to a reasonable approximation

$$f_{j\mathbf{h}} \simeq f_{\mathbf{h}} Z_j \tag{50}$$

where $f_{\mathbf{h}}$ is a universal shape function and Z_j is the atomic number of the jth atom, we have

$$f_{j\mathbf{h}}/s_2^{\frac{1}{2}} \simeq f_{\mathbf{h}} Z_j \bigg/ \left(\sum_{j=1}^{N} f_{\mathbf{h}}^2 Z_j^2\right)^{\frac{1}{2}} = Z_j \bigg/ \left(\sum_{j=1}^{N} Z_j^2\right)^{\frac{1}{2}} = Z_j/\sigma_2^{\frac{1}{2}} \tag{51}$$

Noting that (51) represents a good approximation, and comparing (46) and (49), it can be verified that $E_{\mathbf{h}}$ is essentially the same as $E_{\mathbf{h}}$ when $(m_2^0 + m_0^2)/n$ is equal to unity. It is also apparent that

$$E_{\mathbf{h}} \simeq [(m_2^0 + m_0^2)/n]^{\frac{1}{2}} E_{\mathbf{h}} \tag{52}$$

so that since the magnitudes $|E_{\mathbf{h}}|$ may be obtained from experimental data by an obvious rescaling of the measured $|F_{\mathbf{h}}|$ using (44), it is also evidently possible to obtain values for $|E_{\mathbf{h}}|$ by means of a rescaling.

It follows from (46) that

$$E_k E_{h-k} = \sigma_2^{-1} \sum_{j=1}^{N} Z_j^2 \exp(\pi i \mathbf{h} \cdot \mathbf{r}_j)$$

$$+ \sigma_2^{-1} \sum_{j \neq 1}^{N} \sum_{j' \neq 1}^{N} Z_j Z_{j'} \exp\{2\pi i[\mathbf{k} \cdot \mathbf{r}_j + (\mathbf{h} - \mathbf{k}) \cdot \mathbf{r}_{j'}]\} \quad (53)$$

and taking the average over all values of $\mathbf{k}$ gives, using (47)

$$\langle E_k E_{h-k} \rangle_k = \sigma_2^{-1} \sum_{j=1}^{N} Z_j^2 \exp(2\pi i \mathbf{h} \cdot \mathbf{r}_j) = \sigma_4^{\frac{1}{2}} \sigma_2^{-1} E_h' \quad (54)$$

since the average of the double sum over $\mathbf{k}$ is zero. Rearranging gives the formula

$$E_h' = \sigma_2 \sigma_4^{-\frac{1}{2}} \langle E_k E_{h-k} \rangle_k \quad (55)$$

For structures in which all the atoms are the same $E_h' = E_h$, and these structure factors do not generally differ greatly from one another, even when atoms have widely differing atomic numbers. A relation similar to (55) with E_h replacing E_h' has been found by probability methods[29]

$$E_h \simeq \sigma_2^{\frac{3}{2}} \sigma_3^{-1} \langle E_k E_{h-k} \rangle_k \quad (56)$$

Relations of the type (55) and (56) have played a central role in methods of phase determination and have appeared in many related forms, for example as inequalities in the work of Karle and Hauptman,[30] as probable relations in studies by Sayre,[31] Hauptman and Karle[28, 29] and Cochran[32] and as algebraic relations in the investigations of Hughes[33] and Cochran.[34]

A form of (56), the tangent formula,[29, 7] has considerable practical significance in the determination of non-centrosymmetric crystal structures. It can be obtained from (56) by writing

$$E_h = |E_h|(\cos \phi_h + i \sin \phi_h)$$

which then becomes

$$|E_h| \cos \phi_h \simeq \sigma_2^{\frac{3}{2}} \sigma_3^{-1} \langle |E_k E_{h-k}| \cos(\phi_k + \phi_{h-k}) \rangle_k \quad (57)$$

$$|E_h| \sin \phi_h \simeq \sigma_2^{\frac{3}{2}} \sigma_3^{-1} \langle |E_k E_{h-k}| \sin(\phi_k + \phi_{h-k}) \rangle_k \quad (58)$$

Dividing (58) by (57) gives the tangent formula

$$\tan \phi_h \simeq \frac{\langle |E_k E_{h-k}| \sin(\phi_k + \phi_{h-k}) \rangle_k}{\langle |E_k E_{h-k}| \cos(\phi_k + \phi_{h-k}) \rangle_k} \quad (59)$$

The tangent formula is ordinarily used in practice with normalized structure factors, E, replacing the quasi-normalized ones, E. This is done because it is necessary to approximate averages over infinite sets of data by averages over small samples and the use of E instead of E generally gives a disproportionate weight to certain of the one- and two-dimensional reflections.

Equation (56) can be put into an alternative form[9] which is useful for the initial stages of a phase determination with non-centrosymmetric crystals. Rewriting (56) gives

$$1 \simeq \sigma_2^{\frac{3}{2}} \sigma_3^{-1} \langle | E_h^{-1} E_k E_{h-k} | \exp [i(-\phi_h + \phi_k + \phi_{h-k})] \rangle_k \qquad (60)$$

It has been shown from probability arguments by Cochran[32] that for the largest $|E|$ values $\phi_h - \phi_k - \phi_{h-k}$ is distributed about zero and generally assumes small values. If we limit the average in (60) to include only those vectors $\mathbf{k}$ associated with the larger $|E|$ values, we may write for the imaginary part of (60)

$$\langle | E_k E_{h-k} | \sin(-\phi_h + \phi_k + \phi_{h-k}) \rangle_{k_r} \simeq 0 \qquad (61)$$

where $\mathbf{k}_r$ signifies the restricted range of vectors considered. Since we have further stated that the argument of the sine function is generally small, we may replace (61) by

$$\langle | E_k E_{h-k} | (-\phi_h + \phi_k + \phi_{h-k}) \rangle_{k_r} \equiv$$
$$p^{-1} \sum_{k_r} | E_k E_{h-k} | (-\phi_h + \phi_k + \phi_{h-k}) \simeq 0 \qquad (62)$$

where there are p terms designated by $\mathbf{k}_r$. Equation (62) may be rewritten to give the phase determining formula

$$\phi_h \simeq \frac{\sum\limits_{k_r} | E_k E_{h-k} | (\phi_k + \phi_{h-k})}{\sum\limits_{k_r} | E_k E_{h-k} |} \qquad (63)$$

in which the ϕ_h are defined as weighted averages of the $\phi_k + \phi_{h-k}$ associated with the largest $|E|$ values. If the $|E|$ are all roughly of the same order of magnitude, we obtain the approximation[9]

$$\phi_h \simeq \langle \phi_k + \phi_{h-k} \rangle_{k_r} \qquad (64)$$

This result is called the sum of angles formula.

For centrosymmetric crystals, in which the phases are either zero or π, it is convenient to write (56) in the form

$$s\,E_{\mathbf{h}} \simeq s \sum_{\mathbf{k}} E_{\mathbf{k}}\, E_{\mathbf{h}-\mathbf{k}} \tag{65}$$

where the normalized structure factors have replaced the quasi-normalized ones and s means "sign of", since a phase of zero gives the structure factor a plus sign and a phase of π gives the structure factor a minus sign. Formula (65) is called sigma-2.

The main formulae for carrying out the phase determining procedures, (64) followed by (59) for non-centrosymmetric crystals and (65) for centrosymmetric ones, evidently require certain phases to be known before others can be determined. Probability measures and auxiliary formulae have been introduced in the attempt to overcome this apparent limitation. In the course of applying these formulae it was found that the limitations were not as severe as first appearances might have implied, and routine procedures for phase determination have been developed employing (59), (64), (65) and appropriate probability measures. There are two main reasons why (64) and (65) are quite useful in practice. One is that although they appear to require averages or sums over large numbers of terms, single terms are often quite reliable when the $|E|$ values are large enough, as shown by the probability measures. The second reason is that the basic set of phases, in terms of which the remaining ones are ultimately defined, can be quite limited in number. Therefore such a set can be readily composed of those phases whose values are known, because they may be specified in order to fix the origin in the crystal and a few others to which unknown symbols have been assigned in the course of the phase determination. There will generally be some ambiguity in the phase determining procedure, depending upon what values are assigned finally to the unknown symbols. For example, with centrosymmetric crystals the symbols represent plus or minus signs. If there are three unknown symbols remaining at the end of a phase determination (three or less is quite common), then at most eight possible different Fourier series need to be considered. This is easy to do with modern computing facilities. In actual practice it is generally possible to eliminate most of the eight possibilities from consideration before the Fourier series are computed. There are more possibilities to consider if the unknown symbols represent general non-centrosymmetric reflections, since they must be allowed possible phase values separated by small intervals, about 45°, rather than the 180° separation that applies to centrosymmetric reflections and pure imaginary reflections. It is still possible, however, to keep the number of ambiguities within practical limitations.

The appropriate probability measure to be associated with (65) for centrosymmetric crystals is

$$P_+(\mathbf{h}) \simeq \tfrac{1}{2} + \tfrac{1}{2}\tanh \sigma_3\, \sigma_2^{-\frac{3}{2}} |E_\mathbf{h}| \sum_\mathbf{k} E_\mathbf{k}\, E_{\mathbf{h}-\mathbf{k}} \qquad (66)$$

where $P_+(\mathbf{h})$ represents the probability that the sign of $E_\mathbf{h}$ be positive. The probability $P_+(\mathbf{h})$ has been obtained in the form of a series,[28] but the hyperbolic tangent form given by Woolfson[35] is preferable for practical application.

The probability distribution for the phase $\phi_\mathbf{h}$ of a non-centrosymmetric reflection, $P(\phi_\mathbf{h})$, when a set of $(\phi_\mathbf{k} + \phi_{\mathbf{h}-\mathbf{k}})$ and their associated

$$\kappa = \kappa(\mathbf{h}, \mathbf{k}) = 2\sigma_3\, \sigma_2^{-\frac{3}{2}} |E_\mathbf{h}\, E_\mathbf{k}\, E_{\mathbf{h}-\mathbf{k}}| \qquad (67)$$

are known, has the form[9]

$$P(\phi_\mathbf{h}) = [2\pi I_0(\alpha)]^{-1} \exp\left[\alpha \cos(\phi_\mathbf{h} - \beta)\right] \qquad (68)$$

where

$$\alpha = \left\{ \left[\sum_{\mathbf{k}_r} \kappa(\mathbf{h}, \mathbf{k}) \cos(\phi_\mathbf{k} + \phi_{\mathbf{h}-\mathbf{k}}) \right]^2 \right.$$
$$\left. + \left[\sum_{\mathbf{k}_r} \kappa(\mathbf{h}, \mathbf{k}) \sin(\phi_\mathbf{k} + \phi_{\mathbf{h}-\mathbf{k}}) \right]^2 \right\}^{\frac{1}{2}} \qquad (69)$$

I_n is the Bessel function of imaginary argument[36] of order n and β is equal to the arctangent of the right-hand side of (59). It is apparent from (69) that $P(\phi_\mathbf{h})$ will be sharpest when all the pairs $(\phi_\mathbf{k} + \phi_{\mathbf{h}-\mathbf{k}})$ have the same value. A function of the sharpness of $P(\phi_\mathbf{h})$ and a good probability measure to associate with (64) is the variance defined by the expected value

$$V = \langle (\phi_\mathbf{h} - \langle \phi_\mathbf{h} \rangle)^2 \rangle \qquad (70)$$

The evaluation of (70) using $P(\phi_\mathbf{h})$ defined in (68) gives[9]

$$V = \frac{\pi^2}{3} + [I_0(\alpha)]^{-1} \sum_{n=1}^\infty \frac{I_{2n}(\alpha)}{n^2} - 4[I_0(\alpha)]^{-1} \sum_{n=0}^\infty \frac{I_{2n+1}(\alpha)}{(2n+1)^2} \qquad (71)$$

The series in (71) are readily evaluated and a graph of the result is shown in Fig. 1.

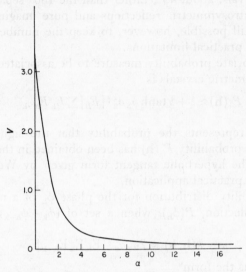

Fig. 1. The variance V (in square radians) of a phase angle determined from known values of other phase angles. The variance is expressed as a function of α defined in (69).[9]

One possible way to determine the values of some of the unknown symbols at the completion of a phase determination, i.e. after the majority of the phases associated with the larger $|E|$ values have been defined in terms of the basic set, is to make use of auxiliary phase determining relations. Some of these, obtained from studies of the joint probability distribution and their associated probabilities, are listed in Table III.

In deriving the formulae appearing in Table III when $E_{\mathbf{h}_\mu}$ is a non-centrosymmetric reflection, it was assumed that $m_{12}^{00} = m_{10}^{02}$ and $m_{112}^{000} = m_{110}^{002}$, where

$$m_{a_1 \dots a_r}^{b_1 \dots b_r} = \int_0^1 \int_0^1 \int_0^1 \xi^{a_1}(x, y, z; \mathbf{h}_1) \dots \xi^{a_r}(x, y, z; \mathbf{h}_r)$$

$$\times \eta^{b_1}(x, y, z; \mathbf{h}_1) \dots \eta^{b_r}(x, y, z; \mathbf{h}_r) \, dx \, dy \, dz \qquad (72)$$

and the functions ξ and η are defined by (7) and (26). The above relations among the moments should be verified before making an application. The various space groups provide a wide variety of formulae Σ_1 and Σ_3 which may be obtained from Table III by making suitable choices of $\mathbf{h}$, $\mathbf{h}_\nu$, and $\mathbf{h}_\mu$ and computing the appropriate moments from (72). Several illustrations of the results of such calculations may be seen in references 28 and 29.

TABLE III. Auxiliary Sign Determining Formulae for $E_\mathbf{h}$ and their Probabilities

Formula	$\Sigma_1 : \sum_\mu \dfrac{m_{12}}{m_{20}^{\frac{1}{2}} m_{02}} (\,|E_{\mathbf{h}\mu}|^2 - 1)$
	$\Sigma_3 : \sum_{\mu,\nu} \dfrac{m_{112}}{m_{200}^{\frac{1}{2}} m_{020}^{\frac{1}{2}} m_{002}} E_{\mathbf{h}\nu}(\,|E_{\mathbf{h}\mu}|^2 - 1)$
Probability sign is positive	$\Sigma_1 : P_+(E_\mathbf{h}) \simeq \frac{1}{2} + \frac{1}{2} \tanh \dfrac{m_{12}\, n^{\frac{1}{3}}\, \sigma_3}{q m_{20}^{\frac{1}{2}} m_{02}\, \sigma_2^{\frac{3}{2}}} \,|E_\mathbf{h}| \sum_\mu (\,|E_{\mu\mathbf{h}}|^2 - 1)$
	$\Sigma_3 : P_+(E_\mathbf{h}) \simeq \frac{1}{2} + \frac{1}{2} \tanh \dfrac{m_{112}\, n\sigma_4}{q m_{200}^{\frac{1}{2}} m_{020}^{\frac{1}{2}} m_{002}\, \sigma_2^2} \,|E_\mathbf{h}| \sum_{\mu,\nu} E_{\mathbf{h}\nu}(\,|E_{\mathbf{h}\mu}|^2 - 1)$
Types of reflections	$E_\mathbf{h}, E_{\mathbf{h}\nu}$: Centrosymmetric
	$E_{\mathbf{h}\mu}$: Centrosymmetric (then $q = 2$)
	$E_{\mathbf{h}\mu}$: Non-centrosymmetric (then $q = 1$)

An approximate formula which has been occasionally found to be useful is

$$E_{\mathbf{h}_1} E_{\mathbf{h}_2} E_{-\mathbf{h}_1-\mathbf{h}_2} |\cos(\phi_{\mathbf{h}_1} + \phi_{\mathbf{h}_2} + \phi_{-\mathbf{h}_1-\mathbf{h}_2})$$
$$\simeq \frac{\sigma_2^3}{c\sigma_4^{\frac{3}{2}}} \langle (\,|E_\mathbf{k}|^2 - 1)(\,|E_{\mathbf{h}_1+\mathbf{k}}|^2 - 1)(\,|E_{\mathbf{h}_2+\mathbf{k}}|^2 - 1)\rangle_\mathbf{k}$$
$$- 2\frac{\sigma_6}{\sigma_4^{\frac{3}{2}}} + \frac{\sigma_8^{\frac{1}{2}}}{\sigma_4} (\,|E_{\mathbf{h}_1}|^2 + |E_{\mathbf{h}_2}|^2 + |E_{\mathbf{h}_1+\mathbf{h}_2}|^2) \ldots \qquad (73)$$

where $c = 8$ for centrosymmetric crystals and $c = 2$ for non-centrosymmetric ones. Equation (73) was derived by Vaughan[37] and Hauptman and Karle.[38] This relation could be used in two different ways. For special choices of $\mathbf{h}_1$ and $\mathbf{h}_2$, depending upon the space group involved, the sum $\phi_{\mathbf{h}_1} + \phi_{\mathbf{h}_2}$ may be a known quantity, even though the separate phases may not be. The remaining unknown phase is $\phi_{-\mathbf{h}_1-\mathbf{h}_2}$, which is defined by (73) in terms of quantities which are all derivable from experiment. The second obvious use of (73) is to obtain $\phi_{-h_1-h_2}$ when the individual values of ϕ_{h_1} and ϕ_{h_2} are known. There are, however, severe limitations on the use of this equation and it is ordinarily restricted to determining the signs associated with pure real and pure imaginary reflections. The limitations arise from the fact that there are additional correction terms whose values depend in a detailed way on the particular structure being studied. Structures having many equal interatomic vectors have large correction terms. An investigation of the modifications to (73), which arise from relations among the interatomic vectors, has been carried out by Hauptman and the modified function has been used to obtain initial phases for the structure determination of a non-centrosymmetric crystal.[39]

B. Procedure for Phase Determination

There were some early attempts to determine phases or atomic positions from the structure factor equations carried out by Ott,[40] Banerjee[41] and Avrami.[42] These procedures were limited in applicability to simple structures. The more recent direct methods for phase determination are deeply rooted in the study of inequality relations among the structure factors. The first such investigation was made by Harker and Kasper[43] and their results were applied by Kasper, Lucht and Harker[44] to determine the structure of orthorhombic decaborane. Gillis[45] discussed the use of inequalities further and applied them to data for monoclinic oxalic acid dihydrate. Several centrosymmetric structures of limited complexity have been solved with the aid of the Harker–Kasper inequalities. The inequalities among the structure factors arise because the electron density function (1), which is expressed as a Fourier series with the structure factors as coefficients, is a nonnegative function. On this basis Karle and Hauptman[30] derived a complete set of inequalities. Written in the order of increasing complexity, the first two inequalities state that F_{000} is non-negative and that the magnitude of any structure factor is less than or equal to F_{000}. The next more complicated relation implies[9] relations (59), (64) and (65) and thus provides a basis for phase determination in both non-centrosymmetric and centrosymmetric crystals. Sayre[31] and Zachariasen[46] advanced the application of this type of formula to centrosymmetric crystals by suggesting procedures and carrying out examples. Since that time the main advances have involved the introduction of the normalized structure factors[28] and probability measures[28, 35] to facilitate and increase the reliability of structure determination for centrosymmetric crystals. A further development has led to a procedure for non-centrosymmetric crystals.[7, 9]

The details for applying the direct method for phase determination, the symbolic addition procedure, have been described recently[7, 9] and the discussion here will be limited to a general outline. In a typical structure investigation the measured X-ray intensities from a crystal are subjected to a series of corrections dictated by the conditions of the experiment and are ultimately expressed as normalized structure factor magnitudes. Attention is then restricted to those normalized structure factors of larger magnitude, the limit often set being $|E| \geqslant 1.5$. In the initial stages of the phase determination it is important to consider only the largest $|E|$ values since they are associated with the highest probabilities. The origin specification[28, 47-49] is made by assigning phases arbitrarily to a properly chosen set of $|E_\mathbf{h}|$. In making the assignment, the largest suitable $|E_\mathbf{h}|$ are used. Suitability is partly determined by the extent to which a particular $\mathbf{h}$ enters into combinations required by formulae (64) and (65). Sometimes a particular vector $\mathbf{h}$ does not form many of the required combinations with other vectors associated with the large $|E|$ values.

For non-centrosymmetric space groups, it is generally also necessary to specify an enantiomorph. An enantiomorphous structure is formed from a given structure by reflection through the origin. In centrosymmetric crystals both enantiomorphous structures so formed are identical. This is not so for most non-centrosymmetric crystals, although both enantiomorphs give the same set of X-ray diffraction intensities. It is therefore not normally possible to distinguish between these two enantiomorphs and so the final structure is given in terms of one arbitrarily chosen one. There are special experimental techniques for distinguishing enantiomorphs in suitably chosen crystals, leading to the determinations of absolute configuration. This will be discussed later on. The manner of specifying the enantiomorph has been described in theoretical papers[29, 41] and in several experimental papers concerning structure determinations of non-centrosymmetric crystals which will be discussed below.

After the specifications which determine the origin and enantio-morph have been made, additional symbols are assigned to other large $|E_h|$ one at a time as required. Formulae (64) and (65) are now employed to determine the phases of the remaining large $|E_h|$ in terms of the phase specifications and unknown symbols. They are used to define as many phases of the largest $|E_h|$ as possible in terms of specified ones and newly determined ones. The phase deter-mining formulae are used with their corresponding probability measures. For centrosymmetric crystals Eq. (66) is used to evaluate the reliability of each new phase determination. The condition that $P_+(\mathbf{h}) \geqslant 0.98$ can normally be readily met. For non-centrosym-metric crystals, the variance as obtained from Fig. 1 may be used as a measure of the reliability. A working rule is $V \leqslant 0.5$ for accepting a phase indication. The basis for assigning additional unknown symbols is the impossibility of continuing the phase determination without violating the preset limitations on the probability measures.

There are several ways in which the number of unknown symbols may be reduced at the end of a phase determination. In the course of the procedure certain symbols are found to be definable in terms of others. Auxiliary formulae as given in Table III and Eq. (73) may also be used. Further restrictions may be based on known structural features or on the fact that a particular assignment of the unknown symbols may lead to a large number of disagreements among the individual contributors to (64) or (65) in the evaluation of the phases. Finally, there will remain a small number of unknown symbols and the alternative assignments to these symbols determines the number of Fourier maps which need to be computed. The initial Fourier map which is computed employs the E_h as coefficients instead of the F_h as shown in (1). Such a Fourier map is called an E-map and has the advantage of clearly resolving the atomic positions. A sufficient number of phases, of the order of ten to twenty per atom in the asymmetric unit, are obtained from (65) for centrosymmetric crystals. In contrast, for non-centrosymmetric crystals (64) is used to obtain of the order of fifty to one hundred phases, after which the alternative values are assigned to the unknown symbols, and the phase determination is continued with each of the alternatives by employing the tangent formula (59). In this way about twenty to thirty phases per atom in the asymmetric unit are obtained for non-centrosymmetric crystals before an E-map is computed.

There is one type of structure for which the procedure as outlined often does not lead to a complete structure determination,
although it generally leads to a partial structure, i.e. a recognizable
fragment of the molecule. For such cases, it has been found[7] that
the tangent formula (59) can be employed with the phase information computed from the partial structure in a procedure which leads
to a complete structure determination. In this procedure, phases
are computed from the known atomic positions substituted into
the definition of the structure factor (3) multiplied by an additional
damping factor $\exp(-Bs^2)$ (with $B \simeq 3\text{--}4$) to account approximately for the damping due to vibrational motion. The quantity
s equals $\sin\theta/\lambda$, where 2θ is the angle between the incident and
scattered ray and λ is the wavelength of the X-rays. The computed
phase is retained if the magnitude of the structure factor F_c so
computed satisfies an acceptance criterion such as $|F_c| \geqslant p|F_0|$, where
p is related to the fraction of the total scattering power contained in
the structural fragment and where the observed magnitude of the
structure factor $|F_0|$ is associated with an $|E_0| \geqslant 1.5$. In this way a
set of phases is obtained which can be further expanded by use of
the tangent formula (59), and the expanded set of phases is used to
compute a new E-map. In successive cycles more of the structure
continues to appear until its determination is completed.

6. APPLICATIONS OF THE SYMBOLIC ADDITION
PROCEDURE

The symbolic addition procedure for direct structure determination has been applied to a considerable variety of problems. The
method is of special significance to organic and biological chemists
since it can be used with crystals containing molecules not only of
unknown configuration but also of unknown chemical composition.
It is the intention of this chapter to demonstrate by means of many
examples the range of problems which can be readily studied by
X-ray analysis employing this procedure.

It should be noted that there are a great number of valuable
structure investigations which continue to be carried out employing
the classical methods of structure determination, such as the
location of heavy atoms by means of the Patterson function and
packing considerations. However, the structures to be described

here, with occasional evident exceptions, are not readily solvable by the classical techniques.

A. A Variety of Applications

A typical example of a moderately complex material is cocarboxylase,[50] a coenzyme which is a key substance in biochemical decarboxylation. After a phase determination was carried out on the X-ray diffraction data from hydrolysed cocarboxylase, (1), an

$$\left[H_3C-\underset{\underset{H}{N^+}}{\overset{N}{\bigvee}}\underset{CH_2}{\overset{NH_2}{\bigvee}}\underset{N^+}{\overset{S}{\bigvee}}\overset{CH_2CH_2O-\overset{\overset{O}{\|}}{\underset{O^-}{P}}-OH}{\underset{CH_3}{}} \right]\quad {}^-O-\overset{\overset{O}{\|}}{\underset{OH}{P}}-OH\cdot 3H_2O$$

(1)

electron density map was computed using Eq. (1). The map was computed in three dimensions and sections were taken through areas of greatest densities, Fig. 2. The sections are projected onto one plane for illustrative purposes forming a composite map.

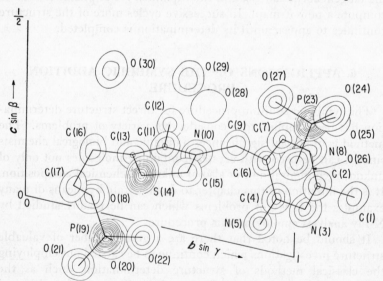

Fig. 2. Composite map of sections of electron density for hydrolysed cocarboxylase.[50]

The types of information which are generally obtained from this type of analysis are atomic positions, vibrational amplitudes, the molecular formula if unknown, the stereoconfiguration of the molecule, bond distances and angles between bonds, the ionic form, the location of hydrogen atoms, the number of waters of hydration and the nature of the hydrogen bonding.

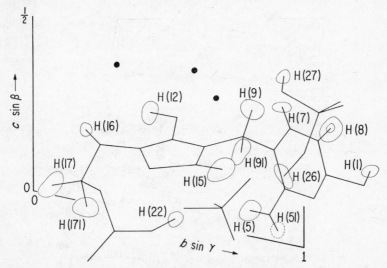

Fig. 3. A difference map for hydrolysed cocarboxylase showing the location of hydrogen atoms.[50]

In order to locate hydrogen atoms, a special type of Fourier map called a difference map is ordinarily computed. Such a map is computed from Fourier coefficients which effectively contain only the scattering power from hydrogen atoms, the scattering power from the heavier atoms having been eliminated. A difference map for hydrated cocarboxylase showing the positions of the hydrogen atoms relative to the locations of the heavier atoms may be seen in Fig. 3.

The nature of the hydrogen bonding in the crystal of hydrated cocarboxylase may be seen in Fig. 4. The hydrogen bonds are indicated by dotted lines. They are seen to occur between oxygen atoms on neighbouring phosphate groups, between a water molecule and a phosphate group and between two water molecules. The black dots represent the positions of water molecules and the hydrogen bonds between them form boat-shaped six-membered rings. In the organic portion a hydrogen bond is formed only between the amino groups on neighbouring molecules.

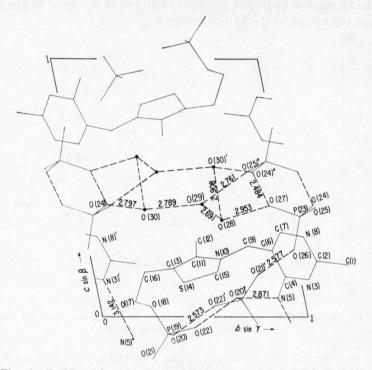

Fig. 4. Packing of molecules and hydrogen bonding in hydrolysed cocarboxylase.[50]

Another kind of result from structure analysis by X-ray diffraction of interest to the organic chemist is the determination of the structural formula and stereoconfiguration of an unknown substance. One example where the structural formula was elucidated by crystal structure analysis was the problem presented by the alkaloids derived from the *Ormosia* plants, some of which have the capability of blood pressure depression. As long ago as 1919 these alkaloids began to be isolated, but their structural formulae had remained unknown in spite of considerable chemical investigation.

The first material examined was jamine[51] (2), $C_{21}H_{35}N_3$. It crystallized in the triclinic space group $P\bar{1}$. The phases for this crystal were obtained in a routine fashion and the first E-map

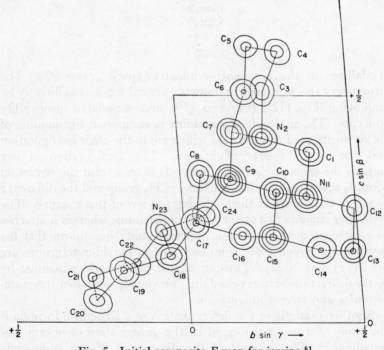

Jamine
$C_{21}H_{35}N_3$
(a)

(2)

computed from a limited number of the experimental data (286 reflections) is illustrated in Fig. 5. The configuration of the molecule was revealed at this point. There are six six-membered

Fig. 5. Initial composite E-map for jamine.[51]

rings, five of which are in the chair configuration and one is in the boat configuration. It still was not known which of the atoms were nitrogen. However, an additional calculation employing Eq. (1) with 1800 independent X-ray reflections indicated from increased electron density and slightly shorter associated distances that atoms 2, 11 and 23 were the nitrogen atoms. The structure determined here was consistent with and helped to clarify the chemical and spectral data which had been collected previously.[52-54]

The second alkaloid in this family which was examined is panamine[55] (3), $C_{20}H_{33}N_3$. A composite density map of the diperchlorate of this material is seen in Fig. 6. This material

Panamine
$C_{20}H_{33}N_3$
(b)
(3)

crystallizes in the non-centrosymmetric space group $P2_1$. The structure of this non-centrosymmetric crystal was solved directly by employing Eqs. (12), (64) and (59) and associated probability concepts. The molecule of panamine is composed, like jamine, of six six-membered rings, five of which are in the chair configuration and one in the boat configuration. The configuration of five rings is the same as that for jamine. It is seen from the structural formula that panamine has one less $-CH_2$ group and the difference concerns the nature of the sixth ring in view of this absence. The sixth ring attaches to a carbon atom in panamine whereas it attaches to a nitrogen atom in jamine. Further study has shown that the two protons which come from each of two perchlorate groups are on N(11) and N(23). Thus studies of jamine and panamine by X-ray diffraction have revealed their previously unknown structural formulae and stereoconfigurations.

Another example of the determination of a structural formula is the study of a material formed by the dimerization of hexafluoro-butadiene.[56] The formula of the dimer is C_8F_{12}, a completely saturated compound which crystallizes in a triclinic space group with one molecule per unit cell. From a statistical analysis, the crystal appeared to have a centre of symmetry. The phases were again determined by means of the symbolic addition procedure and and E-map was computed. Part of the map is shown in Fig. 7. There is a centre of symmetry at the centre of the figure through

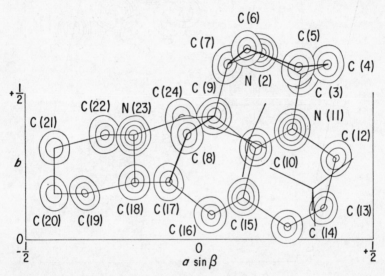

Fig. 6. Composite map of sections of electron density for panamine.[55]

which all peaks should be reflected. Interpretation of the map showed that there was a disorder in the crystal and that the molecule of C_8F_{12} assumed two different orientations which are superimposed in the map. Furthermore, the molecule does not possess a centre of symmetry. The apparent centre exists only as a consequence of an averaging of the two different orientations in which the molecule is found in the disordered crystal. The positions of the fluorine atoms, represented by the heavy peaks, do not change significantly when the molecule is in either orientation. The light peaks are carbon atoms which occur at one-half weight since their positions are quite different in the two orientations. Figure 8 shows how the density map was interpreted with the two superimposed molecules and Fig. 9 shows the geometry of the molecule with a

7

four-membered ring in the middle and five-membered rings on either side. The twelve surrounding atoms are all fluorine atoms.

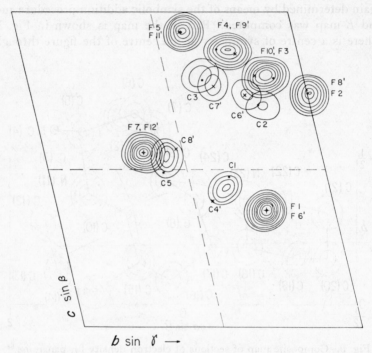

Fig. 7. Part of the initial composite E-map for C_8F_{12}, a dimer of hexafluorobutadiene.[56]

In view of the structure of the molecule, the reaction which formed this substance may probably be represented by (**4**).

$$2 \; \underset{F}{FC}=\underset{F}{C}-\underset{F}{C}=\underset{F}{CF} \xrightarrow{\Delta} \begin{array}{c} F \quad F \quad F \\ FC-C=C-CF \\ \\ FC-C=C-CF \\ F \quad F \quad F \end{array} \xrightarrow{\Delta} \begin{array}{c} F \quad F \quad F \quad F \\ FC-C-C-CF \\ \quad\quad\times \\ FC-C-C-CF \\ F \quad F \quad F \quad F \end{array}$$

(4)

The dimerization of hexachlorocyclopentadiene forms a stable chlorocarbon $C_{10}Cl_{12}$. Physico-chemical methods indicated that the dimer was in the form of a cage with the two most likely alternative

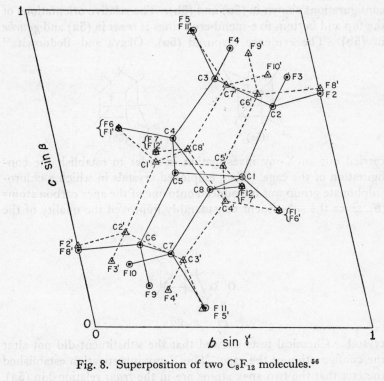

Fig. 8. Superposition of two C_8F_{12} molecules.[56]

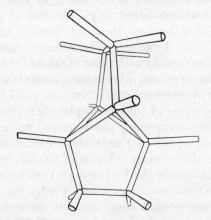

Fig. 9. Model for bonding pattern in C_8F_{12}.[56]

configurations shown in (5a) and (5b). The relative orientations of
the top and bottom five-membered rings is *trans* in (5a) and *gauche*
in (5b). The evidence favoured (5a). Okaya and Bednowitz[57]

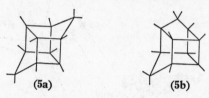

(5a)   (5b)

carried out an X-ray investigation in order to establish the configuration of the cage. They employed crystals in which a chloro-
sulphonate group was introduced onto one of the apex carbon atoms
(6), since the substituent considerably improved the quality of the

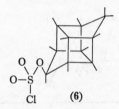

(6)

crystals. Chemical tests showed that the substituent did not alter
the configuration of the cage. The X-ray investigation established
the fact that the two apex atoms are in the *trans* relationship (5a).
The cage can be described as consisting of four five-membered
rings and two four-membered rings. Because of possible strains
introduced in fusing the rings, the bond distances and angles in this
structure are of special interest. The authors[57] have discussed the
comparison of their results with those obtained for free cyclobutane
and cyclopentane rings and other related compounds.

Another type of structural study concerns quasi-racemates. It is
well known that a pair of enantiomorphs often crystallize together
to give a racemic compound, characterized by a special crystal
lattice where the (+) and (−) molecules are often arranged in pairs
having a centre of symmetry. A pair of similar, but not identical
substances which are in the (+) and (−) configurations, respectively, will often form quasi-racemates. Fredga has made use of the
formation of quasi-racemic compounds to determine the absolute
configuration of many unknown substances.[58] The quasi-racemate

has been assumed to be in a regular array with a 1 : 1 correspondence of the two different substances. To verify this hypothesis, a quasi-racemate formed from the two optically active molecules (**7a**) and (**7b**) was investigated. The new compound was optically inactive

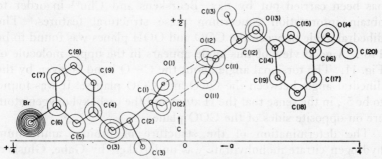

$$H_3C-CH-COOH \qquad H_3C-CH-COOH$$

$$(-) \qquad\qquad (+)$$

(**a**)   (**b**)

(**7**)

and the crystal structure analysis by X-ray diffraction showed that the crystal was ordered with the molecules occurring in pairs.[59] The map in Fig. 10 shows the two different molecules bound together by hydrogen bonds forming a pseudo-dimer. The structure is almost like that of a true racemate, although the crystal does not have a centre of symmetry and all the bromine-substituted molecules have the laevo-form whereas all the methoxy substituted molecules have the dextro-form.

Fig. 10. Composite electron density map for the quasi-racemate.[59]

A material which occurred sporadically in animal feed fat in 1957 caused significant losses to poultry farmers. The chickens suffered an accumulation of fluid in the heart sac and gross kidney and liver damage. Five micrograms could be lethal. In 1961, about 4 mg of the toxic material was isolated from 100 lb of contaminated fat by J. C. Wooten. Two small crystals were grown from some of this material and an X-ray investigation was undertaken by Cantrell, Webb and Mabis.[60] By means of this analysis, the material was identified to be 1,3,7,9-hexachlorodibenzo-p-dioxin (8). This material and similar derivatives are of considerable interest both because of their physiological potency and also their specificity.

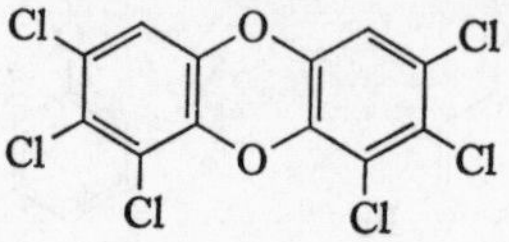

(8)

The organic peroxyacids form a class of compounds which have received little attention in the field of X-ray crystal structure analysis. Of particular interest is the conformation of the peroxy-carboxyl group and the nature of the intermolecular hydrogen bonding. A study of the structure of o-nitroperoxybenzoic acid, (9)

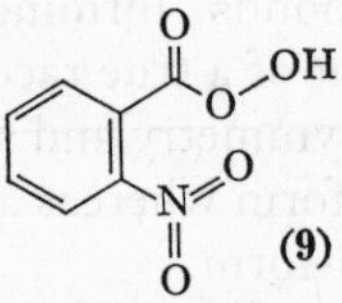

(9)

has been carried out by Sax, Beurskens and Chu[61] in order to obtain information concerning these structural features. The dihedral angle between the COO and OOH planes was found to be 146°. A good view of this angle appears in the upper molecule of Fig. 11. The torsional angle about the C—O bond is given by the dihedral angle between the OCO and COO planes. It was found to be 5°, in the sense that the H atom and the carboxyl oxygen atom are on opposite sides of the COO plane.

The determination of the structure of lithium ammonium hydrogen citrate monohydrate was undertaken by Gabe, Glusker, Minkin and Patterson[62] in order to determine which of the three carboxyl groups are ionized and to establish the stereochemistry of the molecule. They found that the central carboxyl group and one of the terminal carboxyl groups are ionized in the crystal. Four citrate ions surround the ammonium ion as shown in Fig. 12. The dotted lines show the closest approaches. The lithium ion was found to be surrounded by four oxygen atoms in four different citrate ions.

An interesting problem in molecular configuration is presented by the molecule of α-keto-1,5-tetramethyleneferrocene (**10**). In

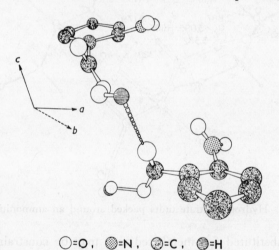

(**10**)

fully-eclipsed configuration.[62] In the fully staggered configuration of ferrocene the angle is 36°. The authors suggest that ferrocene the cyclo-pentadienyl rings are fully staggered, one relative to the other, whereas in other related compounds, they are

O=O, ⊙=N, ⊙=C, ◐=H

Fig. 11. Perspective view of o-nitroperoxybenzoic acid molecules in a hydrogen-bonded chain.[61]

not. The X-ray analysis of the α-keto compound was carried out by Fleischer and Hawkinson.[63] It was found that the cyclo-pentadienyl rings were parallel within experimental error. The dihedral angle between the planes was 1.5°. As is seen in Fig. 13 where the rings

are viewed along a normal to their planes, the rings are almost in
the eclipsed position. The average angle of deviation from the
fully eclipsed configuration is 8.9°. In the fully staggered configuration of ferrocene, the angle is 36°. The authors suggest that

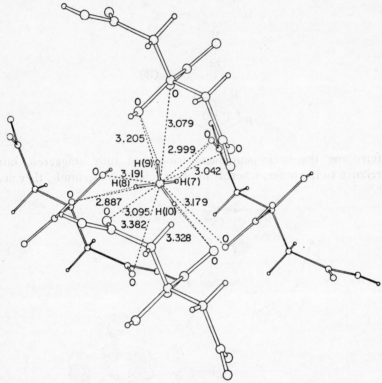

Fig. 12. Hydrogen citrate units packed around an ammonium ion.[62]

the unsubstituted five-membered ring may be constrained to its
conformation because of H ⋯ H contacts. They also report that the
unsubstituted cyclo-pentadienyl ring shows no significant distortion from pentagonal symmetry.

The optical properties of crystals depend upon the nature of the
electron distributions. An interesting example of this is shown by
crystals of tetrahydroxy-p-benzoquinone dihydrate (THQ dihydrate) whose structure was investigated by Klug.[64] The crystals

of THQ dihydrate possess a glistening black colour. In aqueous solution THQ is light red. The presence of a quinoid structure, a chromophoric unit, in THQ might readily explain the light red colour, but not the black colour of the crystals. As pointed out by Klug,[64] a strong absorption in the visible region of the spectrum is characteristic of charge transfer complexes. Klug suggests that

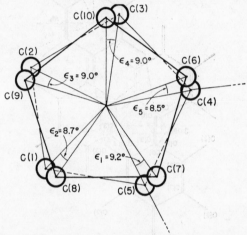

Fig. 13. View perpendicular to the planes of the five-membered rings in α-keto-1,5-tetramethyleneferrocene.[63]

the quinoid structure and self-complexing charge transfer interactions may account for the black colour of the crystals. The quinoid structure is quite planar and forms chains through pairs of hydrogen bonds. There are also hydrogen bonds to form water chains and hydrogen bonds between the quinone and water chains. Figure 14 shows a projection of the quinone-molecule chains perpendicular to their plane.

The configurations of the boron hydride compounds have been of considerable interest to chemists because of the fascinating polyhedra, or fragments of them, which are formed and the possibility of making correlations with theoretical studies of the bonding. The reactions of the boron hydrides represent a field of chemistry of special study. Recently the structure of the dimeric bis(o-dodecacarborane), $H_{11}B_{10}C_2$-$C_2B_{10}H_{11}$ has been studied by Hall, Perloff, Mauer and Block.[65] They found that the C_2B_{10} unit is a slightly distorted icosahedron. Figure 15 shows the molecular packing. The two halves of the molecule are joined by a carbon—carbon single bond and are related by a centre of symmetry.

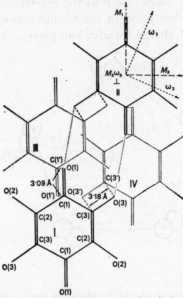

Fig. 14. Projection of the quinone-molecule chains perpendicular to their plane. Molecules I and II are part of a chain in the plane of the paper formed by hydrogen bonding, indicated by the dotted lines. Molecule III is in a chain above the plane and molecule IV is in a chain below the plane of the paper. Abnormally short C $\cdots$ O distances are shown.[64]

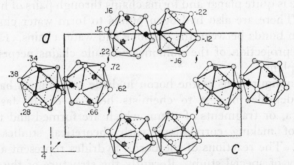

Fig. 15. Molecular packing of bis(o-dodecacarborane). Solid circles are carbon atoms and open circles are boron atoms. The cross-hatched circles are related to models of disorder.[65]

A material called ranunculin, (11a) can be extracted from the buttercup and other anemone flowers. It can be hydrolysed to produce protoanemonin, (11b) which is a disagreeable liquid blistering agent. Protoanemonin, however, dimerizes spontaneously in aqueous solution to yield a crystalline product with no blistering properties. It is remarkable that only one cyclo-dimer, anemonin, is formed from (11b) since at least twelve cyclo-butane structures are possible for various combinations of the monomer. The structural formula of the dimer was known from chemical evidence; however, the stereoconfiguration was unresolved. The *cis*-form, (11c) had been favoured from chemical evidence. An X-ray investigation[66] of the structure, however, showed that the material exists in the *trans*-configuration, (11d).

The structures of cyclo-butane derivatives are of interest because the four-membered rings are under considerable strain. X-ray and electron diffraction structure determinations showed that the four-membered ring was planar in tetraphenylcyclo-butane, but was not planar in cyclo-butane and octafluoro- and octachlorocyclo-butane. The four-membered ring in anemonin, 11d is not planar. An X-ray structure determination has been made of *cis*- and *trans*-1,2-dibromo-1,2-dicarbomethoxycyclo-butane,[67] Figs. 16 and 17, respectively. The latter figures are stereographic drawings and can be viewed with some practice at a comfortable distance of the order of arm's length without the use of a viewer. The drawings were made by a computing machine by means of a programme prepared by C. K. Johnson of the Oak Ridge National Laboratory.

The geometric isomers both possess a folded four-membered ring with a dihedral angle of approximately 150° and internal CCC angles of approximately 88.5°. The *trans*-molecule has an approximate twofold axis of rotation whereas the *cis*-molecule has no symmetry.

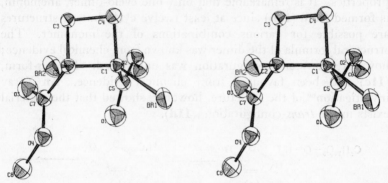

Fig. 16. Stereographic drawing of *cis*-1,2-dibromo-1,2-dicarbo-methoxycyclo-butane.[67]

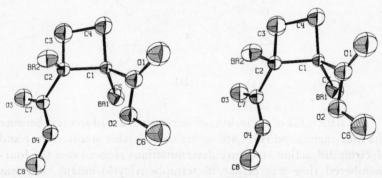

Fig. 17. Stereographic drawing of *trans*-1,2-dibromo-1,2-dicarbo-methoxycyclo-butane.[67]

Serotonin or 5-hydroxytryptamine is an animal hormone concerned with physiological, neurophysiological and psychological processes. It causes smooth muscle contraction, inhibits nerve impulse transmission and produces mental changes. The free base is quite unstable and sensitive to light, heat and pH changes. Biological samples are usually isolated as the monohydrate complex of 5-hydroxytryptamine, creatinine and sulphate. The structure of this complex has been studied by X-ray diffraction[68] and a map is shown in Fig. 18. Among the interesting structural features revealed was an intricate network of hydrogen bonding in all three dimensions. All the hydrogen atoms which are available for hydrogen bonding have been utilized. The hydrogen bonding arrangement suggests an $=NH_2^+$ group on the creatinine and an $-NH_3^+$ group on the serotonin.

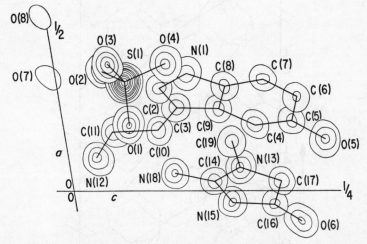

Fig. 18. Composite electron density map for the serotonin–creatinine sulphate complex projected along the b-axis.[68]

The effects of overcrowding by substituents on aromatic systems are quite varied. Poly-substituted aromatic compounds show distortion of the aromatic nucleus, sometimes considerable, and unusual exocyclic bond lengths. As a further study of compounds of this type Dickinson, Stewart and Holden[69] have studied the crystal structure of 2,3,4,6-tetranitroaniline. They found that the benzene ring is distorted, the carbon—carbon bonds at the amine group carbon average 1.43 Å and the remainder 1.37 Å. The nitro groups at C(2), C(3) and C(4) were rotated by 45°, 64° and 19° respectively out of the plane of the benzene ring. The nitrogen atoms on carbons 1, 2, 3 and 4 were also a significant distance away from the plane of the benzene ring, ranging from 0.10 Å to 0.22 Å in magnitude. The longest carbon—nitrogen bond, 1.487 Å for

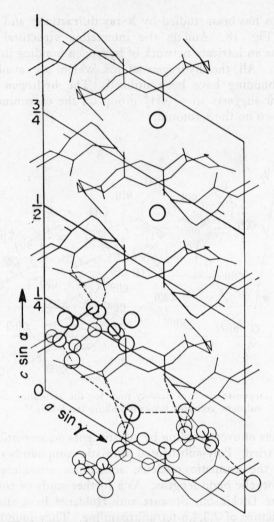

Fig. 19. The contents of the unit cell of cyclo-hexaglycyl hemihydrate
projected along the b-axis.[70]

C(3)—N(3), is correlated with the nitro group which had been
reported to have enhanced chemical reactivity.

Cyclo-hexaglycyl is a synthetic polypeptide in which six glycine
residues form an eighteen-membered ring. The material crystallizes
as a hemihydrate with eight molecules in the unit cell.[70a] Figure 19

shows the contents of the unit cell mainly in the form of a stick model, the corners of which represent the positions of the atoms. A surprising feature of this structure is that the eighteen-membered ring crystallised in four different conformations. Apparently they are all about equally stable. Bryan and Dunitz[70b] have found that some nine-membered rings were stable in two different conformations, but this is the only case with four stable conformations. The cyclo-hexaglycyl crystal is perfectly ordered and has many hydrogen bonds joining the molecular units. In this case all the hydrogen atoms capable of forming hydrogen bonds have been so used.

One of the most active arylphthalamic acids which interferes with the geotropism of vegetable roots is α-naphthyl-4-chlorophthalamic acid. The structures of this material and related materials are currently under study by Mornon.[71] The structure obtained for the α-napththyl compound is seen in projection in Fig. 20.

When α-phenylcinnamoyl chloride is pyrolysed with thionyl chloride and 2% sulphuryl chloride, a complex material (12) is

α-Phenylcinnamoyl   (12)
chloride

formed. The structure of (12) was studied by Bednowitz and Hamilton.[72] They found that the stereoisomers of (12) crystallize in an ordered arrangement, each pair related by a centre of symmetry. The molecule consists essentially of two planes folded along the ϕ—C—CH bonds.

The crystal structure of azulene-1,3-dipropionic acid has been studied by Ammon and Sundaralingam.[73] As seen in Fig. 21, the hydrogen bonding results in infinite ribbons of molecules parallel to the glide directions. The propionic acid residue on one of the carbon atoms is folded in contrast to an extended form on the other carbon atom. The authors relate this observation to the nature of the packing of the molecules.

The dithiocarbamates find a variety of applications, e.g. rubber chemistry and fungicides. The structure of zinc dimethyldithio-carbamate has been investigated by Klug.[74] He found that the

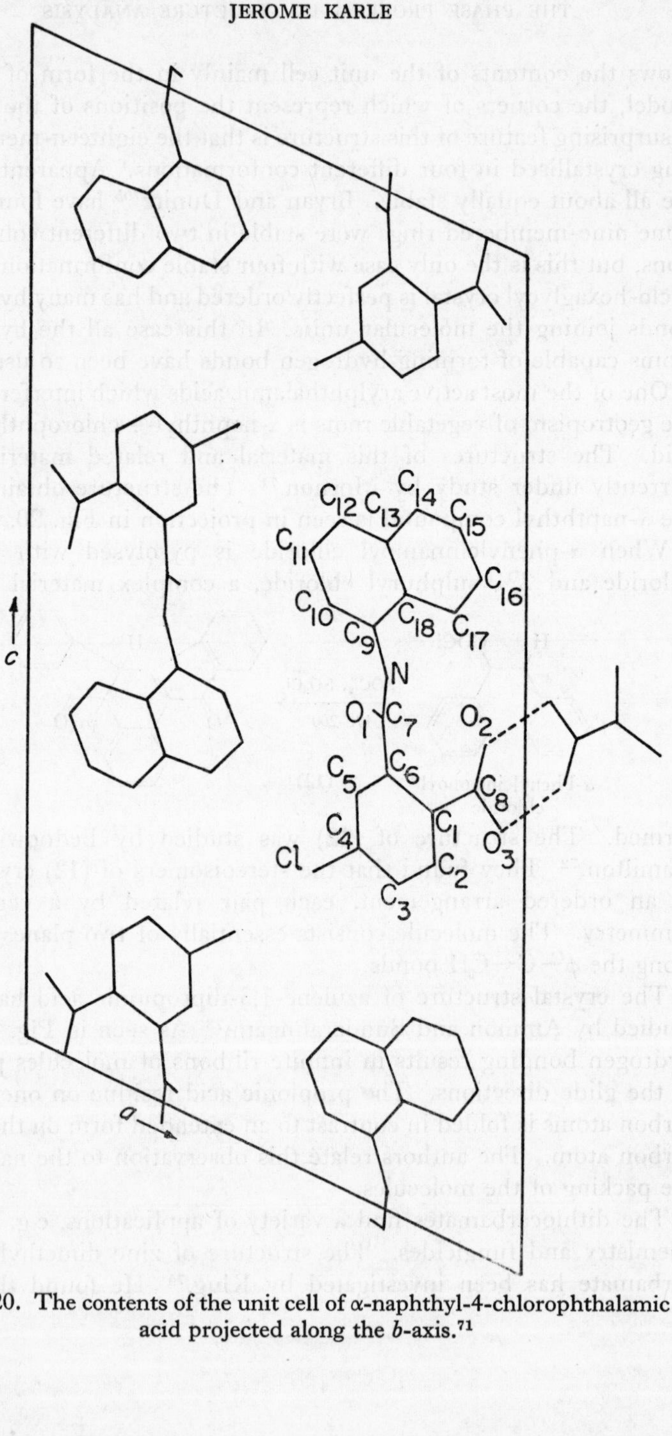

Fig. 20. The contents of the unit cell of α-naphthyl-4-chlorophthalamic acid projected along the b-axis.[71]

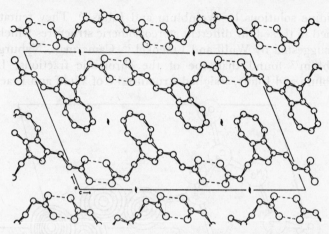

Fig. 21. The contents of the unit cell and parts of surrounding cells of azulene-1,3-dipropionic acid projected along the b-axis. Hydrogen-bonding to form molecular chains is indicated by dashed lines.[73]

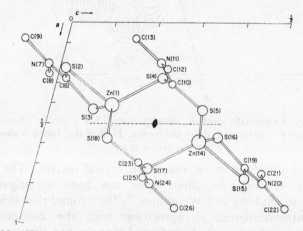

Fig. 22. The binuclear molecule of zinc dimethyldithiocarbamate projected along the b-axis.[74]

molecules occur as dimers in the crystal, as seen in Fig. 22. There is tetrahedral coordination of the sulphur atoms about each zinc atom and the dimethyldithiocarbamate groups are almost planar.

Four different isomers were obtained from u.v. irradiation of aqueous solutions of thymidylthymidine and from frozen aqueous thymidine solutions by Weinblum and Johns.[75] They tentatively assigned to these four different stereoisomeric structures which had been suggested by Wulff and Fraenkel.[76] Camerman, Nyburg and Weinblum[77] found that one of the chromatic fractions, D, of Weinblum and Johns, assigned structure IV of Wulff and Fraenkel,

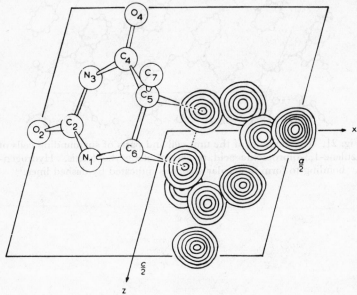

Fig. 23. Composite electron density map for a thymine photodimer ("thymer E") projected along the b-axis. Half of the dimer is shown in the perspective drawing.[77]

contained a triclinic and a monoclinic crystal species. The crystal structure of the monoclinic form has been investigated by Camerman, Nyburg and Weinblum.[77] They found that this dimer was centrosymmetric, in agreement with the assignment of Weinblum and Johns, and confirmed that the two thymine molecules are linked by a cyclo-butane ring formed between their two 5,6 double bonds, as seen in Fig. 23. A comparable molecular structure has been found for a photodimer of 1-methylthymine by Einstein and coworkers.[78] The authors report that their structure refinement thus far has shown that corresponding bond lengths and angles of the two different molecules have no appreciable differences.

Anhydrous borates with greater than 50 mole % B_2O_3 can readily form glasses. As a further study of this type of material, Hyman and coworkers[79] have investigated the crystal structure of the high-temperature form of $Na_2O\cdot4B_2O_3$ grown from a stoichiometric melt on a hot wire loop. They found that the structure

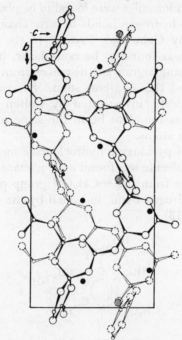

Fig. 24. The two networks of $Na_2O\cdot4B_2O_3$ projected along the *a*-axis. The large open circles represent oxygen atoms and the small open circles represent boron atoms. One network is indicated by dashed circles and open links, the other by unbroken circles and filled links. The solid circles represent sodium atoms. The shaded circles represent atoms from the next unit cell along the *a*-axis. They have been displaced slightly for clarity.[79]

consists of two infinite, independent and interlinking boron–oxygen networks, each containing alternating single and double rings, Fig. 24. The sodium atoms hold the networks together through their coordination with oxygen atoms. A rule is given for determining the relative numbers of triangularly and tetrahedrally coordinated boron atoms as a function of composition. The rule does not hold in the presence of unusual coordinations between boron and oxygen atoms. The authors suggest that density measurements may indicate the presence of unusual coordinations.

Among other interesting studies is the determination of the structure of β-fumaric acid by Bednowitz and Post.[80] Within the limits of error, the molecules were found to be planar and they were linked together by hydrogen bonds to form chains. The glycolate ion was studied by Gabe and Taylor[81] in the form of [6]Li-(S)-glycolate-2-d. It was found to be non-planar, both atoms of the hydroxyl group being significant distances from the plane of the carboxyl group and the α-carbon atom. By contrast, in lithium glycolate monohydrate, investigated by Colton and Henn,[82] the hydroxyl group was found to be in the same plane as the other carbon and oxygen atoms.

The structure of pyridoxine hydrochloride has been studied by Hanic[83] and the molecule was found to be planar with the exception of the oxygen atom from the $-CH_2OH$ group participating in an intramolecular hydrogen bond, indicated by the dotted line in the molecular model (13).

(13)

B. Almost Equal Atom Non-centrosymmetric Crystals

Equal atom and almost equal atom non-centrosymmetric crystals have constituted an area of difficulty in structure determination. At present they cannot be handled with the routine simplicity that pertains to most centrosymmetric crystals or to non-centrosymmetric ones which contain one or more atoms which have a significantly higher atomic number than the majority of the atoms present. Both the analysis of Patterson vector maps[1-6] and the symbolic addition procedure for phase determination,[9] usually used in conjunction with a reiteration procedure employing the tangent formula,[7] have been successfully applied to moderately complex almost equal atom non-centrosymmetric crystals. Such structures often arise among substances of biochemical interest.

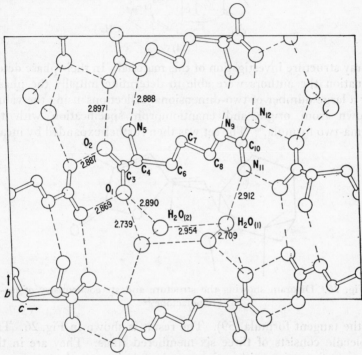

Fig. 25. The contents of the unit cell of L-arginine dihydrate projected along the *a*-axis.[84]

The structure determination of the amino acid L-arginine dihydrate[84] has been carried out by means of the symbolic addition procedure. A stick model of the contents of the unit cell is shown in Fig. 25. There is an extensive amount of hydrogen bonding, as shown by the dotted lines. As a consequence the organic molecules form infinite chains. The water molecules also form hydrogen-bonded infinite chains which are perpendicular to the plane of the figure.

Isoeremolactone is an isomer of eremolactone, a diterpene. It had been suggested that the structure of isoeremolactone could be represented by the structural formula (14). Owing to the uncertainty in the proposed structure Oh and Maslen[85] undertook the

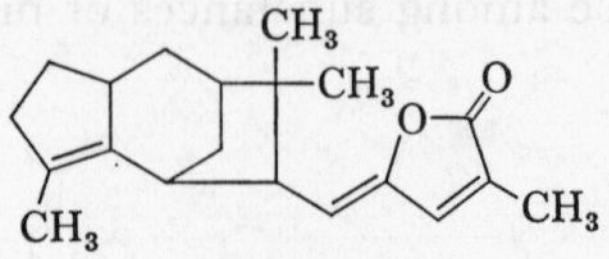

(14)

X-ray structure investigation of this material. In their phase determination the authors were able to determine initially the phases for a large number of two-dimensional reflections using the values known from origin and enantiomorph specification with the sigma-two formula.[65] The set was then further expanded by means

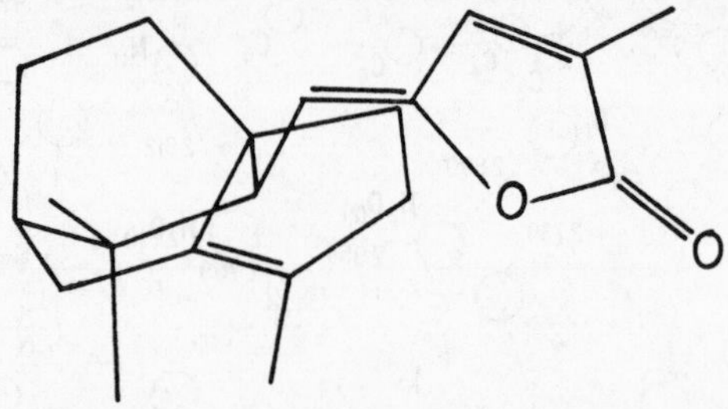

Fig. 26. Diagram showing the structure and stereochemistry of iso-eremolactone.[85]

of the tangent formula (59). The result is shown in Fig. 26. The molecule consists of three six-membered rings. They are in the boat configuration with a five-membered ring attached to one of them. The γ-lactone structure of the side-chain was confirmed.

Reserpine (15) is an important drug for the treatment of hypertension and nervous disorders. The stereochemistry had been

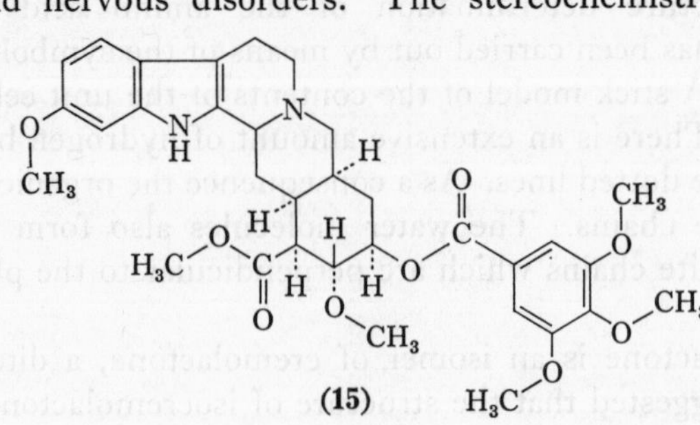

(15)

established by chemical means, but there remained the problem of finding the spatial arrangement of the atoms in the molecule. An X-ray investigation,[86] employing the symbolic addition procedure, revealed a partial structure which was combined with the reiteration procedure, employing the tangent formula, to give the complete structure. Figure 27 shows sections from an electron density map

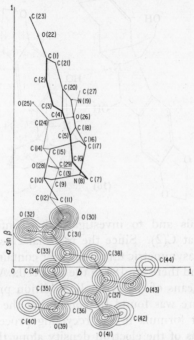

Fig. 27. Sections from part of a three-dimensional electron density map for reserpine projected along the c-axis. The contours for atoms in the remainder of the molecule are omitted for purposes of clarity.[86]

projected down the c-axis. The bonds are outlined and the top half is not contoured since the overlap would obscure the details. The trimethoxybenzoxy group is nearly perpendicular to the remainder of the reserpine molecule. The indole group is planar and only C(7) and N(8) are significantly out of the plane in the adjacent ring. The next two rings are puckered and in the chair configuration. The dihedral angle between least-squares planes for the indole group and the benzoxy group is 82°.

The structure of the 6-deoxy-derivative of 6-hydroxycrinamine
(16) has been derived from chemical and spectral evidence by
Fales and Wildman.[87] The X-ray diffraction analysis[88] was undertaken to confirm the structure and stereoconfiguration of the

(16)

crinamine nucleus and to investigate the configuration of the
hydroxyl group at C(2). Since the material crystallizes with two
molecules in the asymmetric unit, the latter contains 46 atoms other
than hydrogen and there are 184 in the unit cell. A partial structure
was found by means of the symbolic addition procedure and the
complete structure was found by combining the partial structure
with the tangent formula in the recycling procedure. The projection of sections of the electron density along the b-axis may be
seen in Fig. 28. The X-ray study confirmed the configuration of the
crinamine nucleus deduced from degradative studies.[87] It also
established that the C(2) hydroxyl group is *trans* to the pyrollidine
ring. The geometry of the two molecules in the asymmetric unit
is quite similar and they are associated into dimers by a pair of
hydrogen bonds between N(4)—O(21)* and N(4)*—O(21).

Digitalis is an important material in heart therapy. It acts by
increasing the ability of the heart muscle to contract and diminishing the heart rate. The active components of digitalis are glycosides
of three materials, one of which is the glycoside of digitoxigenin.
The individual glycosides are all quite potent, but the free aglycones

are fairly weak. The structural formula of digitoxigen (**17**) has been worked out by chemical means. It was of interest to examine the

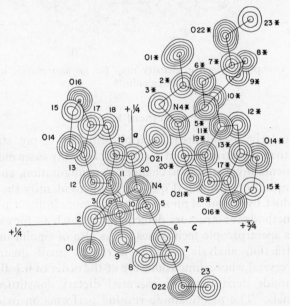

(17)

structure of the molecule by X-ray analysis and obtain the detailed structural parameters.[89] A partial structure obtained by means of the symbolic addition procedure was combined with the tangent

Fig. 28. Composite electron density map for an asymmetric unit of 6-hydroxycrinamine.[88]

formula in the recycling procedure to yield the complete structure, Fig. 29. The result was in agreement with the structure derived by chemical means and the molecular parameters were normal. The hydrogen atoms on the methyl group attached to the central portion of the molecule apparently had low thermal motion since they were readily found in a difference map. It has been reported that the unsaturated lactone ring and the 14-hydroxyl group are both essential to activity. Their configurations in the crystal were not unusual.

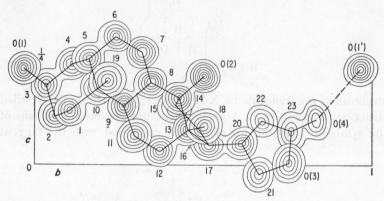

Fig. 29. Composite electron density map for an asymmetric unit of digitoxigenin.[89]

C. Rearrangement Products

A particularly useful area of application of X-ray structure analysis is that of rearrangement reactions. In many cases molecular rearrangements involve drastic changes in configuration, and often also in composition, making it quite difficult to identify the nature of the product by chemical means or by the more indirect physicochemical methods for structure determination such as, for example, the various spectroscopic techniques. It is also of significance that X-ray diffraction analysis requires a rather small quantity of material, a crystal whose dimensions are of the order of 0.1–0.2 mm.

Nicotinamide derivatives are essential dietary constituents for many animals. The nicotinamide residue performs an oxidation-reduction function (18). When R is ribose pyrophosphate adenosine, the above substances are the coenzymes denoted by DPN and DPNH, respectively. N-benzyl-1,4-dihydronicotinamide, an analogue of DPNH, serves as a good model for the DPN–DPNH transhydrogenase system. It was of particular interest to know if the dihydronicotine ring is planar and therefore an investigation of the structure was carried out[90] by the direct method for phase determination[28] that preceded the symbolic addition procedure. It was found that the dihydronicotine ring is planar and that the molecules

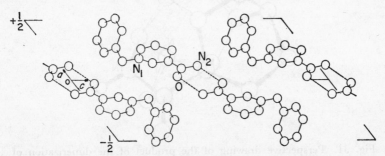

$$+ 2H \rightleftharpoons + H^+$$

(18)

occur as dimers in the crystal, Fig. 30, held together by hydrogen bonding. It was therefore reasonable to assume that the N-methyl derivative of dihydronicotinamide also forms a dimer by means of hydrogen bonding. However Ammon and Jensen[91] found from

Fig. 30. The contents of the unit cell of N-benzyl-1,4-dihydronico-tinamide projected along the b-axis. Dimers are formed by hydrogen bonds across symmetry centres.[90]

an X-ray analysis that the structure of the dimer was quite unusual, Fig. 31. In the dimerization both double bonds open to form the cage compound. An amide group is involved in a ring closure by the transfer of a hydrogen atom from an —NH_2 group. Further discussion of the reaction mechanism is given by the authors.[91] The dimer possesses seven six-membered rings, six of which have the distorted boat shape.

A novel oxygen-transfer reaction in a photosensitized autoxidation has been reported by Wasserman, Doumaux and Davis.[92] A methanolic solution of furano-*para*-cyclo-phane (**19**) was irradiated in the presence of oxygen and then was concentrated and hydrogenated. Three products were obtained by chromatography of

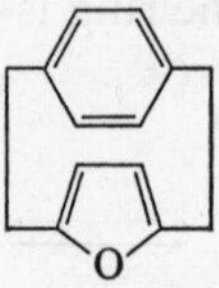

(19)

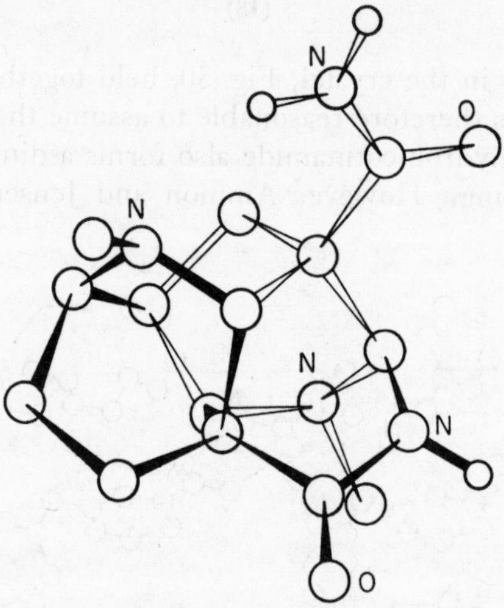

Fig. 31. Perspective drawing of the product of the dimerization of 1-methyl-1,4-dihydronicotinamide.[91]

which two were identified by chemical and physico-chemical methods and the third was characterized by X-ray diffraction. The third product was found to have the configuration (20). The

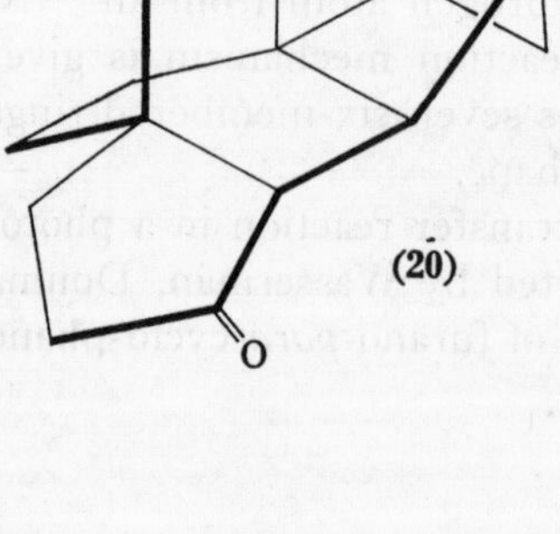

(20)

authors have proposed a reaction mechanism for the formation of this material which involves the intermediate formation of a benzene epoxide followed by an intramolecular Diels–Alder reaction and hydrogenation to form the final product.

The photosensitized autoxidation of (2,2)-*para*-cyclo-naphthane has been reported by Wasserman and Keehn.[93] A solution of anti-(2,2)-*para*-cyclo-naphthane (**21**) in methanol in the presence

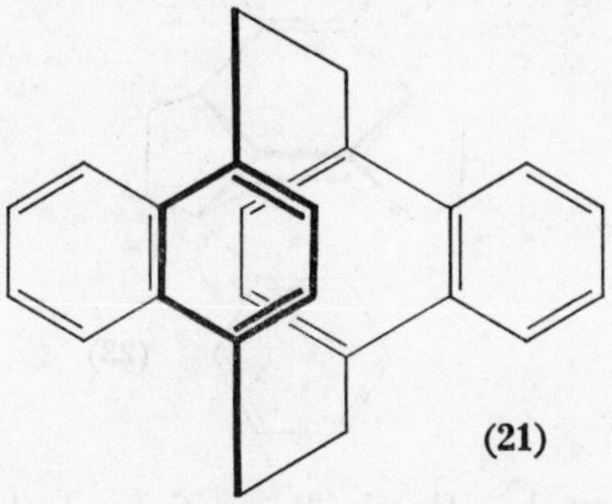

(21)

of methylene blue was aerated and subjected to irradiation by a 150 W floodlamp. The oxidation product was recovered and subjected to analysis by chemical and physico-chemical methods. The structure consistent with this analysis is (**22**). The latter

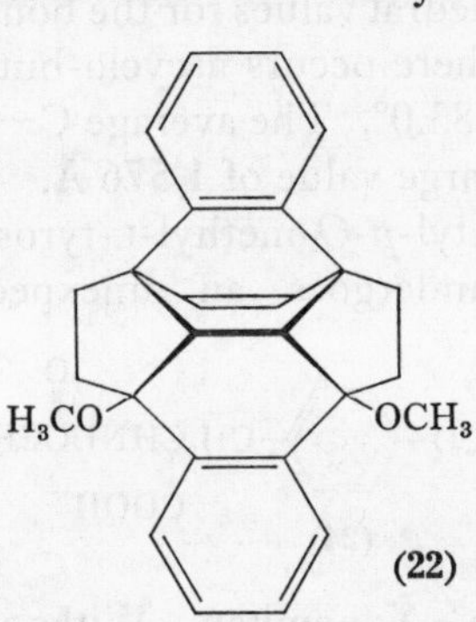

H₃CO   OCH₃

(22)

structure was confirmed by an X-ray analysis carried out by Fratini.[94] Wasserman and Keehn[93] have suggested a reaction mechanism which involves the intermediate formation of a transannular peroxide, followed by an internal Diels–Alder reaction and solvolysis in methanol to form the final product.

If instead of irradiating (2,2)-*para*-cyclo-naphthane (**21**) by means of a floodlamp, u.v. irradiation is used in the absence of oxygen, it has been reported by Wasserman and Keehn[95] that another new rearrangement product is produced. The authors described a series of physico-chemical tests with which the structure (23), termed dibenzoequinene, is consistent. An X-ray

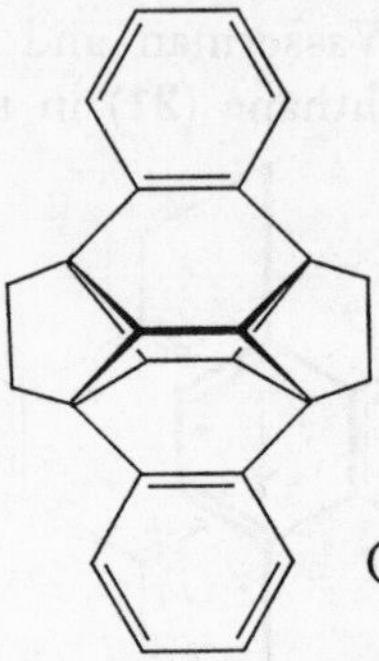

(23)

analysis carried out by Fratini[94] confirmed this structure. The parameters of the inner ring systems in the rearrangement products from (2,2)-*para*-cyclo-naphthane are of particular interest owing to the constraints imposed by the configuration. There are large deviations from tetrahedral values for the bond angles. For example in dibenzoequinene there occurs a cyclo-butane ring in which the average CCC angle is 83.0°. The average C—C distance in this ring is the comparatively large value of 1.576 Å.

When N-chloroacetyl-p-O-methyl-L-tyrosine (24) is subjected to irradiation, it undergoes an unexpected photocyclization

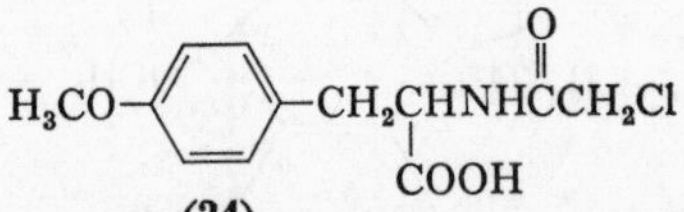

(24)

reaction reported by Yonemitsu, Witkop and Karle.[96] The identification and the determination of the structure of the methyl ester of the product were carried out by X-ray analysis.[97] The material crystallized in a non-centrosymmetric space group. A partial structure was obtained by means of the symbolic addition procedure and the use of the partial structure with the tangent formula in the recycling procedure yielded the complete structure shown in Fig. 32. It is seen that a five- and a seven-membered ring are formed, an aldehyde appears and a methoxy group is lost. With the aid of the identification of the final product, a detailed mechanism for the reaction has been proposed.[96]

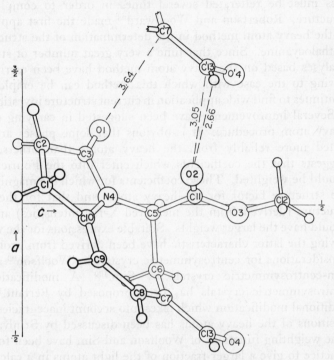

Fig. 32. View along the *b*-axis of the molecule formed from the rearrangement of *N*-chloroacetyl-*p*-*O*-methyl-L-tyrosine. Closest molecular approaches are indicated.[97]

7. GENERALIZED HEAVY ATOM TECHNIQUES

A. Phases from Knowledge of Partial Structure

If there are few heavy atoms in a crystal, it is often a very simple matter to overcome the phase problem. The positions of the heavy atoms may be found from analysis of the Patterson function (5) and approximate values for the phases may then be obtained by computing the expression for the structure factor (3) using the heavy atom positions. In the simplest procedure, the phases are then used to compute a Fourier map from (1). The positions of the lighter atoms may be determined by examining this map. Generally only a fraction of the light atom positions is obtained and the process must be reiterated several times in order to complete the structure. Robertson and Woodward[98] made the first application of the heavy atom method in the determination of the structure of phthalocyanine. Since that time a very great number of structure analyses based on the heavy atom method have been carried out. Owing to the ease with which this method can be employed, it continues to find wide application in current structure investigations.

Several improvements have been suggested in carrying out the heavy atom procedure. It is obvious that some phases are computed more reliably from the heavy atoms than others. This suggests that the coefficients which enter into the Fourier series should be weighted. Those coefficients for which the magnitude of the structure factor for the heavy atoms and that for the entire structure (derived from the measured X-ray intensities) are large should have the larger weights. Suitable expressions for the weights having the latter characteristic have been derived from probability considerations for centrosymmetric crystals by Woolfson[99] and for non-centrosymmetric crystals by Sim.[100] A modification for centrosymmetric crystals has been proposed by Bertaut.[101] An additional modification which takes into account inaccuracies in the positions of the heavy atoms has been discussed by Srinivasan.[102] The weighting functions of Woolfson and Sim have been found in practice to give a larger fraction of the light atoms in a calculation of a Fourier map from a partial structure, thus decreasing the total number of reiterations required.

Another alteration of the heavy atom procedure of practical significance is the use of phase determining formulae with the approximate phases computed from the positions of the heavy atoms or the heavy atoms and a partial structure of light atoms. It was seen in the previous section that the tangent formula (59) combined with phases computed from a partial structure of light atoms can play a valuable role in generating additional phases. In this way complete structures have been obtained for equal atom non-centrosymmetric crystals. Similarly, the starting partial structure could be composed of the heavy atoms in a crystal. Interposing the application of the tangent formula to the phases from the partial structure before computing the Fourier map can reduce the number of iterations required and occasionally yields a complete structure determination in one step without the need for reiteration. This has been observed in three recent structure investigations, that of two anti-radiation agents[103, 104] (25), (26) and of the anti-heroin

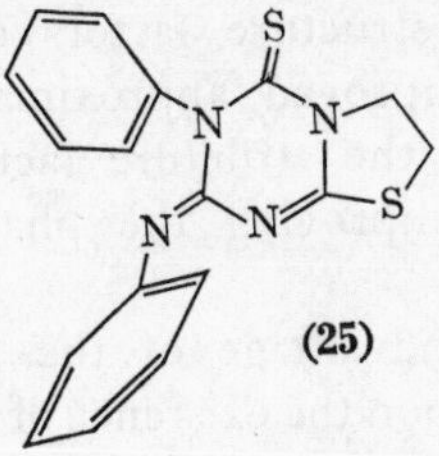

(25)

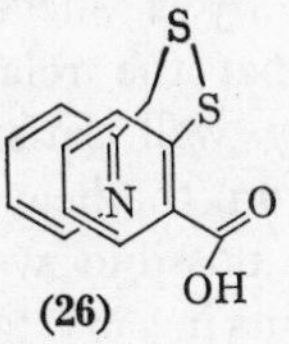

(26)

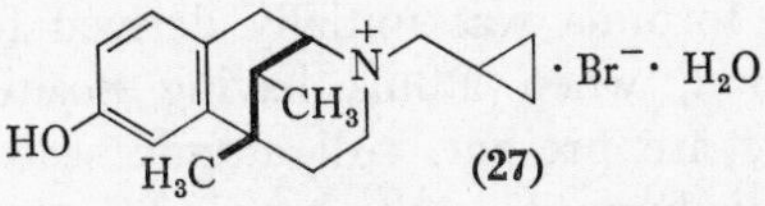

(27)

agent, cyclazocine, in the form of the bromide[105] (27). The structure determination for the first two substances was begun with a knowledge of the positions of the two sulphur atoms in each. The use of the tangent formula as an intermediate step gave the complete structure in one step. For the cyclazocine bromide, which crystallized in space group $P2_1$, the positions of two light atoms in addition to that of the bromine atom were used for the starting partial structure, because a single atom of bromine in this non-centrosymmetric space group is centrosymmetrically placed and cannot properly define non-centrosymmetric phases. In this case one reiterative step was required.

It is apparent that the use of the tangent formula in reducing the number of iterations for non-centrosymmetric crystals should find a parallel in the use of the sigma-2 relation (65) with centro-symmetric crystals. A somewhat similar recommendation for application with heavy atom or light atom partial structures in centrosymmetric crystals has been made by Hoppe[106] and by Hoppe and Huber.[107] The authors use the Sayre equation[32] which is expressed in terms of observed structure factors. Their procedure should be improved by the direct application of the sigma-2 relation (65) instead, which is expressed in terms of normalized

8

structure factors. It is merely necessary to employ the approximate phases obtained from the partial structure as a basic set for further expansion of the phases employing (65).

Rossman[108] has described an alternative method for refining the phases obtained from knowledge of a partial structure, e.g. heavy atoms. He derived a relation between the structure factors of the known partial structure and the structure factors of the total structure. By a succession of improved approximations, it is expected that the relation among the structure factors will be increasingly well satisfied, thus improving the phases initially obtained from the heavy atoms.

Another possible way to take advantage of the presence of heavier atoms in a crystal depends upon the existence of a formula[109] which exaggerates the contribution of the heavier atoms to the scattering. The formula was initially derived for application in neutron diffraction, when atoms having positive and negative scattering factors are present. All atomic scattering factors for X-rays are essentially positive numbers, except near an absorption edge where they may become complex numbers. The latter case will be discussed below under the subject of anomalous dispersion.

For X-ray scattering, the formula comparable to that for neutron scattering is

$$|E'_{\mathbf{h}}|^2 - 1 \simeq \frac{(\sigma_2^2 \sigma_4^{-1} - 1)}{\langle(|E_{\mathbf{k}}|^2 - 1)^2\rangle_{\mathbf{k}}} \langle(|E_{\mathbf{k}}|^2 - 1)(|E_{\mathbf{h}-\mathbf{k}}|^2 - 1)\rangle_{\mathbf{k}} \quad (74)$$

where the quasi-normalized structure factors for the original and for the squared structure are defined by (46) and (47) and the σ_n are defined by (48). The value of this formula lies in the fact that quasi-normalized structure factors for the squared structure may be computed from those obtained experimentally for the original structure. Since the squared structure is one in which the atomic scattering factors for the original structure are squared, it is apparent that the heavier atoms will make a proportionately larger contribution. For example, a chlorine atom in a squared structure would make almost nine times the contribution to a structure factor on the average that a carbon atom would make, whereas in the original structure the ratio would be about three to one. Given sufficient accuracy in the calculation of (74), a set of structure factors may be obtained in suitable cases which essentially represent the heavy atom structure with little contribution from the lighter atoms. This may facilitate direct phase determination or the interpretation of Patterson syntheses (5). In cases where there may be difficulty in locating the positions of moderately heavy atoms from a Patterson synthesis, it may be useful to employ (74) to transform the data to that for the squared structure and then recompute the Patterson map. If the data for the squared structure were computed with sufficient accuracy, a useful map would be obtained.

B. Isomorphous Replacement

A particularly useful application of heavy atom techniques arises when such atoms are present in isomorphous crystals. Crystals having the same unit cell geometry, but differing in chemical composition are called isomorphous. Often, isomorphous crystals contain many light atoms in a fixed configuration and some heavy ones which are added or replaced. The method of phase determination based on isomorphous structures is called the isomorphous replacement method. It has the advantage that the replaceable atoms need not have proportionately as great an electron content as is required in the heavy atom method.

If the two structures of an isomorphous pair are centrosymmetric the signs of the structure factors for both crystals can be found. In practice certain of the signs would not be determinate because of experimental errors and small magnitudes for some of the structure factors. The appropriate equation is

$$F_{R+X} + F_{Y-X} = F_{R+Y} \qquad (75)$$

where F_{R+X} is the structure factor for the structure consisting of invariant atoms R and replaceable atoms X, F_{R+Y} concerns the invariant atoms R and replaceable atoms Y, and F_{Y+X} is the structure factor for the configuration of the difference between atoms Y and atoms X. The quantities $|F_{R+X}|$ and $|F_{R+Y}|$ are known and F_{Y-X} is determined from, for example, Patterson maps from which the positions of atoms X and Y are obtained. With this information the assignment of signs to $|F_{R+X}|$ and $|F_{R+Y}|$ to satisfy (75) is unique.

In the case of two non-centrosymmetric crystals which form an isomorphous pair, it can be seen from the construction in Fig. 33 that a twofold ambiguity exists in the assignment of the phases.

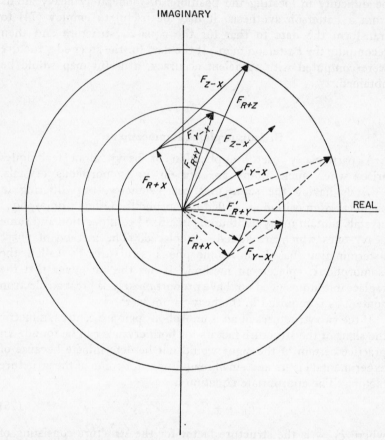

Fig. 33. Construction for multiple isomorphous replacement, showing the manner of choosing between alternative values for the phase angle associated with F_{R+X}. The dashed lines indicate the unsuitable alternatives.

Equation (75) holds and the known quantities are again $|F_{R+X}|$, $|F_{R+Y}|$ and F_{Y-Y}. The construction is made by drawing two circles of radii $|F_{R+X}|$ and $|F_{R+Y}|$. The vector F_{Y-X} is placed so that its tail touches the circle of radius $|F_{R+X}|$ and its point the circle of radius $|F_{R+Y}|$. There are then two closed triangles which satisfy (75) and the known information. They are distinguished from each other by the use of primed and unprimed symbols. Bijvoet[111] discussed the application of isomorphous substitution to non-centrosymmetric crystals and noted the existence of the twofold ambiguity. One method of overcoming the ambiguity was suggested by Bokhoven, Schoone and Bijvoet[112] and is now termed the method of multiple isomorphous replacements. In this method a second isomorphous replacement is made involving $|F_{R+X}|$, $|F_{R+Z}|$ and F_{Z-X} as shown in Fig. 33. It is clear from the figure that the ambiguity has been resolved by choosing the pair of triangles which have the same F_{R+X}.

It is commonplace in the study of protein structures to make several isomorphous replacements in order to overcome difficulties which arise from experimental error, lack of exact isomorphy, insufficient differences for some reflections in the directions of F_{Z-X} and F_{Y-X} and as a further verification of the results. A study of the treatment of errors in isomorphous replacement has been carried out by Blow and Crick.[113] Probability formulae were presented which gave the probability of a correct sign in the centrosymmetric case and the relative probability of different values for a phase in the non-centrosymmetric case, when the lack of isomorphism and observational errors were taken into account. For proteins, it is the probability formula for the non-centro-symmetric case which is of particular significance. Ideally, for a given reflection Eq. (75) holds and if we assume that heavy atoms are added to an unsubstituted crystal

$$F_R + F_Y = F_{R+Y} \qquad (76)$$

where F_R is the structure factor for the unsubstituted crystal, F_Y is the replaceable or heavy atom contribution and F_{R+Y} is the structure factor for the substituted crystal. Blow and Crick[113] defined a function of the phase angle which represents the lack of closure of the vector triangle (76) owing to experimental difficulties

$$x(\phi) = F_{R+Y} - F_R - F_Y \qquad (77)$$

where the discrepancy $x(\phi)$ is assumed to be in the direction of F_{R+Y}. They presented an expression which affords a relative probability of different values for a phase based on the value of $x(\phi)$

$$P(\phi) = \exp\left(-x^2(\phi)/2\varepsilon^2\right) \qquad (78)$$

where ε is the root mean square error associated with the measurements including errors from all sources. When measurements from several isomorphous replacements are included, the joint probability was formed by multiplying the various probability functions given by (78). Rossmann and Blow[114] continued the study of optimizing the information available from multiple isomorphous replacement for phase determination. They formed joint probability distributions composed of the probability distribution for the phase from isomorphous replacement (78) multiplied by the probability distribution for the phase based on the fact that the structure is partially known since the heavy atoms have been located.

A test to determine whether a pair of crystals is isomorphous or non-isomorphous, based on the use of the higher moments of the normalized intensities, has been developed by Parthasarathy and Ramachandran.[115] A difficulty may arise in practical application owing to the fact that higher moments of the intensities obtained from experimental data are very sensitive to the errors in the data.

Another procedure for using a single isomorphous pair for non-centrosymmetric crystals, suggested by Coulter,[116] is to combine the information obtained from the pair with the tangent formula (59). For those large normalized structure factors, having two ambiguous indications which do not differ by more than a specified range of angle, the average value is accepted. Such phases form the basic set from which additional phases may be computed and the original ones recomputed. A further study[117] of the combination of isomorphous replacement and phase determining formulae has shown that when the substituent atoms are centrosymmetrically placed, the ambiguity can be readily resolved by means of a simple sign determination on the magnitudes of the imaginary parts of the structure factor. This may be seen by referring to Fig. 33 and noting that in the case that F_{Y-X} refers to centrosymmetric substituents, it would be placed along the real axis and the ambiguous triangles would be symmetric about the real axis. The real parts of the structure factors would then be known and only the signs of the imaginary parts would be in doubt. A phase determining procedure[117] similar to that for centrosymmetric crystals could then be carried out to determine the signs of the imaginary parts.

The use of Eq. (74) can provide an artificial way of producing an isomorphous pair. In a structure containing light and heavy atoms, the ratio of their scattering powers is different in the squared structure from that in the original structure. If the structure factors for the squared structure could be computed exactly from (74), the original and squared structures would make an ideal isomorphous pair. In practice, using a finite set of data, inaccuracies do arise, but they are smaller for the larger structure factors[109] and with suitable crystals useful information should be obtainable.

C. Anomalous Dispersion

In the discussions up to this point, it has been assumed that the X-ray scattering factor for atoms is a positive number. For heavier atoms and incident photon energy, sufficient or somewhat more than sufficient to remove electrons from the deep-lying shells, this is not a good assumption. Significant changes take place in the values of the scattering factors and the phenomenon is called anomalous dispersion. The scattering factor becomes a complex number defined as

$$f = f_0 + f' + if'' \tag{79}$$

where f_0 is the contribution from the normal or non-anomalous scattering and the quantities f' and f'' are the real and imaginary contributions from the anomalous scattering, respectively. Bijvoet[118] recognized that anomalous dispersion could be used to determine the absolute configuration of a molecule containing an asymmetric carbon atom. This is so because, when significant anomalous dispersion is present, the intensity for a reflection having indices h, k, l is in general different from one with indices $\bar{h}, \bar{k}, \bar{l}$, an effect which does not occur in the absence of anomalous dispersion. It is possible to relate the differences in intensity, called Bijvoet differences, to the handedness in a particular molecule. The mathematical expression for the different intensities associated with the Miller indices h, k, l and $\bar{h}, \bar{k}, \bar{l}$ can be obtained by substituting formula (79), describing the atomic scattering factor, into the expression for the structure factor (3) and then forming the absolute value squared of the structure factor. The latter is proportional to the measured intensity.

It is apparent that the Miller indices have to be assigned with care in the indexing of diffraction patterns obtained by photographic techniques. A discussion of this subject has been given by Bijvoet and Peerdeman[119] and was further extended by Vaidya and Ramaseshan[120] to include the trigonal, monoclinic and triclinic systems. The latter authors also discussed the double-layer screen technique with half slits for Weissenberg photographs of Hanson,[121] which permits reflections for both h, k, l and $\bar{h}, \bar{k}, \bar{l}$ to be recorded on the top and bottom halves of the film. It is quite probable that the increasingly widespread use of automatic diffractometers and the requirement of high accuracy in anomalous dispersion experiments will make the use of film methods fairly rare.

The statistical distribution of Bijvoet differences for non-centrosymmetric crystals has been studied by Parthasarathy and Srinivasan[122] and Parthasarathy.[123] The theory was worked out for one, two and many anomalous scatterers, all of the same type, in the presence of a large number of non-anomalous scatterers. It was found, using the average value of the Bijvoet differences suitably normalized as a measure, that a single anomalous scatterer gave the largest differences followed, usually in order, by two anomalous scatterers, many such scatterers with a non-centrosymmetric distribution and many with a centrosymmetric distribution. Generally the distinction among these cases is not great. However, when the anomalous scatterer contributes of the order of one-half the scattered intensity on the average, and the ratio of the imaginary to the real part of the total atomic scattering factor is large, somewhat significant differences appear among the different categories studied. As is pointed out by Parthasarathy,[123] the average values of the Bijvoet differences are only approximate indicators, because the theory assumes that the non-anomalous atoms are randomly distributed. Since in real structures this is rarely the case, structural features can be expected to play a significant role in determining the magnitudes of the Bijvoet differences and hence their ease of measurement.

In the initial application of anomalous dispersion to the determination of absolute configuration, Bijvoet, Peerdeman and van Bommel[124, 125] determined the absolute configuration of D-tartaric acid in the form of its sodium rubidium double salt and confirmed the chemical convention of Emil Fischer. Many determinations of absolute configuration have been carried out since this original work and the number of such investigations is increasing rapidly at the present time. They concern a variety of inorganic and organic materials. Among the more recent studies is the determination of the absolute configuration of the alkaloid cleavamine by Camerman and Trotter,[126] establishing the absolute stereochemistry at all the asymmetric centres in the molecule. Moncrief and Lipscomb[127] determined the stereochemistry and absolute configuration of leurocristine methiodide dihydrate and thereby established the absolute configuration of the antileukemia agent leurocristine. Tulinsky and van den Hende[128] determined the absolute configuration of N-brosylmitomycin A, a derivative of an anticancer antibiotic extracted from soil isolates of *Streptomyces verticillatus* strains. Okaya[129] established the absolute configuration of (+)alpha-naphthylphenylmethylgermane and Stevenson and Dahl[130] determined the absolute configuration of the cation in a crystal containing a trigonal-bipyramidal nickel (II) cyanide complex with tris(3-dimethylarsinopropyl)phosphine. Beurskens-Kerssen and coworkers[131] found that specimens of sodium chlorate and sodium bromate which rotate polarized light with the same rotation sign have opposite absolute configurations.

Given sufficient accuracy, it is possible to determine phases from an anomalous dispersion experiment with a twofold ambiguity in the case of non-centrosymmetric crystals. The geometric features are illustrated in Figs. 34 and 35 and are similar to constructions of Ramachandran and Raman.[132] The quantities shown in Fig. 34 are the structure factors F and $\bar{F}$ corresponding to the reflections **h** and $-$**h**, respectively, whose magnitudes may be obtained from experiment. The quantity F can be written

$$F = F' + F_A'' \tag{80}$$

where F_A'' is the contribution to the structure factor from the imaginary part of the atomic scattering factors of the anomalously scattering atoms and F' is the remainder of the contribution. The quantity F' in turn may be written

$$F' = F_A' + F_R \tag{81}$$

where F_R is the contribution from the non-anomalously scattering atoms and F'_A is the contribution to the scattering from the real part of the atomic scattering factors of the anomalously scattering atoms.

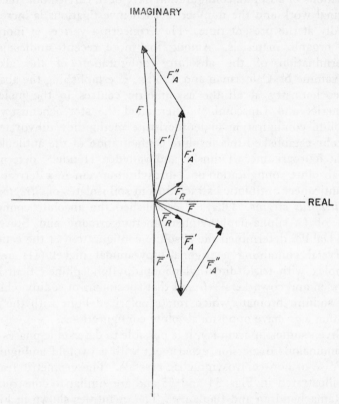

Fig. 34. Diagram for anomalous dispersion, showing the components of the structure factors for the reflections **h** and $-$**h**.

F'_A is therefore composed of the normal scattering and that from the real part of the anomalous correction to the atomic scattering factors of the anomalously scattering atoms. Several authors have termed the quantities F'_A the contribution from the normal dispersion and F''_A the contribution from the anomalous dispersion, ignoring the fact that the real part of the anomalous atomic scattering factors contribute to the quantity F'_A. This terminology is therefore not desirable.

Figure 34 shows the relationships among the various quantities for the reflections $\mathbf{h}$ and $-\mathbf{h}$. It is seen that F_R and $\bar{F}_R$, F'_A and $\bar{F}'_A$ and F' and $\bar{F}'$ are symmetric about the real axis whereas F''_A and

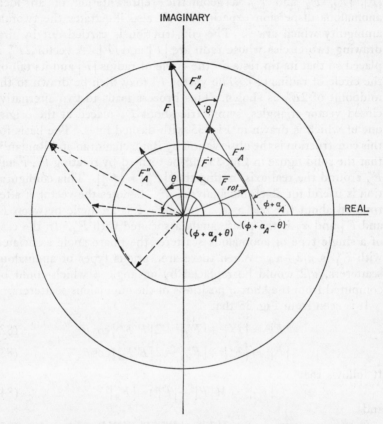

Fig. 35. Construction based on Fig. 34, showing in solid lines a diagram which is convenient for algebraic analysis and in dashed lines the ambiguous alternative which arises from the known experimental quantities. When all the anomalous scatterers are of the same type, $\phi = \pi/2$.

$\bar{F}''_A$ are symmetric about the imaginary axis. It is also apparent that $F \neq \bar{F}$ in general. When all the anomalously scattering atoms are of the same type, F'_A and F''_A are perpendicular to each other. These many relationships follow from the definition of the structure factor (3) and the definition of the atomic scattering factor (79) for anomalously scattering atoms.

Figure 35 illustrates the construction from the known quantities, $|F|$, $|\bar{F}|$, F_A'' and $\bar{F}_A''$, a geometric representation of an ideal anomalous dispersion experiment. It also illustrates the twofold ambiguity which arises. The construction is carried out by first drawing two circles whose radii are $|F|$ and $|\bar{F}|$. A vector $2F_A''$ is placed so that its tip rests on the circle of radius $|F|$ and its tail on the circle of radius $|\bar{F}|$. The vector F' may then be drawn to the midpoint of $2F_A''$ as shown. This process leads to two alternative closed vector triangles, symmetric about F_A placed at the origin, one of which is drawn in Fig. 35 with dashed lines. The basis for this construction is the observation by Ramachandran and Raman[132] that the solid figure in Fig. 35 can be formed by rotating $\bar{F}'$, $\bar{F}$ and $\bar{F}_A''$ around the real axis, noting that $|\bar{F}_A''| = |F_A''|$. This configuration is useful for algebraic analysis. $\bar{F}_{\rm rot}$ denotes the vector $\bar{F}$ after rotation about the real axis. The angle θ is an angle between F' and F_A'' and α_A is the phase angle associated with F_A'. In the case of a single type of anomalous scatterer the phase angle associated with F_A'' is $\pi/2 + \alpha_A$. When there are various types of anomalous scatterers, $\pi/2$ would be replaced by an angle ϕ which could be computed from the known positions of the anomalous scatterers.

It is seen from Fig. 35 that

$$|F|^2 = |F'|^2 + |F_A''|^2 + 2|F'F_A''|\cos\theta \tag{82}$$

$$|\bar{F}|^2 = |F'|^2 + |F_A''|^2 - 2|F'F_A''|\cos\theta \tag{83}$$

It follows that

$$|F'|^2 = \tfrac{1}{2}(|F|^2 + |\bar{F}|^2) - |F_A''|^2 \tag{84}$$

and

$$\cos\theta = (|F|^2 - |\bar{F}|^2)/4|F'F_A''| \tag{85}$$

The ambiguity derives from (85) in which $\pm\theta$ are both solutions. It is seen in Fig. 35 that the phase angle associated with F' is $\phi + \alpha_A - \theta$ or $\phi + \alpha_A + \theta$ for the ambiguous alternative to F' drawn with dashed lines. This is in agreement with the algebraic indications from (85).

An analysis of the twofold ambiguity in the phase arising from an anomalous dispersion experiment has been carried out by Bijvoet and Peerdemann,[133] Ramachandran and Raman[132] and Okaya and

Pepinsky.[134] It was apparent to Bijvoet[111] that since the ambiguity occurring with anomalous dispersion was different from the ambiguity occurring with isomorphous replacement, the combined experiment could yield one result in common, thus resolving the ambiguity of each method. Bijvoet[111] considered the case of the centrosymmetrically placed heavy atom substituents and Ramachandran and Raman[132] generalized this to arbitrary placement in the non-centrosymmetric cell. An alternative suggestion for resolving the ambiguity in phase arising from anomalous dispersion was made by Bijvoet and Peerdeman[133] and Ramachandran and Raman.[132] It is to choose, between the two alternatives, that phase which is closest to the phase which can be computed from the heavy atom substituents. This will give a correct choice in a majority of instances. Still another method for resolving the ambiguity by means of a two-wavelength experiment has been suggested by Okaya and Pepinsky.[134] The detailed analysis of this type of experiment has been described by Mitchell.[135] In all these instances the positions of the heavy atoms were assumed to be known. An algebraic analysis of an anomalous dispersion experiment in which the positions of the heavy atoms are not known has also been carried out.[136] Further experimental details for performing a multiple-wavelength experiment have been discussed by Herzenberg and Lau.[137] They described the use of the characteristic L-multiplet of heavier atoms to excite the K-edge of lighter atoms, thus obtaining information from several wavelengths in a single experiment. As an example they have suggested the possibility of studying protein structure by exciting the K-edge of sulphur atoms by means of Mo L-radiation.

The main applications of anomalous dispersion experiments so far have been to determine the absolute configurations of substances, whose structures were determined by other means, and also to combine the anomalous dispersion information with that from isomorphous replacement in the study of proteins. There have been a few studies involving the direct determination of phase from an anomalous dispersion experiment alone. Raman[138] has determined the structure and absolute configuration of L(+)-lysine hydrochloride dihydrate, taking advantage of the excitation of the chlorine atoms by Cu K_α-radiation. In this case the anomalous scattering effect is quite small and a Geiger-counter spectrometer was employed to measure the relative intensity differences, i.e. $|F_{\mathbf{h}}|^2 - |F_{-\mathbf{h}}|^2$, on a relative scale. The formula used to obtain the phase angle of F', which is the structure factor when the contribution of the imaginary part of the anomalous scattering is negligible, was

$$\sin \alpha = \frac{|F_{\mathbf{h}}|^2 - |F_{-\mathbf{h}}|^2}{4 |F'| F_{\mathrm{Cl}}''} \tag{86}$$

This formula follows directly from the general formula (85), since the anomalous scatterers were of the same type and centrosymmetrically placed in the unit cell. Under these circumstances we have from the previous discussion $\pm \theta = \alpha - \pi/2 - \alpha_A$, or

$$\cos \theta = \cos \alpha_A \sin \alpha \tag{87}$$

The angle α_A is the phase angle associated with the contribution from the total real part of the atomic scattering factor for the anomalously scattering atoms. Since the chlorine atoms are centrosymmetrically placed, α_A is either 0 or π. The contribution from the imaginary part F_A'' is $\pi/2$ out of phase and will have a positive or negative sign depending upon whether F_A' has a positive or negative sign, respectively, i.e. whether $\alpha = 0$ or π. Therefore

$$|F_A''| \cos \alpha_A = |F_{\mathrm{Cl}}''| \cos \alpha_{\mathrm{Cl}} = F_{\mathrm{Cl}}'' \tag{88}$$

completing the derivation of (86) from the general formula (85) for this special example. Formula (86) was originally derived by Bijvoet and Peerdeman.[133] All the quantities on the right-hand side of (86) are known from experiment. There remains the twofold ambiguity of α and $\pi - \alpha$. This was resolved in the study of the structure of the derivative of lysine by taking that phase, α or $\pi - \alpha$, which was closest to the phase computed from the known positions of the chlorine atoms.

In another phase determination employing anomalous dispersion, Dale, Hodgkin and Venkatesan[139] studied the structure of an aquo cyanide of the natural vitamin B_{12} nucleus containing cobalt. Cu K_α-radiation was employed and the anomalous dispersion effect was readily measured by visual estimates of the intensities recorded on film. The ambiguity in the phase determination was again resolved by choosing the phase nearer to that of the cobalt atom.

Okaya, Saito and Pepinsky[140] have suggested a generalization of the Patterson synthesis employing data from anomalous scattering which avoids direct confrontation with the phase problem in a manner which is similar to that of the Patterson function. The synthesis of particular interest is

$$P_s(\mathbf{u}) = \sum_{\substack{\mathbf{h} \\ -\infty}}^{\infty} |F_{\mathbf{h}}|^2 \sin 2\pi \mathbf{h} \cdot \mathbf{u} \qquad (89)$$

Equation (89) may be compared with the Patterson synthesis (5). Were it not for anomalous scattering $P_s(\mathbf{u})$ would be equal to zero, since then $|F_{\mathbf{h}}|^2 = |F_{-\mathbf{h}}|^2$ and it is recalled that $\sin(-x) = -\sin x$. The peaks of $P_s(\mathbf{u})$ represent vectors between unlike anomalous scatterers, between anomalous and non-anomalous scatterers and the sense of the vectors between them. It is often the case that there is only one type of anomalous scatterer in the crystal. In that case $P_s(\mathbf{u})$ is simplified. Several structures and absolute configurations have been determined by this method and a short review of this work and other studies employing anomalous dispersion has been written by Okaya and Pepinsky.[141]

Geurtz, Peerdeman and Bijvoet[142] have determined the structure of cytisine hydrobromide, both by means of the phase formula (86) and the P_s function (89). They found that both methods led to an unambiguous solution although the P_s function gave somewhat less clear results. The authors explain this observation on the basis that the P_s function is computed with $|F_{\mathbf{h}}|^2 - |F_{-\mathbf{h}}|^2$ as coefficients and does not make use of $|F'|$. On the other hand, the latter quantity coupled with the appropriate phase, α, is the coefficient employed in the direct Fourier series calculation of the structure.

Kartha and Parthasarathy[143] have described a Fourier synthesis composed of information from isomorphous replacement combined with that from anomalous dispersion which gives only the interatomic vectors among the heavy atoms. A further study of this synthesis has been made by Matthews.[144] Kartha and Parthasarathy[145] have also described correlation functions based on information from isomorphous replacement and anomalous dispersion which can establish the positions of the heavy atoms in different isomorphous heavy atom derivatives with respect to the same origin. This is a basic problem in the multiple isomorphous replacement method which is generally employed in the study of protein crystals.

The representations of anomalous dispersion, given in Figs. 34 and 35 and their accompanying discussions, concern ideal experiments in which it is assumed that there are no experimental errors and no limitations in the theoretical interpretation. The question of errors has received particular attention from investigators in the field of protein structures and the studies have embraced both isomorphous replacement and anomalous dispersion experiments. Blow and Crick[113] defined the "best" Fourier as that having the minimum expected mean-square error in electron density, as averaged over the entire unit cell. On the basis of certain assumtions regarding randomness of errors, they showed that the "best" Fourier is obtained by taking for each coefficient the centroid of the probability distribution, $p(|F|, \phi)$

$$F_c = \int_0^{2\pi} \int_0^{\infty} |F| \exp(i\phi) p(|F|, \phi) |F| \, d|F| \, d\phi \qquad (90)$$

where the joint probability distribution $p(|F|, \phi)$ for the magnitude and phase of a structure factor is the resultant distribution for a parent crystal and a number of isomorphous derivatives. Rossmann and Blow,[114] North,[146] Matthews[147] and Einstein[148] have extended the treatment of Blow and Crick to the case in which the anomalous dispersion of the heavy atom derivatives is also included by measuring both $|F_h|^2$ and $|F_{-h}|^2$.

A probability function similar to (78) has been suggested by North[146] to represent the probability of a phase angle obtained from anomalous dispersion. It was further suggested that the probability distributions for isomorphous replacement and anomalous dispersion could be combined by taking their product. North discussed the importance of properly estimating the values for the errors to be used in the various probability functions in order to achieve an appropriate weighting of the two types of information. A further investigation of the combined probability distribution has been carried out by Matthews,[147] and the analysis has been generalized to include the possibility of having any combination of replacement atoms in the isomorphous derivatives rather than all of the same type. A main problem in making the joint probability distribution a reliable measure for combining the information from isomorphous replacement and anomalous dispersion concerns the requirement of making reasonably accurate estimates of the errors which enter into the experimental procedures. These estimates naturally govern the relative weighting of the individual data.

Einstein,[148] working from the same assumptions as Blow and Crick,[113] has carried out a more exact analysis of the combination of isomorphous replacement and anomalous dispersion, both in the manner of the use of probability theory and the nature of the approximations made. For example, the probability is considered a function of amplitude and phase, not just phase, and only probabilities related to independent variables are multiplied, those related to variables which are interdependent are convoluted. Overlooking the latter point has led to an overweighting of the measurement of the amplitude for the structure factor for the unsubstituted crystal. Einstein has also derived an expression for the standard error for the probability expression involved in anomalous scattering and has not assumed that the ratio of the heavy atom structure factor magnitude to that for the complete structure is very small, but has retained terms of the first order. The resultant probability expression of Einstein generally would lead to different values for the centroids than the previous expressions. It will be of interest to see if a material improvement occurs in the resulting Fourier maps with the use of this new function for computing the coefficients.

D. Identical Molecular Units

Another approach to the problem of determining phases for protein structures has been developed by Main and Rossmann,[149] who considered chemically identical molecules in different crystallographic environments and the consequent restraints on the phases. When a molecule crystallizes in different crystal forms or occurs more than once in an asymmetric unit, a relation[149] exists among the structure factors which restricts the values of the phases

$$|F_\mathbf{p}| \exp(i\alpha_\mathbf{p}) = UV^{-1} \sum_{\substack{\mathbf{h} \\ -\infty}}^{\infty} |F_\mathbf{h}| \exp(i\alpha_\mathbf{h}) \sum_{n=1}^{N} G_{\mathbf{hp}n} \exp(i\phi_{\mathbf{hp}n}) \quad (91)$$

where N represents the number of identical molecules in the "p" crystal unit cell, $|F_\mathbf{p}|$, $\alpha_\mathbf{p}$, $|F_\mathbf{h}|$ and $\alpha_\mathbf{h}$ are the structure factors and their phases having Miller indices $\mathbf{p}$ and $\mathbf{h}$ in either the same or different crystals, and U is the volume occupied by the basic molecular unit. The quantities $G_{\mathbf{hp}n}$ and $\phi_{\mathbf{hp}n}$ are functions of the rotation and translation parameters relating the various molecules and it is necessary to know these quantities, as well as the magnitudes of the structure factors, in order to implement (91). This implementation, which essentially involves a succession of approximations designed to improve the agreement between the left- and right-hand sides of (91), has been described by Main[150] utilizing a hypothetical structure.

An alternative method for solving Eqs. (91) has been presented by Crowther.[151] He rewrites (91) in the form

$$F_{\mathbf{p}}^{*} = \sum_{\mathbf{h}} B_{\mathbf{ph}} F_{\mathbf{h}} \tag{92}$$

where the $B_{\mathbf{ph}}$ are assumed to be calculable in terms of the shape and the rotational and translational parameters of the subunits within the asymmetric unit. Equations (92) are then put into matrix formulation having solutions which are expressible as eigenvectors of a Hermitian matrix. The virtue of this procedure is that the amplitudes and phases are treated together, composing the unknown complex structure factors, thus permitting the use of the techniques of linear analysis. In the iterative procedures for solving (91) mentioned above, the objective is to determine phases separately from the magnitudes of the structure factors, resulting in a non-linear problem.

8. CONCLUDING REMARKS

The considerable progress which has recently been made in the facility with which structure analyses can be performed by X-ray diffraction has been largely due to the continued discovery and implementation of useful mathematical properties of the structure factor equations and their Fourier transforms, the Fourier series. It is possible now to approach moderately complex unknown structures, having up to several hundred atoms per unit cell, with considerable confidence that a successful analysis can be carried out, provided that one-half or more of the reflections from the sphere of reflection from a copper target are available for measurement. This includes the most difficult type of problem, the approximately equal atom non-centrosymmetric structure of unknown molecular formula which arises, for example, in photorearrangement reactions.

An outstanding area of difficulty which still remains concerns the complex biochemical structures, such as proteins, in which the unit cell may contain thousands of atoms and sufficient irregularities to limit the experimental data to about one-eighth of the copper sphere of reflection or less. The preparation and analysis of large numbers of suitable isomorphous derivatives have generally been a major undertaking, covering several years and involving many investigators for just one type of protein. The direct results from the X-ray analysis have yielded the overall configurations of the protein structure (tertiary structure), but not generally the amino acid sequences nor the atomic positions. There is a great opportunity and need in the application of X-rays to protein structure analysis to simplify the procedures and enhance the resolution and accuracy of the final results. Efforts in these directions are at present gaining considerable momentum.

Another problem which will probably be receiving increasing attention is the determination of electron distributions in crystals by X-ray diffraction. The opportunity to make such studies is increasing rapidly because of the greater accuracy in the measurement of the X-ray intensities afforded by automatic diffractometers, continued improvements in the theoretical understanding of the scattering process as, for example, given in the work of Zachariasen[152] and the labour saving afforded by computers. A major problem in obtaining accurate electron distributions concerns the removal of the effect of the vibrational motion of the atoms, which on the average tends to smear out the distributions.

At the present time, it is generally not possible to predict molecular and crystal structures. Some beginnings in the latter area have been made with the use of packing considerations. Significant progress in these fields would mark a major advance in the understanding of the forces which account for molecular configuration and the formation of the solid state.

References

1. Hoppe, W., *Z. Elektrochemie* **61**, 1076 (1957).
2. Hoppe, W., and Will, G., *Z. Kristallogr.* **113**, 104 (1960).

218 JEROME KARLE

3. Huber, R., and Hoppe, W., *Chem. Ber.* **98**, 2403 (1965).
4. Nordman, C. E., and Nakatsu, K., *J. Am. Chem. Soc.* **85**, 353 (1963).
5. Nordman, C. E., and Kumra, S. K., *J. Am. Chem. Soc.* **85**, 2059 (1965).
6. Papers by Buerger, M. J., Raman, S., Corfield, P. W. R., Rosenstein, R. D., Nordman, C. E., and Jacobson, R. A., in *Transactions of the American Crystallographic Association*, Vol. II, Bradley, W. F., and Hanson, H. P., Eds., Polycrystal Book Service, Pittsburgh, 1966, p. 29.
7. J. Karle, *Acta Cryst.* **B24**, 182 (1968).
8. Karle, J., "The Determination of Phase Angles", in *Advances in Structure Research by Diffraction Methods*, Vol. I, Brill, R., Ed., Interscience, New York–London, 1964, p. 55.
9. Karle, J. and Karle, I. L., *Acta Cryst.* **21**, 849 (1966).
10. Dickerson, R. E., "X-Ray Analysis and Protein Structure", in *The Proteins*, Vol. II, Neurath, H., Ed., Academic Press, New York, 1964, p. 603.
11. Holmes, K. C., and Blow, D. M., *The Use of X-ray Diffraction in the Study of Protein and Nucleic Acid Structure*, Interscience, New York–London, 1966.
12. Phillips, D. C., "Advances in Protein Crystallography", in *Advances in Structure Research by Diffraction Methods*, Vol. II, Brill, R., and Mason, R., Eds., Interscience, New York–London, 1966, p. 75.
13. Vainshtein, B. K., *Soviet Physics Uspekhi* **9**, 251 (1966); *Usp. Fiz. Nauk* **88**, 527 (1966).
14. Patterson, A. L., *Phys. Rev.* **46**, 372 (1935) *Z. Kristallogr.* **90**, 517 (1935).
15. Bokhoven, C., Schoone, J. C., and Bijvoet, J. M., *Acta Cryst.* **4**, 275 (1951).
16. Bijvoet, J. M., Peerdeman, A. F., and van Bommel, A. J., *Nature* **168**, 271 (1951).
17. Bijvoet, J. M., *Nature* **173**, 888 (1954).
18. *International Tables for X-ray Crystallography*, Vol. III, The Kynoch Press, Birmingham, 1962, p. 201.
19. Wilson, A. J. C., *Nature* **150**, 152 (1942).
20. Wilson, A. J. C., *Acta Cryst.* **2**, 318 (1949).
21. *International Tables for X-ray Crystallography*, Vol. I, The Kynoch Press, Birmingham, 1952.
22. Wilson, A. J. C., *Acta Cryst.* **9**, 143 (1956).
23. Lipson, H., and Woolfson, M. M., *Acta Cryst.* **5**, 680 (1952).
24. Hargreaves, A., *Acta Cryst.* **8**, 12 (1955).
25. Rogers, D., and Wilson, A. J. C., *Acta Cryst.* **6**, 439 (1953).
26. Parthasarathy, S., *Z. Kristallogr.* **123**, 27 (1966).
27. Parthasarathy, S., *Z. Kristallogr.* **123**, 77 (1966).
28. Hauptman, H., and Karle, J., *Solution of the Phase Problem. I. The Centrosymmetric Crystal*, A.C.A. Monograph No. 3, Polycrystal Book Service, Pittsburgh, 1953.
29. Karle, J., and Hauptman, H., *Acta Cryst.* **9**, 635 (1956).
30. Karle, J., and Hauptman, H., *Acta Cryst.* **3**, 181 (1950).

31. Sayre, D., *Acta Cryst.* **5**, 60 (1952).
32. Cochran, W., *Acta Cryst.* **8**, 473 (1955).
33. Hughes, E. W., *Acta Cryst.* **6**, 871 (1953).
34. Cochran, W., *Acta Cryst.* **6**, 810 (1953).
35. Woolfson, M. M., *Acta Cryst.* **7**, 61 (1954).
36. Watson, G. N., *Theory of Bessel Functions*, Cambridge University Press, 1945, p. 77.
37. Vaughan, P. A., *Acta Cryst.* **11**, 111 (1958).
38. Hauptman, H., and Karle, J., *Acta Cryst.* **10**, 267, 515 (1957).
39. Hauptman, H., *Acta Cryst.* **17**, 1421 (1964); Hauptman, H., Norton, D., Fisher, J., and Hancock, H., Meeting of American Crystallographic Association, Tucson, February, 1968, Abstract J1, p. 43.
40. Ott, H., *Z. Kristallogr.* **66**, 136 (1928).
41. Banerjee, K., *Proc. Roy. Soc.* **A141**, 188 (1933).
42. Avrami, M., *Phys. Rev.* **54**, 300 (1938).
43. Harker, D., and Kasper, J. S., *Acta Cryst.* **1**, 70 (1948).
44. Kaspar, J. S., Lucht, C. M., and Harker, D., *Acta Cryst.* **3**, 436 (1950).
45. Gillis, J., *Acta Cryst.* **1**, 174 (1948).
46. Zachariasen, W. H., *Acta Cryst.* **5**, 68 (1952).
47. Hauptman, H., and Karle, J., *Acta Cryst.* **9**, 45 (1956).
48. Hauptman, H., and Karle, J., *Acta Cryst.* **12**, 93 (1959).
49. Karle, J., and Hauptman, H., *Acta Cryst.* **14**, 217 (1961).
50. Karle, I. L., and Britts, K., *Acta Cryst.* **20**, 118 (1966).
51. Karle, I. L., and Karle, J., *Acta Cryst.* **17**, 1356 (1964).
52. Lloyd, H. A., and Horning, E. C., *J. Am. Chem. Soc.* **80**, 1506 (1958).
53. Lloyd, H. A., and Horning, E. C., *J. Org. Chem.* **26**, 2143 (1961).
54. Clarke, R. T., and Grundon, M. F., *J. Chem. Soc.* 535 (1963).
55. Karle, I. L., and Karle, J., *Acta Cryst.* **21**, 860 (1966).
56. Karle, I. L., Karle, J., Owen, T. B., and Hoard, J. L., *Acta Cryst.* **18**, 345 (1965).
57. Okaya, Y., and Bednowitz, A., *Acta Cryst.* **22**, 111 (1967).
58. Fredga, A., *Tetrahedron* **8**, 126 (1960).
59. Karle, I. L., and Karle, J., *J. Am. Chem. Soc.* **88**, 24 (1966).
60. Cantrell, J. S., Webb, N. C., and Mabis, A. J., Meeting of American Crystallographic Association, Atlanta, January, 1967, Abstract B10, p. 27; *Acta Cryst.* **B25**, 150 (1969).
61. Sax, M., Beurskens, P., and Chu, S., *Acta Cryst.* **18**, 252 (1965).
62. Gabe, E. J., Pickworth Glusker, J., Minkin, J. A., and Patterson, A. L., *Acta Cryst.* **22**, 366 (1967).
63. Fleischer, E. B., and Hawkinson, S. W., *Acta Cryst.* **22**, 376 (1967).
64. Klug, H. P., *Acta Cryst.* **19**, 983 (1965).
65. Hall, L. H., Perloff, A., Mauer, F. A., and Block, S., *J. Chem. Phys.* **43**, 3911 (1965).
66. Moriarty, R. M., Romain, C. R., Karle, I. L., and Karle, J., *J. Am. Chem. Soc.*, **87**, 3251 (1965); Karle, I. L., and Karle, J., *Acta Cryst.* **20**, 555 (1966).
67. Karle, I. L., Karle, J., and Britts, K., *J. Am. Chem. Soc.* **88**, 2918 (1966).

68. Karle, I. L., Dragonette, K. S., and Brenner, S. A., *Acta Cryst.* **19**, 713 (1965).
69. Dickinson, C., Stewart, J. M., and Holden, J. R., *Acta Cryst.* **21**, 663 (1966).
70a. Karle, I. L., and Karle, J., *Acta Cryst.* **16**, 969 (1963).
70b. Bryan, R. F., and Dunitz, J. D., *Helv. Chim. Acta.* **43**, 3 (1960).
71. Mornon, J. P., *C. R. Acad. Sci. Paris* Série C, **263**, 286 (1966); *Acta Cryst.* **23**, 367 (1967).
72. Bednowitz, A. L., and Hamilton, W. C., Meeting of American Crystallographic Association, Atlanta, January, 1967, Abstract B9, p. 27.
73. Ammon, H. L., and Sundaralingam, M., *J. Am. Chem. Soc.* **88**, 4794 (1966).
74. Klug, H. P., *Acta Cryst.* **21**, 536 (1966).
75. Weinblum, D., and Johns, H. E., *Biochim. Biophys. Acta* **114**, 450 (1966).
76. Wulff, D. L., and Fraenkel, G., *Biochim. Biophys. Acta* **51**, 332 (1961).
77. Camerman, N., Nyburg, S. C., and Weinblum, D., *Tetrahedron Letters* 42, 4127 (1967).
78. Einstein, J. R., Hosszu, J. L., Longworth, J. W., Rahn, R. O., and Wei, C. H., *Chem. Comm.* **20**, 1063 (1967).
79. Hyman, A., Perloff, A., Mauer, F., and Block, S., *Acta Cryst.* **22**, 815 (1967).
80. Bednowitz, A. L., and Post, B., *Acta Cryst.* **21**, 566 (1966).
81. Gabe, E. J., and Taylor, M. R., *Acta Cryst.* **21**, 418 (1966).
82. Colton, R. H., and Henn, D. E., *Acta Cryst.* **18**, 820 (1965).
83. Hanic, F., *Acta Cryst.* **21**, 332 (1966).
84. Karle, I. L., and Karle, J., *Acta Cryst.* **17**, 835 (1964).
85. Oh, Y. L., and Maslen, E. N., *Tetrahedron Letters* 28, 3291 (1966).
86. Karle, I. L., and Karle, J., *Acta Cryst.* **B24**, 81 (1968).
87. Fales, H. M., and Wildman, W. C., *J. Am. Chem. Soc.* **82**, 197 (1960).
88. Karle, J., Estlin, J. A., and Karle, I. L., *J. Am. Chem. Soc.* **89**, 6510 (1967).
89. Karle, I. L., and Karle, J., *Acta Cryst.* to be published.
90. Karle, I. L., *Acta Cryst.* **14**, 497 (1961).
91. Ammon, H. L., and Jensen, L. H., *J. Am. Chem. Soc.* **88**, 613 (1966); *Acta Cryst.* **23**, 805 (1967).
92. Wasserman, H. H., Doumaux, A. R., and Davis, R. E., *J. Am. Chem. Soc.* **88**, 4517 (1966).
93. Wasserman, H. H., and Keehn, P. M., *J. Am. Chem. Soc.* **88**, 4522 (1966).
94. Fratini, A. V., *J. Am. Chem. Soc.* **90**, 1688 (1968).
95. Wasserman, H. H., and Keehn, P. M., *J. Am. Chem. Soc.* **89**, 2770 (1967).
96. Yonemitsu, O., Witkop, B., and Karle, I. L., *J. Am. Chem. Soc.* **89**, 1039 (1967).
97. Karle, I. L., Karle, J., and Estlin, J. A., *Acta Cryst.* **23**, 494 (1967).
98. Robertson, J. M., and Woodward, I., *J. Chem. Soc.* 219 (1937); 36 (1940).

99. Woolfson, M. M., *Acta Cryst.* **9**, 804 (1956).
100. Sim, G. A., *Acta Cryst.* **13**, 511 (1960).
101. Bertaut, E. F., *Acta Cryst.* **10**, 670 (1957).
102. Srinivasan, R., *Acta Cryst.* **20**, 143 (1966).
103. Karle, J., Flippen, J., and Karle, I. L., *Z. Kristallogr.* **125**, 201 (1967).
104. Karle, J., Mitchell, D., and Karle, I. L., *Acta Cryst.*, in press.
105. Karle, I. L., Gilardi, R. D., Fratini, A. V., and Karle, J., *Acta Cryst.*, in press.
106. Hoppe, W., *Naturwiss.* **19**, 536 (1962).
107. Hoppe, W., and Huber, R., "Additional Structure Factor Signs in the Heavy Atom Technique by Use of the Sayre Relations", in *Crystallography and Crystal Perfection*, Ramachandran, G. N., Ed. Academic Press, London–New York, 1963, p. 61.
108. Rossman, M. G., *Acta Cryst.* **23**, 173 (1967).
109. Karle, J., *Acta Cryst.* **20**, 881 (1966).
110. Levy, H. A., and Ellison, R. D., Meeting of American Crystallographic Association, Atlanta, January, 1967, Abstract A9, p. 21.
111. Bijvoet, J. M., "The Isomorphous Substitution Method in the Noncentrosymmetrical Case", in *Conference on Computing Methods and the Phase Problem in X-ray Crystal Analysis*, Pepinsky, R., Ed., Pennsylvania State College, 1952, p. 84.
112. Bokhoven, C., Schoone, J. C., and Bijvoet, J. M., *Acta Cryst.* **4**, 275 (1951).
113. Blow, D. M., and Crick, F. H. C., *Acta Cryst.* **12**, 794 (1959).
114. Rossman, M. G., and Blow, D. M., *Acta Cryst.* **14**, 641 (1961).
115. Parthasarathy, S., and Ramachandran, G. N., *Acta Cryst.* **21**, 163 (1966).
116. Coulter, C. L., *J. Mol. Biol.* **12**, 292 (1965).
117. Karle, J., *Acta Cryst.* **21**, 273 (1966).
118. Bijvoet, J. M., *Nature* **173**, 888 (1954).
119. Bijvoet, J. M., and Peerdeman, A. F., *Acta Cryst.* **9**, 1012 (1956).
120. Vaidya, S. N., and Ramaseshan, S., "Some Procedures in the Determination of the Absolute Configuration of Crystals", in *Crystallography and Crystal Perfection*, Ramachandran, G. N., Ed., Academic Press, London–New York, 1963, p. 243.
121. Hanson, A. W., *J. Sci. Instr.* **35**, 180, 268 (1958).
122. Parthasarathy, S., and Srinivasan, R., *Acta Cryst.* **17**, 1400 (1964).
123. Parthasarathy, S., *Acta Cryst.* **22**, 98 (1967).
124. Bijvoet, J. M., Peerdeman, A. F., and van Bommel, A. J., *Nature* **168**, 271 (1951).
125. Peerdeman, A. F., van Bommel, A. J., Bijvoet, J. M., *Proc. Roy. Soc. Amsterdam* **B54**, 16 (1951).
126. Camerman, N., and Trotter, J., *Acta Cryst.* **17**, 384 (1964).
127. Moncrief, J. W., and Lipscomb, W. N., *Acta Cryst.* **21**, 322 (1966).
128. Tulinsky, A., and van den Hende, J. H., *J. Am. Chem. Soc.* **89**, 2905 (1967).
129. Okaya, Y., Meeting of American Crystallographic Association, Minneapolis, August, 1967, Abstract M1, p. 60.

130. Stevenson, D. L., and Dahl, L. F., *J. Am. Chem. Soc.* **89**, 3424 (1967).
131. Beurskens-Kerssen, G., Kroon, J., Endeman, H. J., van Laar, J., and Bijvoet, J. M., "Absolute Configuration and Rotatory Power of the Crystals of $NaBrO_3$ and $NaClO_3$", in *Crystallography and Crystal Perfection*, Ramachandran, G. N., Ed., Academic Press, London–New York, 1963, p. 225.
132. Ramachandran, G. N., and Raman, S., *Current Sci. (India)* **25**, 348 (1956).
133. Bijvoet, J. M., and Peerdeman, A. F., *Acta Cryst.* **9**, 1012 (1956).
134. Okaya, Y., and Pepinsky, R., *Phys. Rev.* **103**, 1645 (1956).
135. Mitchell, C. M., *Acta Cryst.* **10**, 475 (1957).
136. Karle, J., *J. Appl. Optics* **6**, 2132 (1967).
137. Herzenberg, A., and Lau, H. M. S., *Acta Cryst.* **22**, 24 (1967).
138. Raman, S., *Z. Kristallogr.* **111**, 301 (1959).
139. Dale, D., Crowfoot Hodgkin, D., Venkatesan, K., "The Determination of the Crystal Structure of Factor V 1a", in *Crystallography and Crystal Perfection*, Ramachandran, G. N., Ed., Academic Press, London–New York, 1963, p. 237.
140. Okaya, Y., Saito, Y., and Pepinsky, R., *Phys. Rev.* **98**, 1857 (1955).
141. Okaya, Y., and Pepinsky, R., "New Developments in the Anomalous Dispersion Method for Structure Analysis", in *Computing Methods and the Phase Problem in X-ray Crystal Analysis*, Pepinsky, R., Robertson, J. M., and Speakman, J. C., Eds. Pergamon Press, New York, 1961, p. 273.
142. Geurtz, Th. J. H., Peerdeman, A. F., and Bijvoet, J. M., *Acta Cryst.* **16**, A6 (1963).
143. Kartha, G., and Parthasarathy, R., *Acta Cryst.* **18**, 740 (1965).
144. Matthews, B. W., *Acta Cryst.* **20**, 230 (1966).
145. Kartha, G., and Parthasarathy, R., *Acta Cryst.* **18**, 749 (1965).
146. North, A. C. J., *Acta Cryst.* **18**, 212 (1965).
147. Matthews, B. W., *Acta Cryst.* **20**, 82 (1966).
148. Einstein, J. R., Meeting of American Crystallographic Association, Minneapolis, August, 1967, Abstract G1, p. 42; *Acta Cryst.* to be published; and private communication.
149. Main, P., and Rossmann, M. G., *Acta Cryst.* **21**, 67 (1966).
150. Main, P., *Acta Cryst.* **23**, 50 (1967).
151. Crowther, R. A., *Acta Cryst.* **22**, 758 (1967).
152. Zachariasen, W. H., *Phys. Rev. Letters* **18**, 195 (1967); *Acta Cryst.* **23**, 558 (1967).

THE PRINCIPLE OF CORRESPONDING STATES FOR CHAIN-MOLECULE LIQUIDS AND THEIR MIXTURES

J. HIJMANS* and TH. HOLLEMAN, *Koninklijke/Shell-Laboratorium, Amsterdam (Shell Research N.V.), The Netherlands*

CONTENTS

1. INTRODUCTION

A. The Principle of Corresponding States for Substances consisting of Spherical Molecules

The extreme difficulty of evaluating thermodynamic quantities from first principles, in any but a few simple cases, often forces one to take recourse to more indirect methods for interpreting the

* Present address, Instítuut von Theoretische Fysica, Universiteit van Amsterdam, Valckenierstraat 65, Amsterdam, The Netherlands.

macroscopic behaviour of a system in terms of its molecular properties. The principle of corresponding states provides a valuable method for deducing molecular information from the observed differences in macroscopic properties of mutually similar species.

The first class of substances to which this principle has been applied successfully comprises the noble gases argon, xenon and krypton and a number of other simple species such as nitrogen, oxygen and carbon monoxide which consist of approximately spherically symmetric molecules. Van der Waals[1] and Kamerlingh-Onnes[2] observed that the equations of state of these species can be reduced to a single, universal equation of state by expressing their pressures, volumes and temperatures in terms of the corresponding critical quantities. As was demonstrated by De Boer and Michels,[3] Guggenheim,[4] Pitzer[5] and others, the validity of this principle is a general consequence of statistical mechanics for all molecular systems which satisfy the following requirements.

(a) The Hamiltonian of the system is separable into a contribution from the internal degrees of freedom of the molecules, e.g. the vibration in the case of nitrogen or oxygen, depending only on their internal coordinates and momenta, and a configurational contribution depending only on the coordinates (and momenta) of the centres of mass of the molecules.

(b) The configurational potential energy of the system can be written as a sum of contributions from pairs of molecules.

(c) The interaction of a molecule-pair depends only on its mutual distance, r, and is of the general form

$$\phi(r) = \varepsilon\phi^*(r/\sigma) = \varepsilon\phi^*(r^*) \tag{1}$$

for all species considered, where ϕ^* is a universal function of r^*.

On account of these assumptions the canonical partition function for each of the considered systems will have the general form

$$Z_N(V, T) = z_{\text{int}}^N(T)(1/N!)Q_N(V, T) \tag{2}$$

where the function $z_{\text{int}}(T)$ stems from the internal degrees of freedom and the kinetic part of the translational degrees of freedom, and $Q_N(V, T)$ is the configurational partition function, defined as

$$Q_N(V, T) = \int d\mathbf{r}_1 \int d\mathbf{r}_2 \ldots \int d\mathbf{r}_N \exp(-1/kT)\sum_{(ij)}\phi(r_{ij}) \tag{3}$$

with the sum running over all pairs of molecules (ij). Introducing reduced coordinates $\mathbf{r}_i^* = \mathbf{r}_i/\sigma$, and expressing the potential in terms of the universal function $\phi^*(r^*)$ by means of (1), one can rewrite (3) as

$$Q_N(V, T) = \sigma^{3N} \int d\mathbf{r}_1^* \int d\mathbf{r}_2^* \dots \int d\mathbf{r}_N^* \exp(-1/T^*) \underset{(ij)}{\sum} \phi^*(r_{ij}^*) \quad (4a)$$

$$= \sigma^{3N}[Q^*(V^*, T^*)]^N \quad (4b)$$

Here $Q^*(V^*, T^*)$ is a universal function of the reduced volume

$$V^* = V/N\sigma^3 \quad (5)$$

and the reduced temperature

$$T^* = kT/\varepsilon \quad (6)$$

Thus the thermal equation of state, which is obtained from the ee-energy function

$$F(V, T) = -kT \ln Z_N(V, T) \quad (7)$$

by means of the thermodynamic relation

$$p(V, T) = -\left[\frac{\partial F(V, T)}{\partial V}\right]_T \quad (8)$$

can be written in the reduced form

$$p^*(V^*, T^*) = T^* \left[\frac{\partial \ln Q^*(V^*, T^*)}{\partial V^*}\right]_{T^*} \quad (9)$$

i.e. as a universal relation between V^* and T^* as defined by (5) and (6), and the reduced pressure

$$p^* = p\sigma^3/\varepsilon \quad (10)$$

The values of V^*, T^* and p^* at the critical point defined by the vanishing of $(\partial p^*/\partial V^*)_{T^*}$ and $(\partial^2 p^*/\partial V^{*2})_{T^*}$ are uniquely determined by (9), and consequently the critical quantities $V_{\mathrm{cr}} = N\sigma^3 V_{\mathrm{cr}}^*$, $T_{\mathrm{cr}} = (\varepsilon/k)T_{\mathrm{cr}}^*$ and $p_{\mathrm{cr}} = (\varepsilon/\sigma^3)p_{\mathrm{cr}}^*$ are uniquely expressible in terms of ε and σ, and may therefore also be used for reducing the thermal equation of state to a universal relation connecting p/p_{cr}, V/V_{cr} and T/T_{cr}.

The reduction of the equations of state of the different molecular systems to a universal reduced equation of state directly yields the relative magnitudes of the molecular units for volume, temperature, and pressure, $N\sigma^3$, ε/k, and ε/σ^3, respectively. From these the parameters ε and σ, characterizing the overall strengths and the ranges of the intermolecular potential $\phi(r)$ for the various species can be determined.

B. The Principle of Corresponding States for Chain-molecule liquids, as derived from Prigogine's Cell Model for r-mer Molecules

The statistical-mechanical derivation of the principle of corresponding states for species consisting of spherical molecules has initiated a considerable amount of work to extend its range of applicability both to mixtures consisting of simple molecules and to systems consisting of more complicated molecules.

Formulations of a principle of corresponding states for mixtures of spherical molecules have been worked out in the conformal solutions theory by Longuet-Higgins,[6] in the average potential model by Prigogine, Bellemans and Englert-Chwoles,[7] and in the theories of Scott,[8] Wojtowicz, Salsburg and Kirkwood,[9] Rice[10] and several others.

Trappeniers[11] has developed a generalized principle of corresponding states, which is applicable to systems such as CH_4, CCl_4, CBr_4 and $C(CH_3)_4$, whose molecules are composed of a central atom surrounded by a symmetric arrangement of peripheral groups.

The principle of corresponding states for chain-molecule liquids, which is the main subject of this survey, has been formulated for the first time by Prigogine, Bellemans and Naar-Colin.[12, 13] These authors have combined the ideas of Trappeniers' paper[11] with results obtained by Mathot[14] and by Prigogine, Trappeniers and Mathot[15] using a generalized cell model for r-mer molecules. Their derivation of the principle is based on the cell model by Lennard-Jones[16] and on the quasi-lattice hypothesis.[17] Each chain molecule in the system is thought to be subdivided into r elements, which are placed on the sites of a regular lattice, in such a way that each vertex is occupied by one element. An element is allowed to move around its equilibrium position at a lattice site, in a cell formed by its neighbouring elements. The cell motions are treated as independent motions, each taking place in the potential field which would arise if all neighbours were held fixed at their lattice positions.

As in the case of spherical molecules, the partition function is assumed to be separable into a density-independent contribution, $z_{int}^N(T)$, arising from the internal degrees of freedom and a configurational contribution $Q_N(V, T)$, arising from the external degrees of freedom [see Eq. (2)]. Prigogine and coworkers[12, 13] present arguments showing that the configurational partition function has the general form

$$Q_N(V, T) = g_N \, \Psi^N(V, T) \exp\left[-E_0(V)/kT\right] \tag{11}$$

In this expression g_N is the "combinatorial factor", i.e. the number of distinguishable arrangements of the N r-mer molecules on the sites of the lattice. The lattice energy E_0, i.e. the intermolecular potential energy of the system when all elements are placed in their lattice positions, is assumed to be expressible in the form

$$E_0(V) = Nq_r \, \varepsilon_r E^*(V^*) \tag{12}$$

where $E^*(V^*)$ represents a universal function of the reduced volume

$$V^* = \frac{V}{Nr\sigma_r^3} \tag{13}$$

and where ε_r and σ_r are parameters characterizing the strength and the range of the interaction potential between two r-mer elements. The parameter q_r determines the number of "contacts" of an r-mer molecule (nearest neighbour sites occupied by elements of other molecules)

$$q_r z = rz - 2(r-1) \tag{14}$$

which is a function of the coordination number, z, of the lattice.

The cell-partition function, $\Psi(V, T)$, in (11), arising from the motions of the r elements of a molecule in their respective cells, is approximated by a temperature-independent function of the form

$$\Psi(V, T) \simeq [\sigma_r \psi^*(V^*)]^{3c_r} \tag{15}$$

where $\psi^*(V^*)$ is again a universal function of V^*. [If $\psi(V, T)$ should depend significantly on the temperature, one could always factorize out an appropriate function of T alone and include it in the internal partition function, so as to make the remaining function practically temperature independent.] The exponent $3c_r$ occurring in (15) represents the number of external degrees of freedom of an r-mer molecule.

Substituting Eqs. (12) and (15) into (11) and using as before Eqs. (2), (7) and (8), one again finds a universal reduced equation of state

$$p^*(V^*, T^*) = 3T^* \frac{\partial \ln \psi^*(V^*)}{\partial V^*} - \frac{\partial E^*(V^*)}{\partial V^*} \tag{16}$$

which relates the reduced pressure

$$p^* = \frac{r\sigma_r^3}{q_r \varepsilon_r} p \tag{17}$$

to the reduced volume (13) and the reduced temperature

$$T^* = \frac{c_r k}{q_r \varepsilon} T \tag{18}$$

The principle of corresponding states for chain molecules, as obtained from the preceding arguments, was found to be satisfied by the normal alkanes, $C_n H_{2n+2}$, with quite a remarkable accuracy. This result is rather surprising, since the quasi-lattice model and the cell model can hardly be considered as an adequate basis for a quantitative description of a chain-molecule liquid. It suggests in fact that arguments of a more general nature, independent of the choice of a specific statistical-mechanical model, should be sufficient to arrive at a principle of corresponding states for chain molecules.

This has led one of us[18] to reformulate the principle from a purely phenomenological point of view, without introducing the cell model or the quasi-lattice assumption. This formulation is based mainly on the conclusion, which is implied by the success of the cell-theory treatment in the case of normal alkanes, that a chain molecule can be characterized exhaustively by three dimensionless parameters, r, q and c, having a similar significance as the quantities r, q_r and c_r, which in the formulation by Prigogine and coworkers[12, 13] determine the number of sites, the number of contacts and the number of external degrees of freedom, respectively. The arguments leading to a principle of corresponding states for pure chain-molecule liquids[18] will be presented in the next Section. In Section 3 we investigate whether the principle is applicable to the thermodynamic functions of normal-alkane liquids. The generalization of the principle for application to the excess functions of mixtures of chain molecules[19] is discussed in Section 4. In Section 5 the validity of the principle for excess functions is tested for the excess volumes, the heats of mixing and the excess free energies of normal alkane mixtures at various temperatures.

2. PHENOMENOLOGICAL FORMULATION OF THE PRINCIPLE OF CORRESPONDING STATES FOR PURE CHAIN-MOLECULE LIQUIDS

The main difficulty in applying the principle of corresponding states to chain-molecular species arises from the fact that the large size of the molecules and the freedom of motion of the individual chemical groups relative to each other rules out a description of the molecules as a whole as point centres. On the other hand, a description in terms of the individual chemical groups, e.g. CH_2 and CH_3 groups in the case of normal alkanes, as the basic units also presents difficulties. When the latter picture is adopted two qualitatively different kinds of forces between the units have to be considered: the strong, short-range chemical bonding forces between units of the same molecules and the van der Waals forces between the units of different molecules, which are considerably weaker and of longer range. Since the possibility of choosing between two different ε's and two different σ's would destroy the uniqueness of the reduction procedure for thermodynamic relations, it is clear that a principle of corresponding states can be expected to hold only for purely configurational quantities, i.e. quantities which are independent of the valency forces (or alternatively for purely internal quantities, which, however, are of little interest in the liquid state).

In order to be able to define configurational quantities in a unique way, we have to impose the separability of the partition function, as was done for spherical molecules [assumption (a) in Section 1], but which plays a more essential role in the present case. Thus we assume that the Hamiltonian of the system is separable into a part depending only on a set of internal coordinates and momenta, and a part depending only on external coordinates and momenta, the former coordinates being independent of the relative positions of the molecules. The criterion for a coordinate to be external or internal is whether a change in that coordinate requires a major change in the relative arrangement of the molecules or not. Thus the electronic degrees of freedom, and all stretching and bending vibrations in the molecules, are considered as internal degrees of freedom, whereas the translations and rotations of the molecules as a whole are external motions. The rotational motions in the molecules would be considered as external motions if they were completely free (see Fig. 1), but in most practical cases, such as for the

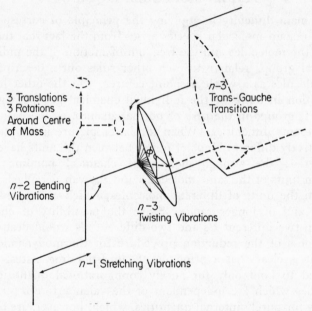

3 Translations
3 Rotations
Around Centre
of Mass

$n-3$
Trans-Gauche
Transitions

$n-2$ Bending
Vibrations

$n-3$
Twisting Vibrations

$n-1$ Stretching Vibrations

Fig. 1. Internal and external degrees of freedom in a normal alkane molecule. Drawn arrows represent internal, dashed arrows external motions.

normal alkanes, the rotations are severely hindered. In this case they can best be considered as twisting vibrations around one of the various possible equilibrium positions (one trans and two gauche positions in the case of normal alkanes), which are superimposed on discrete jumps from one equilibrium position to another. The former are considered as internal, the latter as external motions.

As a result of our assumption the partition function of a chain-molecule liquid will again be of the form

$$Z_N(V, T) = z_{int}^N(T)(1/N!) Q_{conf}(T, V) \qquad (19)$$

where $z_{int}(T)$ is the contribution from the internal degrees of freedom and the kinetic part of the external degrees of freedom, and where the configurational partition function $Q_{conf}(T, V)$ has been defined in such a way that no density-independent factors can be split off from it. From Eq. (19) we obtain by means of (7) expressions for thermodynamic quantities such as the energy

$$U = NkT^2 \frac{d \ln z_{int}(T)}{dT} + kT^2 \left[\frac{\partial \ln Q(V, T)}{\partial T} \right]_V \qquad (20a)$$

$$= U_{int}(T) + U_{conf}(T, V) \qquad (20b)$$

and the entropy

$$S = Nk \frac{d}{dT} T \ln z_{int}(T) + k \left[\frac{\partial}{\partial T} T \ln Q(V, T) \right]_V \qquad (21a)$$

$$= S_{int}(T) + S_{conf}(T, V) \qquad (21b)$$

which are sums of an internal and a configurational contribution. The principle of corresponding states will be applied only to the *configurational parts* of such quantities, as well as to the thermal equation of state

$$p(V, T) = kT \left[\frac{\partial \ln Q(V, T)}{\partial V} \right]_T \qquad (22)$$

which is a purely configurational relation.

Now that we have restricted ourselves to configurational properties only, we may disregard the valency forces and consider our system as one in which only intermolecular van der Waals forces are acting. But even then we are dealing with a system whose behaviour is intermediate between that of a system in which each molecule behaves as a structureless point centre, and one in which the individual chemical groups behave as independently moving and interacting units. Thus, if we compare a liquid of N chain molecules, each consisting of n units, with an assembly of Nn independent units interacting according to the same force law, we observe the following characteristic differences.

9

(a) The total intermolecular interaction energy will be smaller in the former case, since the number of "contacts" of a chain molecule is smaller than that of n independently interacting units.

(b) The volume within which the interaction from a chain molecule is being felt is smaller than the corresponding total volume for n independent units, since in the former case the units are held closely together by chemical bonds.

(c) The conformations of the n units of a chain molecule will be more ordered than those of n independent units, because of the constraints imposed by the chemical bonds.

One may put these observations on a somewhat more formal basis by introducing a hypothetical reference species consisting of structureless spherical particles whose interaction potential is characterized by a strength parameter ε and a range parameter σ. More specifically one might think of liquid argon or liquid methane as the reference compound.

Then three dimensionless parameters q, r and c, all having the significance of an effective number of units per chain molecule, may be defined by requiring that

(a) one chain molecule consisting of n units has the same average interaction with its surroundings as $q(n)$ independently interacting reference molecules;

(b) the volume of the domain of interaction of one chain molecule with n units equals the combined volume of the spheres of action of $r(n)$ reference molecules;

(c) the number of external degrees of freedom of one chain molecule equals the total number of translational degrees of freedom of $c(n)$ reference molecules, i.e. $3c(n)$.

The basic assumption which determines the validity of the principle of corresponding states is that the parameters $q(n)$, $r(n)$ and $c(n)$ can be defined *independently of the thermodynamic state* of a system. For, if this is the case, the only difference, from a macroscopic point of view, between a given chain-molecular compound, on the one hand and the reference species, on the other, is that in the former case the effective number of units contributing to the configurational energy, the volume and the configurational entropy is $Nq(n)$, $Nr(n)$ and $Nc(n)$, respectively, whereas it is always N in the latter case. Accordingly, the fundamental relation between the three extensive thermodynamic variables U_{conf}, V, and S_{conf} of each system can be "translated" into the corresponding relation for the reference system by introducing the reduced quantities

$$U^* = \frac{U_{\text{conf}}}{N\varepsilon q(n)} = \frac{U_{\text{ref}}}{N\varepsilon} \tag{23a}$$

$$V^* = \frac{V}{N\sigma^3 r(n)} = \frac{V_{\text{ref}}}{N\sigma^3} \tag{23b}$$

and

$$S^* = \frac{S_{\text{conf}}}{Nkc(n)} = \frac{S_{\text{ref}}}{Nk} \tag{23c}$$

The relation

$$U^* = U^*(S^*, V^*) \tag{24}$$

represents a universal law, independent of the species.

Taking the derivatives with respect to S^* and V^*

$$T^*(S^*, V^*) = -(\partial U^*/\partial S^*)_{V^*} \tag{25a}$$

and

$$p^*(S^*, V^*) = -(\partial U^*/\partial V^*)_{S^*} \tag{25b}$$

and solving S^* from (25a) as a function of T^* and V^*, we obtain the thermal and caloric equations of state

$$p^* = p^*(V^*, T^*) \tag{26a}$$

and

$$U^* = U^*(V^*, T^*) \tag{26b}$$

respectively, in reduced form.

Here a reduced pressure

$$p^* = \frac{r(n)\,\sigma^3}{q(n)\,\varepsilon}\,p \tag{27a}$$

and a reduced temperature

$$T^* = \frac{c(n)}{q(n)}\frac{kT}{\varepsilon} \tag{27b}$$

have been introduced, as in the theory of Prigogine and co-workers.[12, 13]

The assumption that q, r and c are independent of the thermo-dynamic state seems to be justified for the quantity r, since the region occupied by a molecule depends mainly on the lengths of its chemical bonds. Similarly, the parameter c, which is a measure of the number of external degrees of freedom of a molecule, is almost uniquely determined by its chemical structure. The only way in which it can be affected by the thermodynamic state is through a slight increase in the freedom of the intra-molecular rotations at higher temperatures, which may safely be disregarded. The parameter q may, however, be influenced by the thermodynamic state, since the mean interaction of a molecule with its surroundings is not only determined by its own composition, but also by the mean composition of the surrounding molecules. In systems such as the normal alkanes, where the strengths of the interactions between the different (CH_2- and CH_3-) groups are almost identical, this influence may again be neglected, but in associating liquids, such as the alcohols, a dependence of q on the degree of association and thus on temperature may give rise to deviations from the principle of corresponding states.

An interesting consequence of the present formulation of the principle is that it enables one to investigate the dependence of the parameters q, r and c on the chemical composition of the molecules, i.e. on their chain length, n. In view of their significance as effective numbers of repeating units per molecule, we may expect these quantities to be additively composed of fixed contributions from the individual chemical groups. Thus for chain-molecules we would anticipate that

$$q(n) = 2q_e + (n-2) q_m \qquad (28a)$$

$$r(n) = 2r_e + (n-2) r_m \qquad (28b)$$

$$c(n) = 2c_e + (n-2) c_m \qquad (28c)$$

where q_e, q_m, r_e, r_m, c_e and c_m are characteristic contributions from end (CH_3-) and middle (CH_2-) groups, respectively. We expect r_e and r_m, as well as c_e and c_m to be independent of the chain length, n, but the contributions q_e and q_m to $q(n)$ should in principle depend on the mean distribution of end and middle groups around a given group. For non-associating liquids this distribution should not be far from a random distribution and we may write

$$q_e(n) = \frac{2}{n} q_{ee} + \frac{n-2}{n} q_{em} \qquad (29a)$$

$$q_m(n) = \frac{2}{n} q_{em} + \frac{n-2}{n} q_{mm} \qquad (29b)$$

where q_{ee}, q_{em} and q_{mm} are constants characterizing the relative strengths of end–end, end–middle and middle–middle interactions, respectively. For relatively large molecules we can neglect the dependence of q_e and q_m on n, in which case $q(n)$, $r(n)$ and $c(n)$ are all linear functions of the chain length. By definition all contributions q_e, q_m, r_e, r_m, c_e and c_m must have values between zero and one. Moreover, since the external motions of a chain-molecule consist of three translations, three rotations of the molecule as a whole and a fraction of a degree of freedom for each intramolecular rotation, we have

$$c(n) = 2 + (n-3)\,c_m \tag{30a}$$

or

$$c_e = 1 - \tfrac{1}{2}c_m \tag{30b}$$

where $0 \leqslant c_m \leqslant \tfrac{1}{3}$.

3. APPLICATION OF THE PRINCIPLE TO PURE NORMAL ALKANES

In order to investigate whether the thermodynamic properties of normal alkanes conform to the principle of corresponding states, we have to carry out the following tests.

(a) Verify whether the relations between configurational quantities for different normal alkanes can be superimposed on each other by expressing these quantities in terms of suitably chosen molecular units.

(b) If this is the case, verify whether the molecular units obtained by reducing different thermodynamic relations are mutually consistent.

(c) Investigate whether the parameters $q(n)$, $r(n)$ and $c(n)$ as obtained from the molecular units depend on the chain length n in the predicted way.

Suitable thermodynamic relations which may be used for this purpose are for instance:

(1) the molar volume as a function of temperature at zero pressure or (as is practically the same for liquids) at atmospheric pressure;

(2) the configurational energy as a function of temperature at zero (atmospheric) pressure. This quantity can be obtained

from the heat of vaporization

$$H = (U+pV)_{\text{vap}} - (U+pV)_{\text{liq}} \tag{31}$$

if one assumes that only the configurational part of the energy changes during evaporation, and that

$$(U_{\text{conf}})_{\text{vap}} \simeq -\frac{RT^2}{V}\frac{\mathrm{d}B}{\mathrm{d}T} \tag{32a}$$

and

$$(pV)_{\text{vap}} \simeq RT\left(1+\frac{B}{V}\right) \tag{32b}$$

can be represented with sufficient accuracy by their virial expansions up to the term of order $1/V$. On these assumptions we have

$$-(U_{\text{conf}})_{\text{liq}} = H + (pV)_{\text{liq}} - RT\left[1 + \frac{1}{V}\left(B - T\frac{\mathrm{d}B}{\mathrm{d}T}\right)\right] \tag{33}$$

where $B(T)$ is the second virial coefficient of the species;

(3) the general pressure–volume–temperature relation. The actual pressure–volume–temperature data usually give the volume as a function of pressure at a number of different temperatures, but for carrying out the reduction it is more convenient to convert the data by interpolation into a set of pressure–temperature curves at constant values of the dimensionless relative volume

$$V_{\text{rel}} = \frac{V(p, T)}{V(0, T)} = \frac{V^*(p^*, T^*)}{V^*(0, T^*)} \tag{34}$$

which is already in the form of a reduced thermodynamic variable;

(4) in addition, it is of interest to study the temperature dependence of the adiabatic compressibility

$$\kappa_S = -\frac{1}{V}\left(\frac{\partial V}{\partial p}\right)_S$$

which can be determined accurately from sound velocity data. This quantity, which involves a derivative at constant *total* entropy (and not at constant configurational entropy), is not truly configurational, but the isothermal compressibility

$$\kappa_T = -\frac{1}{V}\left(\frac{\partial V}{\partial p}\right)_T$$

does satisfy this requirement, and the ratio

$$\kappa_S/\kappa_T = c_p/c_V$$

i.e. the specific heat ratio, does not depend very sensitively on the temperature and on the chain length, and therefore the function $\kappa_S(T)$ is expected to satisfy the principle of corresponding states at least approximately.

The test to determine whether the various relations can be reduced to species-independent, universal relations has been carried out graphically. In each case the relation between the logarithms of the two variables was plotted on transparent paper and the curves obtained for different species were shifted relative to each other along the two coordinate axes until they coincided. The ratios of the molecular units for different species are obtained directly from the amounts by which their curves have to be shifted in order to achieve coincidence.

The following experimental data have been analysed in this way:

(1) liquid densities at atmospheric pressure for the normal alkanes from C_2H_6 up to $C_{20}H_{42}$ as recorded in the tables of the American Petroleum Institute;[20]

(2) heats of vaporization for the normal alkanes from C_3H_8 up to $C_{12}H_{26}$ at 25°C and at their normal boiling points, as quoted in the American Petroleum Institute tables,[21] together with some data at intermediate temperatures for hexane, heptane and octane.[22] The correction terms in Eq. (33) were obtained from vapour pressures and liquid densities reported in the American Petroleum Institute tables[23] and second-virial coefficient data by McGlashan and Potter;[24]

(3) Pressure–volume–temperature data for the compounds C_7H_{16}, C_8H_{18}, C_9H_{20}, $C_{12}H_{26}$ and $C_{16}H_{34}$, as published by Boelhouwer;[25]

(4) adiabatic compressibilities for the same alkanes and hexane, calculated from sound velocity data at atmospheric pressure by Boelhouwer.[26]

For all thermodynamic relations investigated, and for all species, a coincidence of curves could be attained to within the experimental accuracy. The results of the reduction procedure are illustrated in Figs. 2–5. Figure 2 shows the reduced volume as a function of the reduced temperature for ethane, heptane and $C_{20}H_{42}$ on a scale relative to heptane as the reference compound. Although the shapes

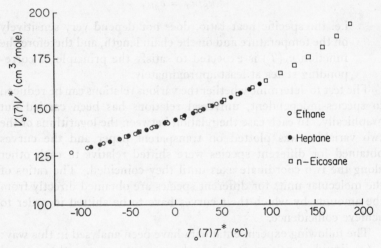

Fig. 2. Reduced molar volume as a function of reduced temperature for chain lengths $n = 2$, $n = 7$ and $n = 20$, relative to heptane as reference species.

of the reduced curves for the various compounds are identical, the end points of the liquid region are seen to be quite different, as could be anticipated, since the melting point and the boiling point are non-configurational properties. In Fig. 3 the reduced configurational energies have been plotted as a function of reduced temperature for the ten investigated compounds. In this figure the temperatures have been reduced by means of the molecular units for temperature, which were deduced from the density data. The same applies to Fig. 4, where sets of p–T curves at constant values of V_{rel} are shown in reduced form for five compounds, and to Fig. 5, showing the reduced adiabatic compressibilities for six normal alkanes. The fact that different functions of temperature could be successfully reduced by means of the same molecular units for temperature, and that in Fig. 4 various curves corresponding to different values of V_{rel} could be made to coincide *simultaneously*, provides additional support for the validity of the principle of corresponding states.

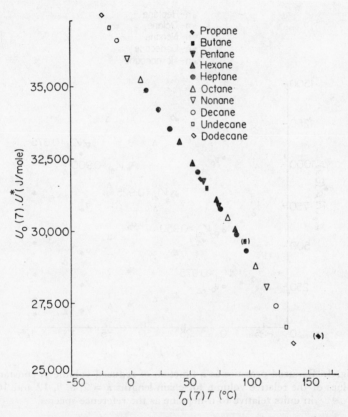

Fig. 3. Reduced molar configurational energy as a function of reduced temperature for chain lengths from $n = 3$ up to $n = 12$, relative to heptane as reference species. The two symbols in parentheses have been obtained from heats of vaporization above atmospheric pressure.

The values of the various reduction parameters relative to those of n-heptane are listed in Table I. The ratios of the volume reduction parameters $r(n)/r(7)$, obtained from density data (Table I, column 1), and the ratios of the molecular units for energy $q(n)/q(7)$, deduced from the heats of vaporization (column 3), have been plotted as functions of the chain length, n, in Figs. 6 and 7, respectively. Figure 8 shows the dependence on n of the ratios $c(n)/c(7)$ obtained by dividing the reduction factors for energy by those for

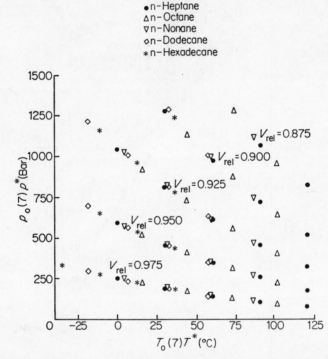

Fig. 4. Reduced pressure as a function of reduced volume at constant values of the relative volume for chain lengths $n = 7, 8, 9, 12$ and 16, in units relative to n-heptane as the reference species.

temperature (Table I, column 2). For all three functions the linear dependence on n, predicted by Eq. (28) is confirmed quite accurately. From the slopes of these straight lines one obtains

$$r_{\mathrm{m}}/r_{\mathrm{e}} = 0.71 \pm 0.02 \tag{35a}$$

$$q_{\mathrm{m}}/q_{\mathrm{e}} = 0.53 \pm 0.02 \tag{35b}$$

$$c_{\mathrm{e}} = 0.92 \pm 0.01; \quad c_{\mathrm{m}} = 0.16 \pm 0.02 \tag{35c}$$

Combining the volume and the temperature reduction factors from density data with the units for pressure deduced from pressure–volume–temperature data an independent set of values

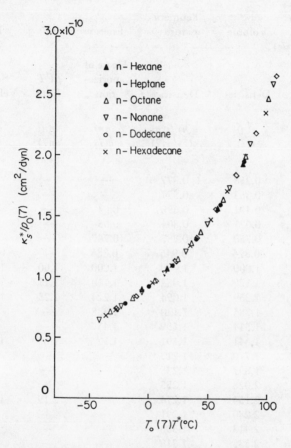

Fig. 5. Reduced adiabatic compressibility as a function of reduced temperature for chain lengths $n = 6, 7, 8, 9, 12$ and 16, in units relative to n-heptane.

of the ratios $q(n)/q(7)$ is obtained (see Table I, column 4). Although these values seem to increase with n faster than would agree with a linear function, their behaviour can roughly be represented by a

TABLE I. Molecular Units for Various Thermodynamic Quantities

	1	2	3	4	5
Reduction factor for:	Volume	Temperature	Energy	pV	V/κ_S
Determined from:	Densities	Densities	Heats of vaporization	pVT Data	Sound Velocities
n	$\dfrac{r(n)}{r(7)}$	$\dfrac{q(n)}{q(7)}\bigg/\dfrac{c(n)}{c(7)}$	$\dfrac{q(n)}{q(7)}$	$\dfrac{q(n)}{q(7)}$	$\dfrac{q(n)}{q(7)}$
1	0.258	0.377	—	—	—
2	0.363	0.577	—	—	—
3	0.491	0.690	0.54	—	—
4	0.621	0.804	0.65	—	—
5	0.750	0.887	0.767	—	—
6	0.874	0.946	0.886	—	0.891
7	1.000	1.000	1.000	1.00	1.000
8	1.129	1.052	1.110	1.11	1.093
9	1.259	1.094	1.224	1.25	1.204
10	1.385	1.130	1.338	—	—
11	1.514	1.159	1.457	—	—
12	1.641	1.191	1.57	1.63	1.517
13	1.770	1.213	—	—	—
14	1.897	1.235	—	—	—
15	2.028	1.255	—	—	—
16	2.154	1.274	—	2.19	1.967
17	2.280	1.291	—	—	—
18	2.413	1.306	—	—	—
19	2.535	1.318	—	—	—
20	2.661	1.331	—	—	—

straight line whose slope corresponds to

$$q_{\mathrm{m}}/q_{\mathrm{e}} = 0.64 \pm 0.11 \tag{35b'}$$

and

$$c_{\mathrm{e}} = 0.90 \pm 0.02; \quad c_{\mathrm{m}} = 0.20 \pm 0.04 \tag{35c'}$$

From the reduction factors for the adiabatic compressibility one obtains in a similar fashion

$$q_m/q_e = 0.43 \pm 0.05 \tag{35b''}$$

and

$$c_e = 0.93 \pm 0.02; \quad c_m = 0.13 \pm 0.03 \tag{35c''}$$

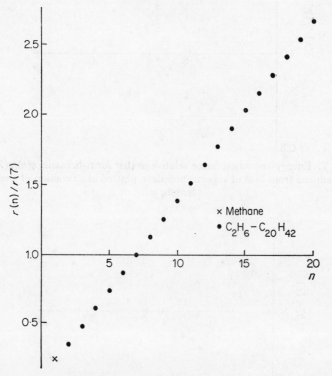

Fig. 6. Volume-reduction factor relative to that for n-heptane, $r(n)/r(7)$, plotted as a function of chain length, n.

The mutual agreement between these results is quite satisfactory if one takes into account the limited accuracy of the experimental data, the non-configurational nature of the adiabatic compressibility and the approximations involved in the calculation of the configurational energy from heat of vaporization data. Thus, apart from the relatively unimportant inconsistencies in the values of the parameters $q(n)$ and $c(n)$ obtained from different thermodynamic

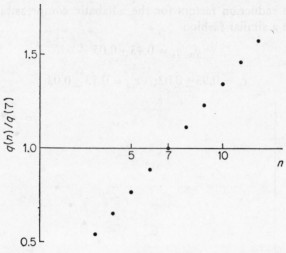

Fig. 7. Energy-reduction factor relative to that for n-heptane, $q(n)/q(7)$, as deduced from heat of vaporization data, plotted as a function of chain length, n.

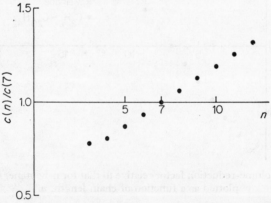

Fig. 8. Entropy-reduction factor relative to that for n-heptane, $c(n)/c(7)$, as deduced from the temperature-reduction factors, and the energy-reduction factors, as a function of chain length.

functions, the validity of the principle of corresponding states as formulated in Section 2 appears to be confirmed in its most important aspects by the thermodynamic data of the normal alkanes.

The same conclusion was drawn in a recent paper by Flory, Orwoll and Vrij,[27] who investigated the principle of corresponding states for chain molecules in a somewhat different form. The principle introduced by these authors is based on arguments and concepts which are essentially equivalent to those used in Section 2 of this paper, but the similarity is sometimes obscured by the difference in notation. (Our quantities r, $q\varepsilon$, c and σ^3 are denoted in their paper by x, $xs\eta/v^*$, cx and v^*, respectively.) Furthermore, Flory and coworkers[27] make use of explicit expressions for the thermal and caloric equations of state, i.e.

$$\frac{p^* V^*}{T^*} = \frac{V^{*\frac{1}{3}}}{V^{*\frac{1}{3}}-1} - \frac{1}{V^* T^*} \tag{36a}$$

and

$$U^* = -1/V^* \tag{36b}$$

respectively. Their analysis of the thermodynamic properties of pure normal alkane liquids is restricted to density and pressure–volume–temperature data only.

4. EXTENSION OF THE PRINCIPLE TO THE EXCESS FUNCTIONS OF LIQUID MIXTURES OF CHAIN MOLECULES

A. The Principle of Congruence

The extension of the principle of corresponding states to liquid mixtures of chain molecules, which has been proposed in a previous publication,[19] is a straightforward consequence of the principle for pure compounds. For, if in a mixture of several components i, the molecules of each species can be characterized exhaustively by three parameters q_i, r_i and c_i, representing the effective numbers of repeating units per molecule, contributing to the three extensive variables U_{conf}, V and S_{conf}, respectively, a system of composition x_i, will be characterized by the arithmetic mean values of these parameters

$$q(x_i) = \sum_i q_i x_i \tag{37a}$$

$$r(x_i) = \sum_i r_i x_i \tag{37b}$$

and

$$c(x_i) = \sum_i c_i x_i \tag{37c}$$

respectively.

As in the case of pure species, the parameters r_i and c_i for a mixture, which are determined by the extent of the domain of interaction and the number of external degrees of freedom of a molecule of type i, are expected to be independent of the composition of the surroundings of such a molecule; however, for the parameter q_i characterizing the overall strength of its interaction with the surroundings this is not necessarily the case. Assuming, as we did in Section 3, that q_i, r_i and c_i are additively composed of contributions from end and middle groups, we would have

$$q_i(x_j) = 2q_e(x_j) + (n_i - 2) q_m(x_j) \tag{38a}$$

$$r_i = 2r_e + (n_i - 2) r_m \tag{38b}$$

and

$$c_i = 2c_e + (n_i - 2) c_m \tag{38c}$$

where r_e, r_m, c_e and c_m are independent of the composition of the mixture, whereas q_e and q_m depend on the average surroundings of an end and of a middle group, respectively. When the strengths of the interactions between the different groups are sufficiently alike to assume a random distribution around a given group, we can write, as in Eq. (29), for pure compounds

$$q_e(x_j) = \frac{2}{\sum_j n_j x_j} q_{ee} + \left(1 - \frac{2}{\sum_j n_j x_j}\right) q_{em} \tag{39a}$$

and

$$q_m(x_j) = \frac{2}{\sum_j n_j x_j} q_{em} + \left(1 - \frac{2}{\sum_j n_j x_j}\right) q_{mm} \tag{39b}$$

Substituting these equations in (38a) and introducing the arithmetic mean of the chain lengths of the components

$$n = \sum_j n_j x_j \tag{40}$$

we see that Eq. (37) can be written

$$q(x_i) = q(n) = \frac{4}{n}q_{ee} + \frac{4(n-2)}{n}q_{em} + \frac{(n-2)^2}{n}q_{mm} \tag{41a}$$

$$r(x_i) = r(n) = 2r_e + (n-2)r_m \tag{41b}$$

$$c(x_i) = c(n) = 2c_e + (n-2)c_m \tag{41c}$$

These equations are equivalent to the so-called "principle of congruence"[28, 29] which was discovered by Brønsted and Koefoed[28] as an empirical rule satisfied by the excess free energies of normal alkane mixtures, according to which a mixture of composition x_i behaves (as far as its configurational properties are concerned) as a hypothetical pure compound whose molecules have a chain length as given by Eq. (40). This principle has been obtained from statistical mechanics by Longuet-Higgins[30] for mixtures of not too short chain molecules. Experiments by Dixon[31] have demonstrated that a similar principle holds also for the molar volumes of mixtures of branched chain molecules.

The implications of the principle of congruence for the thermodynamic excess functions of binary mixtures of chain molecules are two fold. In the first place the principle provides a relation between excess quantities such as the molar energy of mixing

$$u^e(n_1, n_2, x_1, x_2, T) = u(n_1, n_2, x_1, x_2, T) - x_1 u(n_1, T) - x_2 u(n_2, T) \tag{42a}$$

the molar volume change on mixing

$$v^e(n_1, n_2, x_1, x_2, T) = v(n_1, n_2, x_1, x_2, T) - x_1 v(n_1, T) - x_2 v(n_2, T) \tag{42b}$$

the molar excess entropy:

$$s^e(n_1, n_2, x_1, x_2, T) = s(n_1, n_2, x_1, x_2, T)$$
$$- x_1 s(n_1, T) - x_2 s(n_2, T)$$
$$+ x_1 \ln x_1 + x_2 \ln x_2 \tag{42c}$$

and the corresponding molar configurational quantities of the pure compounds. Since the internal contributions to u, v and s are by definition unaffected by the mixing process, i.e.

$$u_{int}(n_1, n_2, x_1, x_2, T) = x_1 u_{int}(n_1, T) + x_2 u_{int}(n_2, T) \tag{43a}$$

and

$$s_{\text{int}}(n_1, n_2, x_1, x_2, T) = x_1 s_{\text{int}}(n_1, T) + x_2 s_{\text{int}}(n_2, T) \qquad (43b)$$

we may express Eqs. (42) in terms of n_1, n_2 and $n = n_1 x_1 + n_2 x_2$ as

$$A^e(n_1, n_2, n, T) = A_{\text{conf}}(n, T) - \frac{n_2 - n}{n_2 - n_1} A_{\text{conf}}(n_1, T)$$

$$- \frac{n - n_1}{n_2 - n_1} A_{\text{conf}}(n_2, T) \qquad (44)$$

where $A^e(n_1, n_2, n, T)$ is one of the excess functions (42) and the function $A_{\text{conf}}(n, T)$ represents the molar configurational energy, the molar volume or the molar configurational entropy of a compound with chain length n, where the entropy is taken exclusive of the ideal entropy of mixing in the case of a mixture.

In the second place the principle of congruence leads to a consistency requirement for the excess functions of different mixtures at the same temperature. Writing Eq. (44) for the four mixtures (n_1, n_2, n), (n_A, n_B, n), (n_A, n_B, n_1) and (n_A, n_B, n_2) and eliminating the function A_{conf} from the equations one obtains

$$A^e(n_1, n_2, n, T) = A^e(n_A, n_B, n, T) - \frac{n_2 - n}{n_2 - n_1} A^e(n_A, n_B, n_1, T)$$

$$- \frac{n - n_1}{n_2 - n_1} A^e(n_A, n_B, n_2, T) \qquad (45)$$

The validity of this relation for different mixtures satisfying $n_A \leqslant n_1 \leqslant n \leqslant n_2 \leqslant n_B$ will be tested in Section 5 for the excess functions of normal alkane mixtures.

B. The Principle of Corresponding States for Excess Functions

Our next aim is to relate the excess functions of different mixtures at different temperatures. For this purpose we return to Eq. (44), expressing an excess function $A^e(n_1, n_2, n, T)$ in terms of the configurational part of the corresponding thermodynamic function $A(n, T)$ at the three values of the chain length, n_1, n_2 and n.

Applying the principle of corresponding states to each of the three terms on the right-hand side of (44), and indicating by $\alpha(n)$ the reduction factor for the quantity A, i.e. $q(n)\,\varepsilon$, $r(n)\,\sigma^3$ or $c(n)\,k$, we can express the excess function

$$A^e(n_1, n_2, n, T) = \alpha(n)\,[A^*(T^*) - X_1 A^*(T_1^*) - X_2 A^*(T_2^*)] \quad (46)$$

as a linear combination of the values of the reduced function

$$A^*(T^*) = \frac{A_{\text{conf}}(n, T)}{\alpha(n)} \quad (47a)$$

at three reduced temperatures

$$T^* = \frac{c(n)\,kT}{q(n)\,\varepsilon}; \quad T_1^* = \frac{c(n_1)\,kT}{q(n_1)\,\varepsilon} \quad \text{and} \quad T_2^* = \frac{c(n_2)\,kT}{q(n_2)\,\varepsilon} \quad (47b)$$

with coefficients

$$X_1 = \frac{n_2 - n}{n_2 - n_1}\frac{\alpha(n_1)}{\alpha(n)} \quad \text{and} \quad X_2 = \frac{n - n_1}{n_2 - n_1}\frac{\alpha(n_2)}{\alpha(n)} \quad (48)$$

From these equations it is seen that the introduction of a reduced excess function

$$A^{e*}(T_1^*, T_2^*, T^*) = \frac{A^e(n_1, n_2, n, T)}{\alpha(n)} \quad (49a)$$

$$= A^*(T^*) - X_1 A^*(T_1^*) - X_2 A^*(T_2^*) \quad (49b)$$

depending only on the three reduced temperatures, (47b), is only possible when X_1 and X_2 can be expressed as functions of T_1^*, T_2^* and T^*. This will in general not be the case, unless the reduction factor for the quantitiy A, $\alpha(n)$ and the one for temperature, $q(n)\,\varepsilon/c(n)\,k$ satisfy certain conditions. These follow from the requirement that the reduced excess function (49a) should have the same value for a mixture (n_1, n_2, n) at temperature T and for a corresponding mixture (n_1', n_2', n') at a different temperature $T' = yT$, where corresponding mixtures are defined in such a way that the reduced temperatures

$$T^* = \frac{c(n')\,kT'}{q(n')\,\varepsilon} = \frac{c(n)\,kT}{q(n)\,\varepsilon} \quad (50a)$$

$$T_1^* = \frac{c(n_1')\,kT}{q(n_1')\,\varepsilon} = \frac{c(n_1)\,kT}{q(n_1)\,\varepsilon} \quad (50b)$$

and

$$T_2^* = \frac{c(n_2')\,kT}{q(n_2')\,\varepsilon} = \frac{c(n_2)\,kT}{q(n_2)\,\varepsilon} \quad (50c)$$

are the same in both cases.

Expressing the relation between corresponding mixtures by the transformation

$$n' = f(n, y); \quad n'_1 = f(n_1, y); \quad n'_2 = f(n_2, y) \tag{51a}$$

$$T' = yT \tag{51b}$$

where the function $f(n, y)$ is such that it leaves invariant the reduced temperatures (50), we must impose the requirement that the ratios

$$X_1 = \frac{n'_2 - n'}{n'_2 - n'_1} \frac{\alpha(n'_1)}{\alpha(n')} = \frac{n_2 - n}{n_2 - n_1} \frac{\alpha(n_1)}{\alpha(n)} \tag{52a}$$

and

$$X_2 = \frac{n' - n'_1}{n'_2 - n'_1} \frac{\alpha(n'_2)}{\alpha(n')} = \frac{n - n_1}{n_2 - n_1} \frac{\alpha(n_2)}{\alpha(n)} \tag{52b}$$

are unaffected by the transformation.

As a consequence of this requirement the cross-ratio of four different n-values

$$\frac{n'_2 - n'_3}{n'_2 - n'_1} : \frac{n'_4 - n'_3}{n'_4 - n'_1} = \frac{n_2 - n_3}{n_2 - n_1} : \frac{n_4 - n_3}{n_4 - n_1} \tag{53}$$

must also be invariant. As is well known, for instance from the theory of conformal mapping,[32] the most general transformation having this property is

$$n' = f(n, y) = \frac{a_1(y) + a_2(y) n}{1 + a_3(y) n} \tag{54}$$

where $a_1(y)$, $a_2(y)$ and $a_3(y)$ are still arbitrary functions of the temperature ratio y. Knowing the general form of the transformation $n' = f(n, y)$ we can now express the invariance requirements for the reduced temperature (50) and for the ratios (52) by means of the functional equations

$$\frac{q[f(n, y)]}{q(n)} \Big/ \frac{c[f(n, y)]}{c(n)} = y \tag{55a}$$

and

$$\frac{\alpha[f(n, y)]}{\alpha(n)} = \frac{a_4(y)}{1 + a_3(y) n} \tag{55b}$$

which determine the reduction factors for the temperature, $q(n)/c(n)$, and for the quantity A, $\alpha(n)$, as functions of n. The function $a_4(y)$ in (55b) can still be chosen arbitrarily.

Equations (55) are easily solved[19] by expanding their right-hand and left-hand sides in powers of $(y-1)$ and equating the coefficients of equal powers. In this way it is found that the molecular units for the temperature and for the quantity A must have the general form

$$\frac{q(n)}{c(n)} = \text{const} \left(\frac{1+\lambda n}{1+\mu n}\right)^u \tag{56a}$$

and

$$\alpha(n) = \text{const}\,(1+\lambda n)\left(\frac{1+\lambda n}{1+\mu n}\right)^v \tag{56b}$$

respectively, where λ, μ, u and v represent arbitrary constants. Finally, the coefficients $a_1(y)$ to $a_4(y)$ may be determined by substituting the solution (56) back into Eqs. (55). In this way the transformation (54) is found to depend on the temperature ratio $y = T'/T$ as

$$n' = f(n, y) = \frac{(y^{1/u}-1)+(\lambda y^{1/u}-\mu)\,n}{(\lambda-\mu y^{1/u})-\lambda\mu(y^{1/u}-1)\,n} \tag{57}$$

The form (56) of the reduction factors for temperature and for the extensive thermodynamic variables, which is required for the applicability of the principle of corresponding states to excess functions (49), is not quite in agreement with the chain-length dependence of the parameters $q(n)$, $r(n)$ and $c(n)$ expressed by Eq. (41), which is expected on the basis of our assumptions. If we disregard the slight non-linearity of the function $q(n)$, i.e. see (41a), the temperature reduction factor is correctly represented by (56a) with $u = 1$, $\lambda = q_m/2(q_e - q_m)$ and $\mu = c_m/2(c_e - c_m)$, and the reduction factors for energy and entropy are given by (56b) with $v = 0$ and $v = -1$, respectively, but the molecular unit for volume is proportional to

$$r(n) = \text{const} \left[1 + \frac{r_m}{2(r_e - r_m)}\,n\right]$$

i.e. to a linear function with a slope which is in general different from both λ and μ. In practice, however, the form of Eqs. (56), which contain four adjustable parameters, is of sufficient generality for representing the molecular units to within the experimental accuracy.

In this way we can apply the principle of corresponding states (49) to any excess function such as the heat of mixing, the volume change on mixing or the excess entropy, by determining in each case a set of values for the parameters λ, μ, u and v, and "translating" an excess function at temperature T into a corresponding function at temperature $T' = yT$ by means of the relation

$$A^e(n_1', n_2', n', T') = \frac{\alpha(n')}{\alpha(n)} A^e(n_1, n_2, n, T) \tag{58}$$

where n_1', n_2' and n' are given by Eq. (57) as functions of n_1, n_2 and n, respectively. In general non-integer values will be obtained for n_1' and n_2'. In this case the principle of congruence should be used for comparing the calculated function $A^e(n_1', n_2', n', T')$ with experimental data for various mixtures at temperature T'.

5. APPLICATION OF THE PRINCIPLE TO THE EXCESS FUNCTIONS OF NORMAL ALKANE MIXTURES

A. Investigation of the Validity of the Principle of Congruence

Since the excess quantities of normal alkane mixtures can be measured much more easily with a high absolute accuracy than the corresponding absolute thermodynamic quantities, they provide a much severer test of the validity of the principle of corresponding states than the thermodynamic relations of the pure normal alkanes.

In the first place we must investigate the validity of the principle of congruence, since it is a prerequisite for the principle of corresponding states for excess functions [Eq. (58)]. The first evidence for the principle of congruence was obtained by Brønsted and Koefoed,[28] who calculated the excess Gibbs free energies of the systems C_6H_{14}–$C_{16}H_{34}$, C_7H_{16}–$C_{16}H_{34}$ and C_6H_{14}–$C_{12}H_{26}$ from vapour pressure data at 20°C. They found that their results could be represented to within the experimental accuracy by the single equation

$$G^e(n_1, n_2, x_1, x_2, 20°C) = -2.7(n_2 - n_1)^2 x_1 x_2 \text{ Joule/mole} \tag{59a}$$

for all three systems.

By introducing the mean chain length $n = n_1 x_1 + n_2 x_2$, this equation can be rewritten in the form

$$G^e(n_1, n_2, n, 20°C) = -2.7(n_2 - n)(n - n_1) \text{ Joule/mole} \tag{59b}$$

which satisfies the principle of congruence identically, as can be seen immediately by substituting it into the different terms of Eq. (45). So far this is the only case for which the principle of congruence has been confirmed for the excess Gibbs function of normal alkane mixtures. It would be of interest to carry out additional vapour pressure measurements, which would allow a verification of this principle at higher temperatures and in a wider range of the chain lengths of the components.

For the volume change on mixing and for the heat of mixing accurate experimental data are available for several different mixtures and over practically the full range of temperatures where the two components are liquids at atmospheric pressure. The validity of the principle of congruence can be tested graphically in the following way. At each temperature the excess function of the mixture with the greatest difference in chain lengths of the components (n_A, n_B) is plotted as a function of the average chain length $n = n_A x_A + n_B x_B$, and a smooth curve is fitted to the experimental data. On this curve the points corresponding to the chain lengths of the two components of another mixture, n_1 and n_2, are selected and a straight line

$$A_{\mathrm{id}}(n_1, n_2, n, T) = \frac{n_2 - n}{n_2 - n_1} A^{\mathrm{e}}(n_A, n_B, n_1, T) + \frac{n - n_1}{n_2 - n_1} A^{\mathrm{e}}(n_A, n_B, n_2, T)$$

representing an ideal (n_1, n_2)-mixture, is drawn. Adding the experimental values of the excess quantity for the mixture (n_1, n_2), i.e. $A^{\mathrm{e}}(n_1, n_2, n, T)$ to the ordinates of this ideal mixing line, one should regain the original excess function for the mixture (n_A, n_B), $A^{\mathrm{e}}(n_A, n_B, n, T)$, if the principle of congruence is satisfied. The results of this procedure for the volume contractions on mixing of several normal alkane mixtures are given in Figs. 9–14. Figure 9 shows the data for the mixtures (5, 16), (6, 16), (7, 16), (8, 16), (10, 16) and (6, 12) at 20°C obtained by Desmyter and van der Waals.[33] The asterisks represent the "excess volumes" determined from the molar volumes of the pure compounds at 20°C by means of the principle of congruence in the form (44). Figures 10–14 present the data obtained by Holleman[34] for various systems at 51, 76, 96, 106 and 126°C in a similar way. A survey of the data which have been analysed is given in Table II. The curves in Figs. 9–14 were obtained by fitting the experimental points to an expression[35] of the form

$$V^e(n_1, n_2, n, T) = -\frac{(n_2 - n)(n - n_1)}{(n_1 - 2)(n_2 - 2)(n - 2)}$$

$$\times \left[a_{-1}(T) + a_{-2}(T)\left(\frac{1}{n_1 - 2} + \frac{1}{n_2 - 2} + \frac{1}{n - 2}\right) + \ldots \right]$$

(60a)

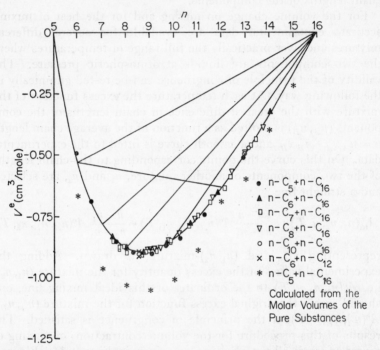

Fig. 9. Excess volumes of various normal alkane mixtures at 20°C, according to Desmyter and van der Waals, plotted as a function of the average chain length, n. Curve represents Eq. (60a) with $a_{-1} = 11.48$ and $a_{-2} = -3.93$ cm³/mole.

which follows from a series expansion of the molar volume

$$V(n, T) = a_1(T)(n - 2) + a_0(T) + a_{-1}(T)\frac{1}{n - 2} + a_{-2}(T)\frac{1}{(n - 2)^2} + \ldots$$

(60b)

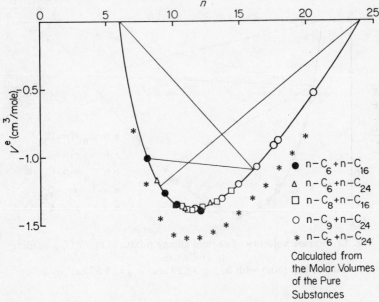

Fig. 10. Excess volumes of normal alkane mixtures at 51°C, according to Holleman.
Curve: Eq. (60a) with $a_{-1} = 17.57$ and $a_{-2} = -2.21$ cm³/mole.

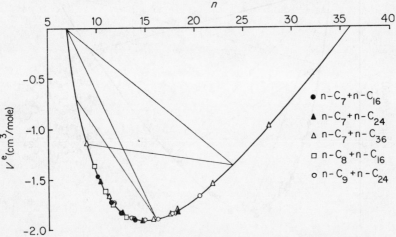

Fig. 11. Excess volumes of normal alkane mixtures at 76°C, according to Holleman.
Curve: Eq. (60a) with $a_{-1} = 24.29$ and $a_{-2} = +2.74$ cm³/mole.

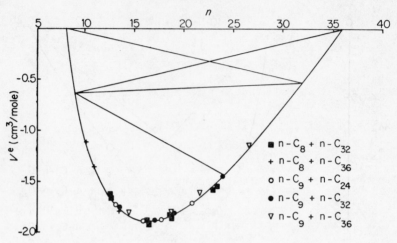

Fig. 12. Excess volumes of normal alkane mixtures at 96°C, according to Holleman.
Curve: Eq. (60a) with $a_{-1} = 32.29$ and $a_{-2} = 4.85$ cm³/mole.

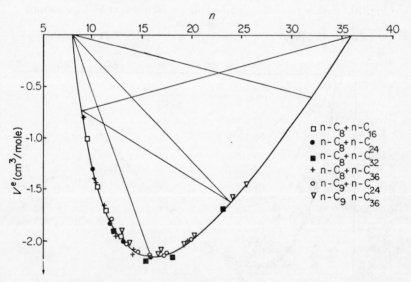

Fig. 13. Excess volumes of normal alkane mixtures at 106°C, according to Holleman.
Curve: Eq. (60a) with $a_{-1} = 35.46$ and $a_{-2} = 10.60$ cm³/mole.

in descending powers of $n - 2$. The values of the first two coefficients in this expression, which provide an excellent representation of all the data, are given in Table III for various temperatures.

As can be seen from Figs. 9–14, the excess volumes for different mixtures are mutually completely consistent with the principle of congruence, and the agreement between the mixing data and the excess volumes derived from the molar volumes of the pure compounds is quite satisfactory if one realizes that the accuracies of the latter are only of the order of 1 cm³/mole.

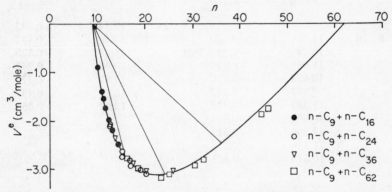

Fig. 14. Excess volumes of normal alkane mixtures at 126°C, according to Holleman.
Curve: Eq. (60a) with $a_{-1} = 48.00$ and $a_{-2} = 11.68$ cm³/mole.

The results of a similar analysis of the heats of mixing are presented in Figs. 15–18. In Fig. 15 data obtained by van der Waals and Hermans[36] and by McGlashan and Morcom[37] for several normal alkane mixtures at 20°C have been plotted as a function of the average chain length n. Apart from some scatter in the points, which may be accounted for completely by experimental inaccuracies, these data conform to the principle of congruence. A comparison with excess enthalpies derived from the molar heat functions of the pure compounds could not be carried out, since these are not known with sufficient accuracy. Bhattacharyya, Patterson and Somcynsky[38] have attempted to translate the reduced configurational energies of pure normal alkanes as a function of reduced temperature (see Fig. 3) into curves for the

TABLE II. Experimental Excess Volumes According to Holleman[34]

n_1	n_2	n $(x_1 n_1 + x_2 n_2)$	V^e exp (cm³/mole)	n_1	n_2	n $(x_1 n_1 + x_2 n_2)$	V^e exp (cm³/mole)
			51°C				
6	16	12.022	−0.730	6	24	12.737	−1.33
		10.268	−0.875			10.901	−1.37
		9.371	−0.892			8.806	−1.16
		8.068	−0.780				
				9	24	17.54	−0.382
8	16	14.37	−0.170			14.84	−0.447
		13.28	−0.264			11.63	−0.370
		11.98	−0.329			20.48	−0.242
		10.675	−0.320			17.87	−0.377
		9.337	−0.242			16.23	−0.437
		11.369	−0.332			13.62	−0.456
		10.692	−0.333				
		12.200	−0.314				
		11.146	−0.336				
		10.318	−0.321				
			76°C				
7	16	13.89	−0.447	8	16	12.984	−0.425
		12.43	−0.678			11.616	−0.501
		11.49	−0.783			10.848	−0.497
		10.03	−0.836			9.720	−0.412
7	24	18.19	−0.928	9	24	17.64	−0.570
		14.57	−1.30			14.87	−0.671
		12.68	−1.37			13.50	−0.661
		10.38	−1.24			11.66	−0.539
						20.48	−0.361
7	36	27.65	−0.974			17.81	−0.563
		21.82	−1.55			16.14	−0.654
		18.17	−1.81			13.88	−0.679
		15.76	−1.89				
		17.41	−1.86				
		11.21	−1.68				
		9.06	−1.13				

TABLE II (*continued*)

n_1	n_2	n $(x_1 n_1 + x_2 n_2)$	V^e exp (cm³/mole)	n_1	n_2	n $(x_1 n_1 + x_2 n_2)$	V^e exp (cm³/mole)
				96°C			
8	32	23.336	−1.22	9	24	20.805	−0.441
		18.632	−1.64			17.700	−0.771
		16.232	−1.73			18.915	−0.879
		12.584	−1.55			13.065	−0.867
		22.887	−1.24				
		18.531	−1.61	9	32	23.895	−0.886
		16.153	−1.71			19.056	−1.21
		12.406	−1.53			16.976	−1.26
						13.480	−1.13
8	36	13.334	−1.80				
		10.954	−1.36	9	36	26.550	−0.927
		10.058	−1.11			21.609	−1.27
		8.995	−0.629			18.720	−1.37
						14.481	−1.29
				106°C			
8	16	12.936	−0.636	9	24	20.010	−0.593
		11.496	−0.765			17.520	−0.865
		10.704	−0.755			15.930	−1.01
		9.520	−0.605			13.605	−1.03
						17.325	−0.915
8	24	15.920	−1.37			14.625	−1.03
		13.248	−1.47			12.015	−0.872
		11.936	−1.43				
		10.176	−1.08	9	36	24.174	−1.25
						19.395	−1.57
8	32	23.288	−1.41			16.965	−1.59
		18.080	−1.92			13.077	−1.30
		15.488	−2.02			25.470	−1.16
		12.296	−1.80			20.378	−1.52
						17.770	−1.60
8	36	14.051	−2.13			13.844	−1.42
		11.349	−1.67				
		10.360	−1.39				
		9.140	−0.802				
		19.480	−2.01				
		14.236	−2.09				
		12.460	−1.95				
		10.458	−1.44				

TABLE II (*continued*)

n_1	n_2	n $(x_1 n_1 + x_2 n_2)$	V^e exp (cm^3/mole)	n_1	n_2	n $(x_1 n_1 + x_2 n_2)$	V^e exp (cm^3/mole)
			126°C				
9	16	13.325	−0.543	9	36	25.956	−1.50
		11.849	−0.692			20.313	−2.02
		11.135	−0.601			17.505	−2.13
		10.211	−0.446			13.887	−1.89
		14.484	−0.345				
		13.296	−0.540	9	62	45.941	−1.75
		12.553	−0.621			32.161	−2.81
		11.367	−0.616			24.953	−3.12
						18.487	−2.98
9	24	20.190	−0.778			44.510	−1.87
		16.995	−1.29			30.412	−2.93
		15.225	−1.47			23.416	−3.19
		12.765	−1.35			16.314	−2.79
		10.049	−0.960				
		16.740	−1.25				
		15.285	−1.35				
		12.804	−1.29				

TABLE III. Values of the Two Coefficients in Expression (60a) for the Excess Volumes at Various Temperatures

Temperature (°C)	$a_{-1}(T)$ (cm^3/mole)	$a_{-2}(T)$ (cm^3/mole)
20	11.48	−3.93
51	17.57	−2.21
76	24.29	+2.74
96	32.29	+4.85
106	35.46	+10.60
126	48.00	+11.68

configurational energy against chain length at constant temperature, using the known chain-length dependence of the reduction factors for energy and temperature. By applying the principle of congruence in the form (44) to the configurational energies obtained in this way it is possible in principle to derive "heats of mixing" from the data of the pure compounds. We tend, however, to be rather sceptical about the physical significance of these results, because of their extreme sensitivity to the precise shape of the reduced configurational energy against reduced temperature curves.

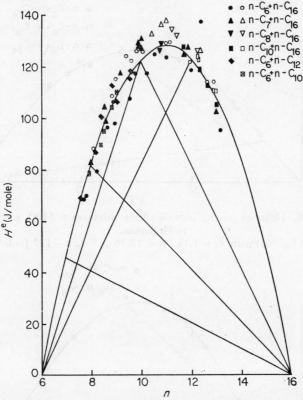

Fig. 15. Heats of mixing of normal alkane mixtures at 20°C, according to van der Waals and Hermans (solid symbols) and McGlashan and Morcom (open symbols).
Curve: Eq. (61) with $A_1 = 5.12$ Joule/mole and $A_2 = A_3 = \ldots = 0$.

We feel that the uncertainties inherent in the determination of configurational energies from heat of vaporization data, and the ambiguities found in the chain-length dependence of the energy reduction factors (cf. Eqs. 35b, 35b′, 35b″) make an analysis such as has been carried out by Bhattacharyya and coworkers[38] somewhat speculative.

Holleman[39] has published accurate heat-of-mixing data at various temperatures above 20°C for several mixtures. The data at 51, 60 and 76°C, as well as more recent unpublished data by the

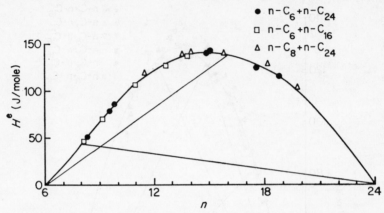

Fig. 16. Heats of mixing normal alkane mixtures at 51°C, according to Holleman.
Curve: Eq. (61) with $A_1 = 1.13$, $A_2 = 17.36$ and $A_3 = -127$ Joule/mole.

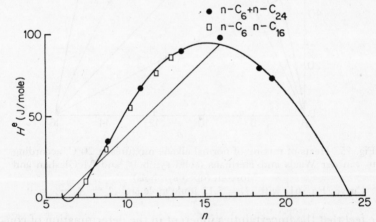

Fig. 17. Heats of mixing at 60°C, according to Holleman. Curve: Eq. (61) with $A_1 = 0.54$, $A_2 = 19.83$ and $A_3 = -160$ Joule/mole.

same author at 106 and 135°C, are shown in Figs. 16, 17, 18, 19 and 20, respectively. The results at 76 and 106°C have been plotted separately for mixtures having the same volatile component

(Figs. 18a, 18b and 19a) and for mixtures of one long-chain component with different short-chain normal alkanes (Figs. 18c and 19b). Although the various mixtures conform qualitatively quite

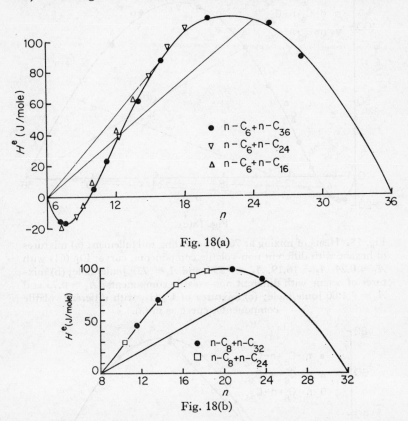

Fig. 18(a)

Fig. 18(b)

well to the principle of congruence, slight systematic deviations exceeding the experimental error are observed for mixtures containing different volatile components.

The investigated data have been collected in Table IV. The values of the coefficients in the series expansion

$$H^e(n_1, n_2, n, T) = (n_2 - n)(n - n_1)\left[A_1 + \frac{A_2}{n} + \frac{A_3}{n^2} + ...\right] \quad (61)$$

which has been used to interpolate the results are listed in Table V.

10

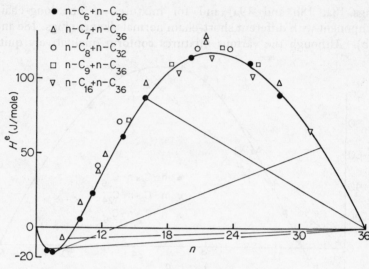

Fig. 18(c)

Fig. 18. Heats of mixing at 76°C, according to Holleman: (a) mixtures of hexane with different non-volatile components; curve: Eq. (61) with $A_1 = 0.24$, $A_2 = 16.19$, $A_3 = -255$, and $A_4 = 739$ Joule/mole; (b) mixtures of octane with different non-volatile components: $A_1 = 0.73$ and $A_2 = -1.00$ Joule/mole; (c) mixtures of $C_{36}H_{74}$ with different volatile components; curve as in (a).

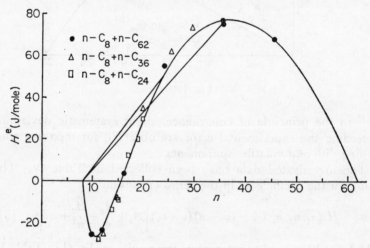

Fig. 19(a)

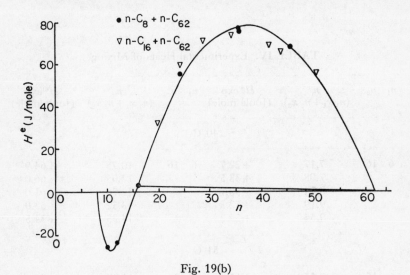

Fig. 19(b)

Fig. 19. Heats of mixing at 106°C, according to Holleman: (a) mixtures of octane with three different long-chain compounds; (b) mixtures of n-$C_{62}H_{126}$ with octane and hexadecane. Curve: Eq. (61) with $A_1 = 0.09$, $A_2 = 2.18$, and $A_3 = -56$ Joule/mole.

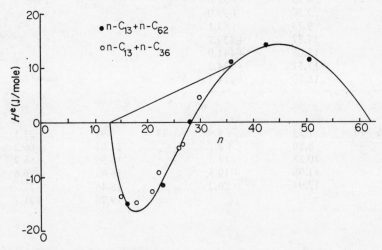

Fig. 20. Heats of mixing at 135°C, according to Holleman. Curve: Eq. (61) with $A_1 = 0.033$, $A_2 = 0.78$, and $A_3 = -49$ Joule/mole.

TABLE IV. Experimental Heats of Mixing

n_1	n_2	$\dfrac{n}{(n_1 x_1 + n_2 x_2)}$	H^e exp (Joule/mole)	n_1	n_2	$\dfrac{n}{(n_1 x_1 + n_2 x_2)}$	H^e exp (Joule/mole)
			40°C				
6	16	7.17	+22.7	6	16	10.75	+64.9
		7.98	+38.5			12.05	+66.0
		8.74	+50.7			12.76	+61.0
		9.19	+52.5			13.52	+53.0
		9.54	+59.2			14.35	+34.5
			51°C				
6	16	8.14	+15.9	8	24	11.45	+84.4
		9.14	+26.5			13.56	+109.3
		10.96	+37.7			13.93	+112.0
		12.60	+36.0			15.74	+117.1
		13.81	+28.5			18.10	+112.5
6	24	8.35	+52.0			19.72	+92.5
		9.56	+79.0				
		9.73	+83.1				
		14.87	+139.7				
		15.05	+141.0				
		17.51	+124.2				
		18.79	+114.3				
			60°C				
6	16	7.54	−4.4	6	24	8.91	+34.1
		8.86	+2.7			11.03	+66.2
		10.35	+14.5			13.59	+88.7
		11.98	+19.5			16.03	+96.6
		12.94	+20.2			18.46	+77.7
						19.20	+71.7

TABLE IV (*continued*)

n_1	n_2	n $(n_1 x_2 + n_1 x_2)$	H^e exp (Joule/mole)	n_1	n_2	n $(n_1 x_1 + n_2 x_2)$	H^e exp (Joule/mole)
			76°C				
6	16	7.11	−28.9	6	36	6.99	−15.6
		8.99	−31.6			7.48	−16.4
		9.73	−24.6			9.99	+5.4
		11.97	−10.2			11.17	+23.0
		13.56	−3.4			14.00	+60.9
						15.92	+87.0
6	24	8.48	−28.3			20.17	+114.1
		12.15	+1.2			25.48	+111.0
		14.91	+21.4			28.11	+88.7
		16.64	+30.7				
		18.22	+32.8				
7	36	8.00	+5.8	8	32	11.60	+45.6
		10.13	+29.4			13.67	+69.5
		11.71	+51.2			20.75	+98.1
						23.63	+88.9
		12.31	+61.0			14.58	+78.0
		15.97	+106.6			18.34	+114.4
		21.46	+132.0	9	36	23.00	+123.9
		21.47	+133.0			26.24	+112.0
		28.12	+100.1				
8	24	10.54	+16.1			18.94	+29.7
		13.82	+36.1	16	36	22.06	+53.7
		15.30	+44.8			25.74	+57.5
		16.93	+43.6			31.05	+42.2
		18.36	+40.6				

TABLE IV (*continued*)

n_1	n_2	n ($n_1 x_1 + n_2 x_2$)	H^e exp (Joule/mole)	n_1	n_2	n ($n_1 x_1 + n_2 x_2$)	H^e exp (Joule/mole)
			96°C				
8	24	11.41	−16.8	8	24	15.66	−5.6
		11.54	−18.0			17.30	−2.1
		12.02	−14.0			19.71	+6.1
		15.01	−7.2			19.86	+2.9
			106°C				
8	24	11.21	−38.0	8	62	9.95	−25.5
		13.95	−32.5			11.95	−23.9
		15.07	−31.0			16.11	+3.4
		17.09	−16.3			24.21	+55.4
		19.14	−14.9			35.73	+76.9
		19.99	−8.1			35.77	+75.5
						45.68	+68.0
8	36	12.04	−36.8				
		14.88	−27.5	16	62	19.92	+29.9
		20.02	+1.3			24.02	+58.0
		25.55	+13.6			28.75	+68.9
		29.63	+13.6			32.46	+73.5
						41.40	+68.0
						43.67	+65.2
						50.39	+54.8
			135°C				
13	36	15.02	−14.3	13	62	16.00	−15.0
		17.92	−16.8			22.89	−11.5
		20.79	−16.1			28.12	0.0
		22.19	−13.3			35.93	+11.0
		25.92	−10.3			42.64	+13.9
		26.65	−10.3			50.68	+11.3
		29.92	−3.0				

TABLE V. Values of the Coefficients in Expression (61) for the Heats of Mixing

n_1	n_2	A_1	A_2	A_3	A_4	Standard error in H^e
						(Joule/mole)
			40°C			
6	16	3.42	−8.26	—	—	1.3
			51°C			
6	16	1.01	22.45	−187	—	0.2
6	24	1.13	17.36	−127	—	2.4
8	24	2.34	−16.24	135	—	1.9
			60°C			
6	16	1.63	−0.31	−112	—	0.9
6	24	0.54	19.83	−160	—	3.3
			76°C			
6	16	0.74	1.89	−187	—	4.4
6	24	0.83	−1.35	−103	—	1.7
6	36	0.24	16.19	−255	739	2.8
7	36	0.40	9.56	−104	—	2.4
8	24	0.99	−5.18	—	—	2.4
8	32	0.73	−1.00	—	—	1.7
9	36	0.68	−0.15	—	—	3.3
16	36	0.47	3.10	—	—	2.4
			96°C			
8	24	0.70	−12.15	−3.2	—	1.9
			106°C			
8	24	−0.11	8.00	−194	—	1.6
8	36	0.17	1.50	−99	—	1.4
8	62	0.09	2.18	−56	—	1.9
16	62	0.09	2.06	—	—	3.2
			135°C			
13	36	−0.085	6.30	−150	—	1.4
13	62	0.033	0.78	−49	—	1.0

B. Test of the Principle of Corresponding States for Excess Functions

For the excess volumes, which were found to satisfy the principle of congruence, the application of the principle of corresponding states for excess functions presents no difficulties. For any mixture of components n_1 and n_2 and average chain length $n = n_1 x_1 + n_2 x_2$ at $T > 20°C$, the appropriate component chain lengths n_1' and n_2' and the average chain length n' of the corresponding mixture at $T' = 20°C$ may be deduced from the transformation (57) with $y = T'/T$. Subsequently the excess volume $V^e(n_1, n_2, n, T)$ may be translated into a corresponding excess volume $V^e(n_1', n_2', n', T')$ at 20°C by means of Eq. (58) with $\alpha(n) = \text{const}\, r(n)$ representing the volume reduction factor. The values of the constants λ and μ appearing in Eq. (57) must be determined from a least-squares fit of the reduction factors for volume and for temperature of the pure normal alkanes, as given in the first two columns of Table I, to expressions of the form (56b) and (56a), respectively. The volume reduction factor $r(n)$ was found in Section 3 to be a linear function of n

$$r(n) = \text{const}\left[1 + \frac{r_m/r_e}{2(1 - r_m/r_e)}n\right] = \text{const}\,(1 + \lambda n) \quad \text{with} \quad \lambda = 1.184$$

(62a)

and a least-squares adjustment of the values of the temperature reduction factors shows that they can be represented by

$$\frac{c(n)}{q(n)} = \text{const}\left[\frac{1 + \lambda n}{1 + \mu n}\right]^u \quad \text{with} \quad \lambda = 1.184; \quad \mu = 0.1805; \quad u = 1.063$$

(62b)

Substituting these values into Eqs. (57) and (58) we have calculated the excess volumes at 20°C, which correspond to the experimental data[34] of the systems (9, 24) at 51°C, (9, 24) at 76°C, (9, 36) at 96°C, (9, 36) at 106°C and (9, 36) at 126°C as given in Table II. The results (see Table VI) are plotted in Fig. 21, together with a least-squares fit to the direct measurements at 20°C by Desmyter and van der Waals.[33] The mutual consistency of the values derived from mixtures at different temperatures shows that the principle of corresponding states (58) is applicable to the volume changes on mixing.

The same procedure may be applied to the heats of mixing given in Table IV. In this case, however, the quantity

$$\alpha(n) = \text{const } q(n)$$

in Eq. (58) represents the energy reduction factor. The values of this quantity, as deduced from the heats of vaporization of the pure

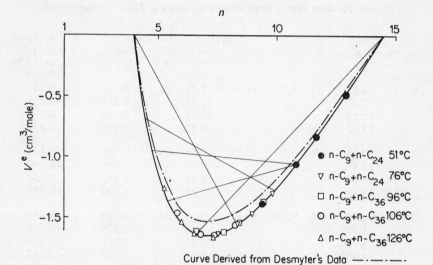

Fig. 21. Excess volumes at 20°C corresponding to the data at higher temperatures. Dashed curve: least-squares fit to data obtained by Desmyter and van der Waals at 20°C.

compounds (see Table I, column 3), can be represented with sufficient accuracy by a linear function

$$q(n) = \text{const}\left[1 + \frac{q_m/q_e}{2(1 - q_m/q_e)}n\right] = \text{const}\,(1 + \lambda'n) \quad \text{with } \lambda' = 0.564$$

(63a)

Accordingly, we represent the temperature reduction factors by the expression

$$\frac{c(n)}{q(n)} = \text{const}\left[\frac{1 + \lambda'n}{1 + \mu'n}\right]^{u'} \quad \text{with } \lambda' = 0.564; \quad \mu' = 0.254; \quad u' = 2.245$$

(63b)

which provides an equally good fit to the data of Table I, column 2, as (62b). Using these values of Eqs. (57) and (58) we have calculated the heats of mixing at 20°C, corresponding to Holleman's[39, 40] data for the systems (6, 16) at 40°C, (6, 16) at 51°C,

TABLE VI. Excess Volumes at 20°C Calculated with the Principle of Corresponding States from Excess Volumes at High Temperatures

n_1	n_2	n	T (°C)	n_1'	n_2'	n'	T' (°C)	$V^e(n_1', n_2', n', T')$	V^e[Eq. (60a)] $a_{-1} = 11.94$ $a_{-2} = -3.71$
9	24	20.48	51	6.70	14.45	12.91	20	−0.161	−0.157
		17.87				11.68		−0.252	−0.255
		16.23				10.87		−0.300	−0.300
		13.62				9.47		−0.325	−0.327
9	24	20.48	76	5.50	10.83	9.88	20	−0.182	−0.171
		17.81				9.03		−0.298	−0.290
		16.14				8.43		−0.357	−0.350
		13.88				7.61		−0.390	−0.384
9	36	26.550	96	4.79	10.87	9.47	20	−0.349	−0.344
		21.609				8.48		−0.525	−0.529
		18.720				7.85		−0.611	−0.600
		14.481				6.69		−0.633	−0.643
9	36	24.174	106	4.50	9.97	8.33	20	−0.459	−0.468
		19.395				7.44		−0.643	−0.639
		16.965				6.87		−0.688	−0.699
		13.077				5.85		−0.623	−0.654
9	36	25.956	126	3.99	8.40	7.41	20	−0.463	−0.433
		20.313				6.61		−0.711	−0.694
		17.505				6.08		−0.803	−0.801
		13.887				5.23		−0.792	−0.813

(6, 24) at 60°C, (6, 36) at 76°C, (8, 24) at 96°C, (8, 62) at 106°C and (13, 62) at 135°C. The results, listed in Table VII, are shown in Fig. 22 together with a least-squares fit of the data obtained by van der Waals and Hermans[26] and by McGlashan and Morcom[37] at 20°C (dashed curve). The drawn curve in this figure, which corresponds to the data of the system (6, 36) at 76°C, has been used as the reference line for applying the principle of congruence. It may be represented by Eq. (61) with $A_1 = 6.12$, $A_2 = -28.7$ and $A_3 = -37.8$ Joule/mole. As can be seen from Fig. 22, the heats of mixing at 20°C, corresponding to the data at various temperatures, are qualitatively rather similar both in magnitude and in chain-length dependence, but quantitatively significant systematic deviations are observed, which increase as the temperature difference becomes larger.

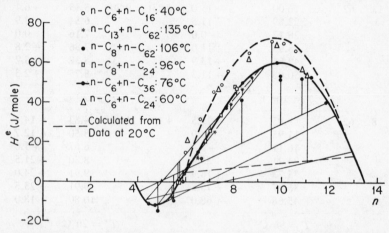

Fig. 22. Heats of mixing at 20°C, corresponding to the data at higher temperatures. Drawn curve: least-squares fit to the data corresponding to C_6-C_{36} at 76°C [Eq. (61) with $A_1 = 6.12$, $A_2 = -28.7$ and $A_3 = -37.8$ Joule/mole]. Dashed curve: least-squares fit to the data at 20°C obtained by van der Waals and Hermans and by McGlashan and Morcom [Eq. (61) with $A_1 = 5.2$ Joule/mole, $A_2 = A_3 = 0$].

For the excess Gibbs free energy only a few data are available which may be used for testing the principle of corresponding states. In analysing these data the question arises whether this principle is applicable to the excess Gibbs function as a whole, or whether a "combinatorial entropy" contribution should first be subtracted from this function, as is done in the cell-model treatment by Prigogine and coworkers.[12, 13] Figure 23(a) shows the result of an application of the principle of corresponding states to excess free energies, when no combinatorial contribution is subtracted. In this figure we have plotted the excess free energy curve at 20°C, which corresponds to the curve deduced by van der Waals[41] from his

TABLE VII. Heats of Mixing at 20°C, Corresponding to the Data at Higher Temperatures

n_1	n_2	n $(n_1 x_1 + n_2 x_2)$	H^e exp (Joule/ mole)	n_1'	n_2'	n'	$H^{e'}$ calc (Joule/ mole)
		$T = 135°C$			$T' = 20°C$		
13	62	16.00	−15.0	4.83	9.20	5.45	−6.1
		22.89	−11.5			6.54	−3.9
		28.12	0.0			7.16	0.0
		35.93	+11.0			7.86	+2.8
		42.64	+13.9			8.32	+3.2
		50.68	+11.3			8.75	+2.3
		$T = 106°C$			$T' = 20°C$		
8	62	9.95	−25.5	4.10	11.89	4.83	−14.4
		11.95	−23.9			5.50	−12.7
		16.11	+3.4			6.67	+1.6
		24.21	+55.4			8.30	+21.5
		35.73	+76.9			9.91	+24.0
		35.77	+75.5			9.91	+23.5
		45.68	+68.0			10.80	+18.0
		$T = 96°C$			$T' = 20°C$		
8	24	11.41	−16.8	4.36	9.00	5.69	−9.5
		11.54	−18.0			5.74	−10.2
		12.02	−14.0			5.91	−6.0
		15.01	−7.2			6.85	−3.7
		15.66	−5.6			7.04	−2.8
		17.30	−2.1			7.48	−1.0
		19.71	+6.1			8.07	+2.8
		19.86	+2.9			8.11	+1.3
		$T = 76°C$			$T' = 20°C$		
6	36	6.99	−15.6	3.91	13.45	4.46	−11.0
		7.48	−16.4			4.72	−11.5
		9.99	+5.4			5.97	+3.6
		11.17	+23.0			6.50	+14.7
		14.00	+60.9			7.67	+36.5
		15.92	+87.0			8.36	+49.9
		20.17	+114.1			9.79	+60.1
		25.48	+111.0			11.21	+52.8
		28.11	+88.7			11.86	+40.4

TABLE VII (*continued*)

n_1	n_2	n $(n_1 x_1 + n_2 x_2)$	H^e exp (Joule/ mole)	n_1'	n_2'	n'	$H^{e'}$ calc (Joule/ mole)
		$T = 60°C$				$T' = 20°C$	
6	24	8.91	+34.1	4.36	12.92	6.15	+25.3
		11.03	+66.2			7.35	+47.2
		13.59	+88.7			8.63	+60.0
		16.03	+96.6			9.79	+62.8
		18.46	+77.7			10.83	+48.4
		19.20	+71.7			11.11	+44.0
		$T = 40°C$				$T' = 20°C$	
6	16	7.17	+22.7	5.08	12.19	5.98	+19.7
		7.98	+38.5			6.61	+33.1
		8.74	+50.7			7.19	+43.2
		9.19	+52.5			7.52	+44.5
		9.54	+59.2			7.76	+49.9
		10.75	+64.9			8.63	+53.9
		12.05	+66.0			9.56	+54.1
		12.76	+61.0			10.06	+49.7
		13.52	+53.0			10.53	+42.7
		14.53	+34.5			11.18	+27.4

vapour pressure data for the system (7, 32) at 73°C, together with a curve corresponding to McGlashan and Williamson's data[42] for the system (6, 16) at 50°C, and the results obtained by Brønsted and Koefoed[28] for the system (6, 12) at 20°C. The curves found in this way from data at different temperatures are seen to be completely inconsistent. This suggests that the principle of corresponding states should be applied to the residual excess function, i.e. excess functions in which the configurational excess entropy is taken *exclusive of the combinatorial contribution*. The "combinatorial entropy" of mixing for a mixture of chain-molecules is given by the Flory–Huggins expression[17]

$$\Delta S_{\text{comb}} = -Nk \sum_j x_j \ln \phi_j \qquad (64)$$

This expression has been derived under quite general assumptions from statistical mechanics by Longuet-Higgins[30] for a mixture in which the components have equal densities. Since in such a mixture the volume fractions ϕ_j are equal to the "reduced volume" fractions

$$\phi_i = \frac{r(n_i)\, x_i}{\sum_j r(n_j)\, x_j} \tag{65}$$

we obtain for the combinatorial excess entropy of a binary mixture the result

$$S^{\mathrm{e}}_{\mathrm{comb}} = -Nk \left[\frac{n_2 - n}{n_2 - n_1} \ln \frac{r(n_1)}{r(n)} + \frac{n - n_1}{n_2 - n_1} \ln \frac{r(n_2)}{r(n)} \right] \tag{66}$$

where the principle of congruence has been used to express x_1 and x_2 in terms of n.

When the reduction procedure, which was applied to the total excess free energies in Fig. 23a, is repeated for the residual Gibbs functions

$$G^{\mathrm{R}} = G^{\mathrm{e}} - G^{\mathrm{e}}_{\mathrm{comb}} = G^{\mathrm{e}} + T S^{\mathrm{e}}_{\mathrm{comb}} \tag{67}$$

one obtains the curves shown in Fig. 23b, which agree remarkably well with the principle of corresponding states. This conclusion confirms the results of Flory, Orwoll and Vrij,[43] who applied the principle of corresponding states in a somewhat different form to the activity coefficients of the same systems.

C. Discussion

The main conclusion emerging from our analysis of the data of normal alkane mixtures is that the excess volumes and the residual Gibbs free energies conform both to the principle of congruence and to the principle of corresponding states for excess functions, whereas the heats of mixing show slight deviations from the former and more significant deviations from the latter principle. It seems plausible to interpret these discrepancies as the result of a slight failure of the principle of congruence, which is manifested much more clearly when the data at different temperatures are compared by means of the principle of corresponding states. In fact, it is not unexpected that the values of the parameters q, r and c for a mixture would show minor deviations from their principle of congruence values (41). For the parameter q such a deviation may

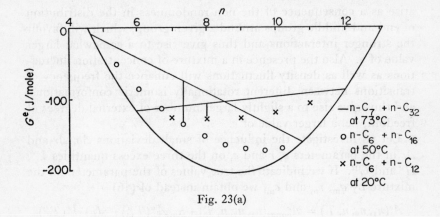

Fig. 23(a)

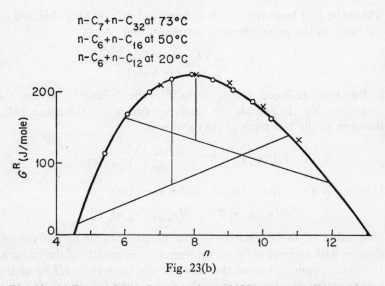

Fig. 23(b)

Fig. 23. (a) Excess Gibbs free energies at 20°C, corresponding to data for C_7–C_{32} at 73°C obtained by van der Waals (curve) and McGlashan and Williamson's data for C_6–C_{16} at 50°C (○). ×: Data at 20°C obtained by Brønsted and Koefoed. (b) Residual Gibbs free energies of the same systems.

arise as a consequence of the non-randomness in the distribution of end and middle groups around a given group, which will favour the stronger interactions and thus give rise to a somewhat larger value of q. Also the presence in a mixture of concentration fluctuations as well as density fluctuations will enhance the frequency of transitions between different rotationally isomeric conformations, and thus give rise to a slightly larger number of external degrees of freedom, i.e. a larger value of c.

Let us investigate the influence of small deviations Δq, Δr and Δc in the parameters q, r and c, on the three excess quantities V^e, H^e and G^R. If we indicate the true values of the parameters in the mixture by q_m, r_m and c_m, we obtain instead of (46)

$$A^e(n_1, n_2, n, T) = A^e_{congr}(n_1, n_2, n, T) + \alpha_m A^*(T^*_m) - \alpha(n) A^*(T^*) \tag{68}$$

where the first term represents the excess function (46) obtained on the basis of the principle of congruence, and

$$T^*_m = \frac{c_m kT}{q_m \varepsilon} = \frac{(c + \Delta c) kT}{(q + \Delta q) \varepsilon} \tag{69}$$

is the true reduced temperature for the mixture. When the deviations Δq, Δr and Δc are small we can apply the mean value theorem to the last term in (68) and write

$$\Delta A^e(n_1, n_2, n, T) \simeq \frac{\partial A(n, T)}{\partial n} (n_m - n) \tag{70}$$

Here n_m is a function of n_1, n_2, and n such that

$$A(n_1, n_2, n, T) = A[n_m(n_1, n_2, n), T] \tag{71}$$

The value of the derivative of the molar volume of the normal alkanes with respect to the chain length, n, is usually of the order of 10–20 cm^3/mole, whereas the value of the derivative $\partial H/\partial n$ of the heat function as estimated from the slope of the function $U^*(T^*)$ in Fig. 3, together with the values of the reduction factors for U and T in Table I, is found to be approximately 4500 Joule/mole. Comparing these values with those of the excess volumes and heats of mixing, respectively, it is seen that for the heats of mixing the deviations from the principle of congruence are significant even when $n_m - n$ is of the order of 0.01, whereas the correction term in the excess volumes will reveal itself only if $n_m - n$ becomes of the order of 0.1. Moreover, the deviations observed when comparing the excess functions of different mixtures at the same temperature may partially compensate each other, but in comparing the excess functions of different mixtures at different temperatures by means of the principle of corresponding states this may not be the case, particularly since the deviations from the principle of congruence are probably non-corresponding quantities.

Summarizing our results, we conclude that (a) the principle of corresponding states appears to be valid for the *configurational* thermodynamic quantities of pure chain-molecule liquids; (b) the principle of congruence in the form (45) of a consistency requirement of the excess functions of different mixtures at the same temperature is satisfied quantitatively by the excess volumes and the excess Gibbs functions and—apart from slight deviations in the case of mixtures containing different volatile components—also for the heats of mixing; (c) the principle of congruence in the form (44) as a relation between the configurational thermodynamic quantities of pure compounds and the excess functions of mixtures, and the principle of corresponding states for excess functions are at least qualitatively valid for the excess volumes, and for the residual Gibbs free energies, but not quite valid for the heats of mixing.

References

1. Van der Waals, J. D., Thesis, Leiden, 1873.
2. Kamerlingh-Onnes, H., *Verhandel. Koninkl. Ned. Akad. Wetenschap.* **21**, 22 (1888); *Arch. Neerl.* **30**, 101 (1897).
3. De Boer, J., and Michels, A., *Physica* **5**, 945 (1938); de Boer, J., *Physica* **14**, 139 (1948).
4. Guggenheim, E. A., *J. Chem. Phys.* **13**, 253 (1945).
5. Pitzer, K. S., *J. Chem. Phys.* **7**, 583 (1939).
6. Longuet-Higgins, H. C., *Proc. Roy. Soc.*, **A205**, 247 (1951).
7. Prigogine, I., Bellemans, A., and Englert-Chwoles, A., *J. Chem. Phys.* **24**, 518 (1956). See also Chapter IX in reference 13, and Bellemans, A., Mathot, V., and Simon, N., *Adv. Chem. Phys.* **9**, 117 (1967).
8. Scott, R. L., *J. Chem. Phys.* **25**, 193 (1956).
9. Wojtowicz, P., Salsburg, Z. W., and Kirkwood, J. G., *J. Chem. Phys.* **27**, 505 (1957).
10. Rice, S. A., *J. Chem. Phys.* **29**, 141 (1958).
11. Trappeniers, N., *Physica* **17**, 501 (1951); *Acad. Roy. Belg., Classe Sci. Mem.* **10**, (1952).

12. Prigogine, I., Bellemans, A., and Naar-Colin, C., *J. Chem. Phys.* **26**, 751 (1957).
13. Prigogine, I., *The Molecular Theory of Solutions*, North-Holland Publ. Comp., 1957, Chapters XVI and XVII.
14. Mathot, V., *Compt. rend. Réunion sur les Changements de Phases*, Paris, 1952, p. 115.
15. Prigogine, I., Trappeniers, N., and Mathot, V., *Discussions Faraday Soc.* **15**, 93 (1953); *J. Chem. Phys.* **21**, 559–60 (1953).
16. Lennard-Jones, J. E., and Devonshire, A. F., *Proc. Roy. Soc.*, **A163**, 63 (1937) and **A164**, 1 (1938).
17. See for instance Guggenheim, E. A., *Mixtures*, Oxford University Press, 1952, Chapter X.
18. Hijmans, J., *Physica*, **27**, 433 (1961).
19. Holleman, Th., and Hijmans, J., *Physica*, **28**, 604 (1962).
20. American Petroleum Institute, *Research Project* 44, Pittsburg, 1953, Selected values of physical and thermodynamic properties of hydrocarbons and related compounds, Table 20d.
21. Reference 20, Table 20m.
22. Waddington, G., and Douslin, D. R., *J. Am. Chem. Soc.* **69**, 2274 (1947); Osborne, N. S., and Ginnings, D. C., *J. Res. Nat. Bur. Stand.*, **39**, 453 (1947); Waddington, G., Todd, S. S., and Huffman, H. M., *J. Am. Chem. Soc.* **69**, 22 (1947).
23. Reference 20, Table 20k.
24. McGlashan, M. L., and Potter, D. J. B., *Joint Conference on Thermal and Transport Properties of Fluids*, London, 1957, session 1, paper 10.
25. Boelhouwer, J. W. M., *Physica* **26**, 1021 (1960).
26. Boelhouwer, J. W. M., *Physica* **34**, 484 (1967).
27. Flory, P. J., Orwoll, R. A., and Vrij, A., *J. Am. Chem. Soc.* **86**, 3507 (1964).
28. Brønsted, J. N., and Koefoed, J., *Kgl. Danske Videnskab. Selskab. Mat. Fys. Medd.* **22**, No. 17, 1 (1946).
29. Hijmans, J., *Mol. Phys.* **1**, 307 (1958).
30. Longuet-Higgins, H. C., *Discussions Faraday Soc.* **15**, 73 (1953).
31. Dixon, J. A., *J. Chem. Eng. Data* **4**, 289 (1959).
32. See for instance: Carathéodory, C., *Conformal Representation*, Cambridge University Press, 1952, Chapters 10 and 11.
33. Desmyter, A., and van der Waals, J. H., *Rec. Trav. Chim.* **77**, 53 (1958); van der Waals, J. H., private communication.
34. Holleman, Th., *Physica* **29**, 585 (1963).
35. Hijmans, J., and Holleman, Th., *Mol. Phys.* **4**, 91 (1961).
36. Van der Waals, J. H., and Hermans, J. J., *Rec. Trav. Chim.* **69**, 949 (1950); van der Waals, J. H., Thesis, Groningen, 1950.
37. McGlashan, M. L., and Morcom, K. W., *Trans. Faraday Soc.* **57**, 907 (1961).
38. Bhattacharyya, S. N., Patterson, D., and Somcynsky, T., *Physica* **30**, 1276 (1964).
39. Holleman, Th., *Physica* **31**, 49 (1965).

40. Holleman, Th., and Hijmans, J., *Physica* **31**, 64 (1965).
41. Van der Waals, J. H., and Hermans, J. J., *Rev. Trav. Chim.* **69**, 971 (1950).
42. McGlashan, M. L., and Williamson, A. G., *Trans. Faraday Soc.* **57**, 588 (1961).
43. Flory, P. J., Orwoll, R. A., and Vrij, A., *J. Am. Chem. Soc.* **86**, 3515 (1964).

ON THE USE OF PSEUDOPOTENTIALS IN THE QUANTUM THEORY OF ATOMS AND MOLECULES

JOHN D. WEEKS,* ANDREW HAZI† and STUART A. RICE,

*Department of Chemistry and The James Franck Institute,
University of Chicago, Chicago, Illinois 60637*

CONTENTS

* N.A.S.A. Predoctoral Fellow.
† Present address: Department of Chemistry, U.C.L.A.

1. INTRODUCTION

It is sometimes said, in condescending tones, that all the non-relativistic problems of chemistry and physics are solved. This comment is presumably intended to mean that all such problems boil down to finding a suitable solution of the Schrödinger equation. Since almost all the problems of interest in atomic and molecular physics involve the coupled motion of many particles, such solutions are difficult to find but, so the comment implies, that is merely a technical problem. It is our contention that the function of theory in chemistry is not primarily that of obtaining exact solutions to the N-body problem. Rather, we believe that theory should shoulder the burden of formulating concepts, of defining constructions which are useful in the description of systems of interacting particles, and of providing algorithms for the calculation of specific properties. Of course, wherever possible, approximations should be avoided. However, given our inability to solve the N-body problem, approximations are necessary in all but the most formal aspects of the atomic and molecular descriptions of matter.

In this review we comment on the development of some simple and conceptually appealing models that have proved useful in a variety of problems. The general technique employed has come to be known as the method of pseudopotentials. As will be seen, the theoretical approach used is to:

(a) Transform the N-electron Schrödinger equation to a new representation, and

(b) Make approximations suggested by the formal structure of the transformed equation and by physical intuition.

The first step in this process can be carried out exactly and in a wide variety of ways. It is the second step that generates the kind of insights and models so useful in the day-to-day work of chemistry. The more accurate and appealing we can make the models of step (b), the better will be our approximation to a solution of the Schrödinger equation.

Although semi-empirical pseudopotentials (model potentials) were introduced into quantum chemistry many years ago,[1] it is the recent work of Phillips and Kleinman[2] and the subsequent extensive exploitation of pseudopotential methods in solid state physics that has revived interest in this subject. The use of pseudo-potentials in solid state physics has been extensively reviewed elsewhere,[3] and will not be commented on here. Instead we shall, using an author's prerogative, focus attention on those applications of pseudopotential theory which have interested us most. These include the study of electron–atom scattering, the nature of the Rydberg spectra of atoms and molecules and the calculation of the total energy, including correlation energy, of atoms. Through-out we shall emphasize the relationship between model potentials and pseudopotentials, the problems that arise and the directions that future developments may take. In following this programme we start in Sections 2 and 3 with a general description of the one-electron pseudopotential formalism and its N-electron generalization. In Section 4 we report briefly the results of calculations, based on one-electron pseudopotential methods, for the zero-energy elastic scattering cross-section of electrons on helium atoms, and for the Rydberg states of helium and beryllium. Section 5 resumes the theoretical development with a general discussion of one-electron model potentials, and these are applied in Section 6 to the study of the Rydberg states of hydrogen, nitrogen and benzene. In Section 7 we discuss the relations between model potentials and pseudopotentials and the model potentials suggested by this analysis are used in Section 8 in the study of the Rydberg states of Be+ and Mg+.

In Section 9 we examine the pseudopotential formalism as applied to the calculation of the ground state energy of atoms with two valence-electrons, e.g. beryllium and magnesium. The relationship between the exact pseudopotential formalism and the theories of Sinanoğlu and of Szasz is delineated and it is shown how use of the one-electron model potentials can simplify such calculations. In Section 10 the results of calculations of the energies of the beryllium and magnesium atoms are reported and analysed. Section 11 deals with other, more approximate, model potential representations of the two-electron problem. Finally, in Section 12 we discuss the utility of the pseudopotential method of analysis, summarize what we understand about its shortcomings and strengths, and indicate possible directions for future research.

2. SOME GENERAL REMARKS

A wide class of problems in atomic and molecular physics can be reduced to finding the extreme value of the functional

$$\frac{\langle \Phi^0 | \mathscr{H} | \Phi^0 \rangle}{\langle \Phi^0 | \Phi^0 \rangle} \tag{1}$$

where $\mathscr{H}$ is a Hermitian operator and Φ^0 satisfies orthogonality constraints of the form

$$\langle \Phi^0 | \phi_c \rangle = 0 \tag{2}$$

for some set of functions ϕ_c lying within a given subspace of the total function space. Some examples are worth citing: (a) in the scattering of an electron by a closed shell atom, the wave function of the scattered electron must be orthogonal to the occupied atomic orbitals; (b) the wave function of an excited valence-electron in, say a Rydberg level, must be orthogonal to the core orbitals; (c) in the many-electron theory of Sinanoğlu[4] it is shown that the correlation energy of an electron pair can be closely approximated by a functional of the type (1) where Φ^0 is a pair function which is one electron orthogonal to the N-electron Hartree–Fock "sea". The total correlation energy of the N-electron system is then closely approximated by the sum of the separate pair correlation energies. Szasz[5] has developed a many-electron theory along rather similar lines.

Löwdin[6] has published a clear discussion of the general problem of variation in a restricted subspace and has emphasized some of the technical difficulties that arise in the use of (1) and (2). In particular, in example (b) above one must require orthogonality to the exact lower eigenfunctions of $\mathscr{H}$ if the minimum of (1) subject to (2) is to be an upper bound to the valence-state energy.

In addition to the case just cited there are other difficulties encountered when the orthogonality constraint must be imposed, and considerable effort has been expended in the attempt to replace (2) by an equivalent constraint. The most successful work along these lines, first placed on a sound theoretical basis by Phillips and Kleinman,[2] has come to be known as the method of pseudopotentials. Phillips and Kleinman consider the upper "valence" eigenfunctions of an effective one-electron Hamiltonian $\mathscr{H}$ which has N low-lying "core" solutions ϕ_c, $c = 1, ..., N$. The orthogonality of the valence eigenfunction ψ_v to the N core eigen-functions ϕ_c requires that ψ_v have several oscillations in the core region. They show[2] that one can transform the eigenvalue equation

$$\mathcal{H}\psi_v \equiv (T + V)\psi_v = E_v\psi_v \tag{3}$$

where

$$\langle\psi_v|\phi_c\rangle = 0, \quad c = 1, ..., N \tag{4}$$

and

$$\mathcal{H}\phi_c = E_c\phi_c \tag{5}$$

to the form

$$[\mathcal{H} + V_R^{\text{PK}}]\chi_v = E_v\chi_v \tag{6}$$

where V_R^{PK} is a repulsive non-local potential defined by

$$V_R^{\text{PK}}\chi = \sum_{c=1}^{N} \langle\phi_c|\chi\rangle [E_v - E_c]\phi_c \tag{7}$$

and χ_v, the (non-orthogonal) pseudowave function, is of the form

$$\chi_v = \psi_v + \sum_{c=1}^{N} a_c\phi_c \tag{8}$$

for arbitrary choices[7] of the core coefficients a_c. We call the term $(V + V_R)$ the effective potential, and shall refer to the term V_R as the pseudopotential.* Cohen and Heine[7] suggest several physically appealing ways of removing the arbitrariness in the core coefficients; for example, one can choose the coefficients to cancel out the oscillations of ψ_v inside the core, thereby generating a smooth function χ_v.

What Phillips and Kleinman have shown is that one can replace orthogonality constraints in an eigenvalue equation by adding a non-local pseudopotential V_R^{PK} to $\mathcal{H}$ without changing the valence eigenvalues. The main point here is that a conceptual shift has been made, from the study of a function satisfying certain ortho-gonality constraints and an unmodified Hamiltonian, to the study of an unconstrained function satisfying a modified Hamiltonian. The advantage of the new viewpoint is that the pseudowave

---

* The term $V + V_R$ is often called the pseudopotential. It is convenient to have a name for V_R alone and we feel the phrase "effective potential" better describes the physical meaning of $(V + V_R)$.

function can take several physically appealing forms. The physical interpretation[7] that the effect of adding the repulsive potential to the Hamiltonian leaves a weak effective potential $(V + V_R)$ is very suggestive, and this viewpoint has proved extremely useful in many problems.

One should note an important feature of the Phillips–Kleinman pseudopotential introduced in Eq. (6). Consider the lowest valence solution ψ_v. The ϕ_c are all solutions of (6) with eigenvalue E_v; it is this degeneracy which allows for an arbitrary core part in the pseudowave function. Thus, the lowest eigenvalue of (6) is E_v and since V_R^{PK} is Hermitian (for this fixed E_v), we may use an arbitrary trial function χ in the functional

$$\frac{\langle \chi | \mathscr{H} + V_R^{\mathrm{PK}} | \chi \rangle}{\langle \chi | \chi \rangle} \tag{9}$$

without fear of "variational collapse" to the core levels. The extreme value of this functional is the lowest valence eigenvalue E_v. We will discuss these points in more detail later in this paper. To summarize, the Phillips–Kleinman pseudopotential has two important properties:

(a) It allows the use of a non-orthogonal *eigenfunction* in the eigenvalue equation (6).

(b) One can use an arbitrary trial function in (9). The constraints $\langle \chi | \phi_c \rangle = 0$, which would normally have to be imposed for E_v to be the extreme value of this functional, have been formally removed by the addition of the term V_R^{PK} to the Hamiltonian.

In this paper we review two applications of the ideas cited and describe a generalization to the cases where the ϕ_c are not eigenfunctions of $\mathscr{H}$ and we have many-electron wave functions. The generalized pseudopotential formalism has been applied to the calculation of the electronic energy of many-electron atoms, a subject we also describe in Section 9. In general we shall examine the use of the pseudopotential in atomic and molecular problems for which use of the variation principle is the most powerful method of solution. The generalized pseudopotential equations we derive have both properties (a) and (b). It will be shown that there exists a large class of repulsive potentials one can add to the Hamiltonian which satisfy (b), i.e. prevent variational collapse, but do not satisfy (a). These we call variation potentials.

3. THE GENERALIZED PHILLIPS–KLEINMAN PSEUDOPOTENTIAL

Consider a Hermitian operator, $\mathscr{H}$, and a set of orthonormal functions, ϕ_c, which span some subspace (core space) of the function space. We wish to determine the minimum value of the functional

$$\langle \Phi^0 | \mathscr{H} | \Phi^0 \rangle \tag{10}$$

for a function Φ^0 subject to the constraints of normalization

$$\langle \Phi^0 | \Phi^0 \rangle = 1 \tag{11a}$$

and orthogonality to the core functions

$$\langle \Phi^0 | \phi_c \rangle = 0 \tag{11b}$$

We call this minimum value $\bar{E}$.

The constraints defined by (11b) describe a core subspace within which Φ^0 cannot lie. Let the projection operator $\mathscr{P}$ be defined by the requirement that it project from any function that part lying within the core space. In formal terms

$$\mathscr{P} = \sum_c | \phi_c \rangle \langle \phi_c | \tag{12}$$

As usual, $\mathscr{P}$ is Hermitian and idempotent and satisfies the relations

$$\mathscr{P}^2 = \mathscr{P} \tag{13}$$

and

$$\mathscr{P}\phi_c = \phi_c \tag{14a}$$

$$(1 - \mathscr{P})\phi_c = 0 \tag{14b}$$

Consider, now, the function

$$\Phi^0 = (1 - \mathscr{P})\Phi \tag{15}$$

and note that Φ^0 always satisfies the constraints (11b) for arbitrary Φ and arbitrary variations of Φ. We can therefore transform the problem of finding the minimum value of (10) subject to (11a) and (11b) into the equivalent problem of minimizing

$$\langle (1 - \mathscr{P})\Phi | \mathscr{H} | (1 - \mathscr{P})\Phi \rangle \tag{16}$$

subject to the normalization constraint

$$\langle (1 - \mathscr{P})\Phi | (1 - \mathscr{P})\Phi \rangle = 1 \tag{17}$$

Arbitrary variation of Φ^* now leads to the equation

$$\langle \delta\Phi | (1-\mathscr{P})\,\mathscr{H}(1-\mathscr{P})\,|\,\Phi\rangle - E\langle \delta\Phi | (1-\mathscr{P})\,\Phi\rangle = 0 \qquad (18)$$

where E is a Lagrange multiplier. To obtain (18) we have made use of the Hermitian and idempotent properties of $\mathscr{P}$ and its complement $(1-\mathscr{P})$. An equivalent relation describes the results of varying Φ. Since $\delta\Phi$ is arbitrary it is necessary that

$$[(1-\mathscr{P})\,\mathscr{H}(1-\mathscr{P}) - E(1-\mathscr{P})]\,\Phi = 0 \qquad (19)$$

or

$$[\mathscr{H} + V_R^{\mathrm{GPK}}]\,\Phi = E\Phi \qquad (20)$$

where

$$V_R^{\mathrm{GPK}} = -\mathscr{H}\mathscr{P} - \mathscr{P}\mathscr{H} + \mathscr{P}\mathscr{H}\mathscr{P} + E\mathscr{P} \qquad (21)$$

is the generalized Phillips–Kleinman pseudopotential. We note that

$$[\mathscr{H} + V_R^{\mathrm{GPK}} - E]\,\Phi = 0 \qquad (22)$$

may be rewritten in the symmetrical form

$$(1-\mathscr{P})\,[\mathscr{H} - E]\,(1-\mathscr{P})\,\Phi = 0 \qquad (23)$$

The generalized pseudopotential V_R^{GPK} has the following two important properties: V_R^{GPK} is non-local and it depends on the eigenvalue E. We fix E in (21) to be the lowest eigenvalue, denoted $\bar{E}$, of (20). By construction $\bar{E}$ is then the extreme value of (10) subject to the constraints (11). As a practical note we observe that it is often possible to calculate $\bar{E}$ self-consistently, or to evaluate it from experimental data. Now, the pseudowave function Φ may have an arbitrary core component. By construction, the core functions satisfy (20) as an identity, and they are all degenerate with eigenvalue $\bar{E}$. That this is true becomes evident when (20) is rewritten in the form (23) and (14b) is used. Having fixed E in (21) at the value $\bar{E}$, we now note that V_R^{GPK} is Hermitian. We may therefore use an arbitrary trial function in a variational solution of (20) without danger of collapse past the lowest eigenvalue $\bar{E}$.

The fact that V_R^{GPK} is basically a repulsive potential becomes clear when it is realized that the minimum of the constrained functional

$$\frac{\langle \Phi^0 | \mathscr{H} | \Phi^0\rangle}{\langle \Phi^0 | \Phi^0\rangle}$$

must lie above that of the unconstrained functional

$$\frac{\langle \Phi | \mathscr{H} | \Phi \rangle}{\langle \Phi | \Phi \rangle}$$

since in the latter case we consider a trial function of greater generality whose additional freedom can only serve to lower the energy.[6] If we insist that these two functionals have the same minimum value it is necessary to add a repulsive (positive) term to the unconstrained functional to raise the minimum value the required amount. The term added, V_R, is repulsive because of the constrained nature of the variation. To use a figurative phrase, V_R prevents collapse when Φ falls into the core space.

It is possible to show that, with the assumptions originally made, our generalized potential V_R^{GPK} reduces to that given by Phillips and Kleinman, Eq. (7). In this case $\mathscr{H}$ is the Hartree–Fock operator and the ϕ_c are the N lowest eigenfunctions of $\mathscr{H}$. Consider the lowest "valence" solution, $\psi_v \equiv \phi_{N+1}$, with eigenvalue $E = E_v$. We define the projection operator on the cth core electron by $\mathscr{P}_c = |\phi_c\rangle \langle \phi_c|$ and note that

$$\mathscr{P} = \sum_c^{\text{core}} \mathscr{P}_c$$

When the ϕ_c are eigenfunctions of $\mathscr{H}$, each with energy E_c, it is easy to show that $\mathscr{H}$ and $\mathscr{P}$ commute. In this case, using (13), Eq. (21) assumes the form

$$V_R^{\text{GPK}} = (E_v - \mathscr{H}) \mathscr{P}$$

$$= \sum_c^{\text{core}} (E_v - \mathscr{H}) \mathscr{P}_c \qquad (24)$$

For an arbitrary function χ,

$$(V_R^{\text{GPK}} \chi) = \sum_c^{\text{core}} [E_v - \mathscr{H}] \langle \phi_c | \chi \rangle \phi_c$$

$$= \sum_c^{\text{core}} [E_v - E_c] \langle \phi_c | \chi \rangle \phi_c \qquad (25)$$

which is just Eq. (7), the original form given by Phillips and Kleinman.

It is instructive to consider this simple case in more detail. We write

$$[\mathscr{H} + \sum_c (E_v - E_c)\mathscr{P}_c]\chi = E\chi \tag{26}$$

and note that we have fixed E in V_R^{GPK} at a particular valence energy E_v. What is the eigenvalue spectrum of (26)? Clearly, the valence spectrum is exactly that of the original Hamiltonian operator,

$$\mathscr{H}\psi = E\psi \tag{27}$$

since

$$\mathscr{P}_c\psi_v = 0 \tag{28}$$

because ψ_v is an upper solution which is orthogonal to all core solutions. The core solutions ϕ_c, $c = 1, ..., N$, are all eigenfunctions of $(\mathscr{H} + V_R^{\mathrm{GPK}})$ with eigenvalue E_v, and they are also eigenfunctions of V_R^{GPK} since

$$\mathscr{P}_c\phi_c = \phi_c \tag{29}$$

Thus, the effect of V_R^{GPK} in (26) is to shift the eigenvalues of the core eigenfunctions to a particular value, E_v. In general, by adding the projection operators $\mathscr{P}_c$ one can shift the core eigenvalues to any desired point on the energy scale. In our case all the core functions have been shifted such that all the core eigenvalues of $\mathscr{H} + V_R^{\mathrm{GPK}}$ are equal to E_v. This results in a degeneracy with the valence solution ψ_v and allows any linear combination of core solutions to be used along with ψ_v to give an eigenvalue E_v. Using the properties of $\mathscr{P}_c$ displayed in (28) and (29) we may write

$$[\mathscr{H} + \sum_c (\tilde{E}_c - E_c)\mathscr{P}_c]\chi = E\chi \tag{30}$$

where the $\tilde{E}_c$ are any numbers we choose. The ϕ_c are then eigenfunctions of (30) with eigenvalues $\tilde{E}_c$. We can, for example, make $n-1$ of our core functions degenerate with a particular valence level E_v and choose the last core function to be degenerate with some other level E'_v. Or we can shift all the core solutions to some large positive number E'. In this latter case the lowest eigenvalue of (30) is the lowest value of E_v. The lowest eigenfunction is then ψ_v and we no longer have an arbitrary core component of the wave function. However, one may use an arbitrary trial function in a variational solution of (30) with this choice of the $\tilde{E}_c$ without collapsing into the core levels, since the core levels are now moved far up to E'. The function to which the solution will converge in a variation calculation is then the true orthogonal function ψ_v. Thus, for the choice of $\tilde{E}_c$ cited we have what might be called a variation potential. We may now use an arbitrary trial function without danger of collapse into core levels. We do not have a pseudo-potential since the use of a pseudopotential allows the mixture of core states in the eigenfunction χ of (30).

A variation potential can arise quite naturally in a given problem. For example, the arguments of Cohen and Heine[7] might lead one to try an equation of the type

$$[\mathscr{H} - \sum_c \mathscr{P}_c V \mathscr{P}_c] \chi = \tilde{E} \chi \tag{31}$$

where we remove the core projection of the potential. This equation has unchanged valence eigenvalues and eigenfunctions and core solutions with eigenvalue

$$\tilde{E}_c = E_c - \langle \phi_c | V | \phi_c \rangle$$
$$= -E_c \tag{32}$$

if we assume the validity of the virial theorem. Thus the core levels are as high above the zero of energy in the variation-potential equation as they are below in the original equation (27). Solving (31) variationally one will converge to the true lowest valence eigenvalue E_v and the orthogonal valence eigenfunction ψ_v.

In all the above we have been careful to keep the term V_R^{GPK} Hermitian. Essentially the only Hermitian pseudopotential which satisfies both criteria (a) and (b) of Section 2 is the Phillips–Kleinman potential.

Austin, Heine and Sham[8] noted that under the same assumptions made by Phillips and Kleinman one could replace the term V_R^{PK} in (6) by a non-Hermitian operator V_R^A having the symbolic form

$$V_R^A = \sum_c | \phi_c \rangle \langle F_c | \tag{33}$$

where the F_c are arbitrary functions. This operator projects any function onto the core space; the amplitudes of the projections are functions of the arbitrary F's. The Austin–Heine–Sham proof depends explicitly on the non-Hermitian nature of Eq. (33). If one attempts to generate a Hermitian form of (33), for example by setting

$$V_R = \sum_c b_c |\phi_c\rangle \langle \phi_c| \tag{34}$$

then one has the general form of a variation potential as in Eq. (30). For the particular choices of the $b_c = E_v - E_c$ one obtains the Phillips–Kleinman pseudopotential (25), and only this choice permits the use of a smooth pseudowave function in Eq. (6).

In the general non-Hermitian case (arbitrary F_i's) one is no longer guaranteed that there do not exist "core" solutions with energies lower than the lowest eigenvalue. It can be shown that it is possible to collapse past this lowest eigenvalue[9] if one assumes a variation principle of the standard form

$$\frac{\langle \tilde{\chi} | \mathscr{H} + V_R^A | \tilde{\chi} \rangle}{\langle \tilde{\chi} | \tilde{\chi} \rangle} = \text{minimum} \tag{35}$$

Since this is the most convenient form of the variation principle for use in calculations of the properties of atoms and molecules, we do not consider this form of the pseudopotential appropriate for further generalization for our purposes. Indeed, one should keep firmly in mind the fact that Eq. (35) does not apply when using it for one-electron problems. It satisfies property (a) but not property (b) of our set of requirements on generalized pseudopotentials.

4. TWO APPLICATIONS OF THE ONE-ELECTRON PSEUDOPOTENTIAL THEORY

In this Section we review two applications of the one-electron pseudopotential theory described in Sections 2 and 3. Our purpose in reviewing these applications is to provide examples of how the theory may be used. That is, we wish to demonstrate that if the pseudopotential is properly calculated the results obtained are very good. On the other hand, because of our basic interest in concepts and model potentials, the calculations cited are not designed to provide maximum accuracy (both could be refined considerably), but rather to show how the pseudopotential formalism lends itself to easy visualization and physical interpretation, while being of comparable accuracy with the results of more tedious and less instructive approaches.

A. Scattering of an Electron from a Helium Atom in the Limit of Zero Incident Energy

The calculation of the cross-section for low energy electron–atom elastic scattering is of considerable theoretical interest. What features of the problem motivate us to examine a pseudopotential representation of this process? The answer is that even the scattering of an electron from a helium atom requires solution of a three-electron problem. The analysis may be simplified on physical grounds by noting that the scattered electron is, to a large extent, independent of the two tightly bound electrons. But in contrast with electron–ion scattering, there is no obvious simple potential which accurately describes the scattering. There are primarily three sources of difficulty; these arise from the Pauli exclusion principle and the fact that electrons are indistinguishable, the requirement that the wave function of the scattered electron be orthogonal to the orbitals of the atom, and the necessity to account for distortion of the atom (represented primarily by polarization effects). Of course, it is always possible in principle to solve the atom plus electron problem completely, without making the atom-scattered electron separation, by using a large computer. However, exact solutions of the type cited are not known even for small atoms and molecules. In addition to the difficult computational problems which arise in such an approach, there are other reasons for searching for simple methods. It is easier to understand a scattering process when the interaction can be expressed in terms of a potential. This feeling leads to the search for some sort of quasipotential (possibly energy-dependent) with which to describe the scattering. It is obviously desirable for these quasipotentials to have simple forms and to correspond to well-defined model systems. This last point is of considerable importance since, for electron–atom or electron–molecule scattering in larger systems, the present prospects for obtaining "exact" solutions are very small.

Kestner, Jortner, Cohen and Rice[10] have described the low energy elastic scattering[14] of an electron by a helium atom in terms of the potential

$$V_R^S \chi = -\sum_c \phi_c \langle \phi_c | V | \chi \rangle \tag{36}$$

where V is the full electron–atom interaction potential and, as before, ϕ_c is a core orbital wave function and χ the pseudowave function. Equation (36) arises when the Hartree–Fock equation for the electron plus atom system is transformed to the pseudo-potential representation. Until approximations are made the transformed pseudopotential form of the equation is equivalent to the standard Hartree–Fock equation. Now the form (36) is one that minimizes the kinetic energy corresponding to χ.[7] The reader should note that the replacement of the Hartree–Fock equation by the pseudopotential equation has the immediate computational advantage that the calculation of V_R^S is relatively simple. Although the calculation of V_R^S involves a self-consistency condition, since V_R^S is used to determine χ but V_R^S is calculated using χ, the smoother the wave function the more rapid is the convergence to the correct wave function. The pseudopotential is influenced by χ in three ways. Because of the self-consistency conditions implicit in the Hartree–Fock equation (of which the pseudopotential equation is merely a transcription) the core orbitals depend on the valence-electron orbitals or the scattered-electron wave function, as well as on each other. However, a comparison of the wave functions of Na with those of Na$^+$, or of K with K$^+$, shows that the core orbitals are insensitive to the presence of the outer electron, much less to the exact form of the wave function. Thus, $V + V_R^S$ depends on χ explicitly in two places: in the exchange potential and in V_R^S.

Because the wave function obtained in this way is smooth, we can by-pass the complete self-consistent calculation of $V + V_R^S$ by simply assuming a form for χ in the calculation of the potential. The resulting energy or scattering length is then even more insensitive to the exact form of χ. Indeed, since χ resembles a simple plane wave, $V + V_R^S$ is calculated for the case of a plane wave. The total effective Hamiltonian is then used to determine χ. The self-consistency of the calculation can easily be carried further, if that proves necessary.

The effective core potentials for electron–helium atom scattering are constructed from a potential V which is the sum of the nuclear, Coulomb, and exchange potentials, plus the polarization potential arising from the small distortion of the atom by the electronic charge:

$$V = -\frac{2}{r_3} + 2\int \frac{\phi^2(r_1)}{r_{13}}\, d\tau_1 - \int \frac{\phi(r_1)\,\phi(r_3)}{r_{13}}\, d\tau_1\, P_{13} + V_{\text{pol}}(r_3) \qquad (37)$$

where r_3 is the distance of the scattered electron from the nucleus,

r_{13} is the distance between two electrons, ϕ is a Hartree–Fock atomic wave function (very similar to the correct self-consistent field core orbital), P_{13} is an operator which permutes electrons one and three in functions to its right, and V_{pol} is the polarization potential evaluated below.

For the study of low energy electron scattering, it is sufficiently accurate to use the adiabatic approximation to calculate the polarization potential. By this we mean that at each distance r_3 the scattered electron is assumed to be at rest relative to the core electrons of the atom, i.e. the core electrons of the atom adjust to any movement of the scattered electron before appreciable relative displacement occurs. The adiabatic assumption is poorest for very small electron–atom separations, when the electron is[11] "well within the atom" but even in this domain two factors tend to limit the error made. First, the potential itself approaches zero, so that any error cannot be very large. Second, in calculations by Temkin and Lamkin on electron–hydrogen atom scattering using various potentials all of which differ at short distances, similar results are obtained in all cases.[11, 12]

To study the two-electron atom using the one-electron Hartree–Fock scheme, or any transformation of it, it is necessary to make another assumption before core polarization potentials may be used directly. To be specific, it must be assumed that the electrons polarize independently, so that each electron reacts as if it were in the field of an effective nuclear charge, z_{eff}. This approximation can be tested using the calculations of Dalgarno[13] and others. Dalgarno has computed the polarizability both in the uncoupled (independent electrons) and the coupled approximations. The difference between the results is less than 10% for the helium atom.

Kestner, Jortner, Cohen and Rice adopted a form for the polarization potential which partially included the effects of penetration of the core of the atom by the scattered electron. Specifically, the form adopted by them is

$$V_{pol} = V_0 + c_1 V_1 + V_2 \tag{38}$$

$$V_1 = -\frac{4.5}{x^4}[1 - \tfrac{1}{3}\exp(-2x)$$

$$\times (1 + 2x + 6x^2 + \tfrac{20}{3}x^3 + \tfrac{4}{3}x^4) - \tfrac{2}{3}\exp(-4x)(1+x)^4] \tag{39}$$

$$V_2 = -\frac{15}{x^6}[1 - \exp(-2x)$$

$$\times (1 + 2x + 2x^2 + \tfrac{4}{3}x^3 + \tfrac{2}{3}x^4 + \tfrac{4}{15}x^5 + \tfrac{4}{45}x^6 + \tfrac{4}{225}x^7)] \tag{40}$$

$$V_0 = 2\exp(-2x)\left[-\frac{2}{x^2} + \frac{1}{2x} + \frac{5}{2} + x + 2(\ln 2 + c)\left(\frac{1}{x^2} + \frac{1}{x}\right)\right.$$

$$+ Ei(-2x)\left(1 - \frac{1}{x^2}\right) + \ln x\left(-2x + 1 + \frac{2}{x} + \frac{1}{x^2}\right)$$

$$\left. - \alpha\left(\frac{2}{x} + 4 + 4x\right) + \beta(2 + 4x)\right] + \frac{4\alpha}{x} + 2\left(1 - \frac{1}{x}\right)Ei(-2x)$$

$$+ 2\exp(-4x)\left[\frac{2}{x^2} + \frac{7}{2x} + \frac{5}{2} + x - \bar{E}i(2x)\right.$$

$$\left. \times \left(\frac{1}{x^2} + \frac{3}{x} + 4 + 2x\right) + \left(\frac{1}{x^2} + \frac{2}{x} + 1\right)\ln x\right] \tag{41}$$

where

$$\alpha = -\left(1 + \frac{1}{x}\right)\exp(-2x)(\ln 2 + c + \tfrac{1}{2}\ln x) + \frac{1}{2}\left(\frac{1}{x} - 1\right)$$

$$\times Ei(-2x) + \exp(-2x)\left(\frac{1}{4x} + \frac{1}{2} - \frac{1}{2}x\right)$$

$$\beta = \left(1 + \frac{1}{x}\right)\{\exp(-2x)[\tfrac{1}{2}\bar{E}i(2x) - \tfrac{1}{2}\ln x - \ln 2 - c] - \tfrac{1}{4}\}$$

$$+ \frac{1}{2}\left(\frac{1}{x} - 1\right)$$

$$\times \left\{[Ei(-2x) - \ln x] + \exp(-2x)\left(\frac{1}{4x} + \frac{1}{2} - \frac{1}{2}x\right)\right\}$$

$$c = 0.5772\ldots (\text{Euler's constant})$$

$$Ei(-x) = -\int_x^\infty \frac{1}{t}\exp(-t)\,dt, \quad \bar{E}i(x) = \mathscr{P}\int_{-\infty}^x \frac{1}{t}\exp t\,dt$$

and in each case $x = r_3 z_{eff}$, the product of the distance of the scattered electron from the nucleus and the effective nuclear charge. For helium a value of z_{eff} of 1.6875 was used, since this corresponds to the best single Slater orbital approximation to the wave function. It might be argued that z_{eff} should be selected to give the correct asymptotic behaviour of the wave function. However, the asymptotic behaviour of the wave function has little influence on the polarization potential, since that potential varies as r_3^{-4}. Kestner, Jortner, Cohen and Rice therefore forced the polarization potential to be correct at large distances by adjusting the constant C_1. Moreover, all parts of the polarization including the pseudopotential part were treated as though they corresponded to a wave function with effective nuclear charge 1.6875. This approximation can be shown to introduce little error.

It is now possible to construct the total effective potential for scattering of an electron by a helium atom. As Hartree-Fock orbitals in Eqs. (36) and (37), except for the polarization calculation, Kestner and coworkers chose

$$\chi = 0.18159 \, \phi_{1s}(2.906) + 0.84289 \, \phi_{1s}(1.453) \qquad (42)$$

where $\phi_{1s}(z)$ is a normalized Slater 1s function with orbital exponent z. The exchange part of the potential was evaluated in the limit of zero electron energy, assuming that the scattered-electron wave function in the pseudopotential formalism is sufficiently smooth that the spatial variation contributions to the exchange are small. Thus

$$\int \frac{\phi(r_1)\chi(r_1)}{r_{13}} \, d\tau_1 \phi(r_3) \to \int \frac{\phi(r_1)}{r_{13}} \, d\tau_1 \, \phi(r_3)\chi(r_1) \qquad (43)$$

in the limit as $k \to 0$, where k is the magnitude of the wave vector of the scattered electron. There is then obtained the local potential

$$V_A = V - V_{pol} = -6.116632 \frac{1}{r}\exp(-\Lambda r) + 0.535625 \frac{1}{r}\exp(-2\Lambda r)$$

$$+ 3.3831894 \frac{1}{r}\exp(-3\Lambda r) + 0.1978492 \frac{1}{r}\exp(-4\Lambda r)$$

$$+ 2.2064609 \exp(-2\Lambda r) + 2.026883 \exp(-3\Lambda r)$$

$$+ 0.191650 \exp(-4\Lambda r) \qquad (44)$$

with $\Lambda = 1.453$. The local pseudopotential approximation implies that the variation of the pseudowave function over the atomic core region is small.

To complete the potential,

$$V_R^S \chi = -\int \phi V \chi \, d^3 r_1 \, \phi(r_3) \tag{45}$$

must be computed. Again taking the limit $k \to 0$ so as to obtain a local potential, it is found that

$$V_R = -\int \phi(r_1) \, V(r_1) \, d^3 r_1 \, \phi(r_3) \tag{46}$$

Substituting $V = V + V_{\text{pol}}$, then,

$$V_R^S = 5.634646 \exp(-1.453 r_3) + 3.434869 \exp(-2.0906 r_3)$$
$$+ 1.107442 \exp(-1.6875 r_3) \tag{47}$$

The total effective potential is, of course,

$$V_T = V_A + V_{\text{pol}} + V_R^S \tag{48}$$

which is easily obtained from Eqs. (44), (38) and (47). Because of the assumptions made in calculating this local potential, it is valid only near $k = 0$.

Fig. 1 displays the total potential for the electron–helium atom scattering. Note that, overall, the potential is strongly repulsive. The potential does not become negative until 6.7 a.u., and has a well depth of -0.0001109 a.u. at 7.8 a.u. From these observations alone, it is obvious that the polarization interaction is not very important in electron–helium scattering.

The interaction potential (48) is so large that the Born approximation cannot be used to calculate the scattering cross-section. Two alternative procedures may be used to calculate the zero energy cross-section: the integrodifferential equation, which reduces to a one-dimensional equation because of the local potential assumption, may be integrated, or standard variational procedures may be used to determine directly the scattering length. In keeping with the philosophy of Section 2 we examine the variational calculation made by Kestner and coworkers.

Now, given the asymptotic conditions

$$\lim_{r_3 \to \infty} \Phi(r_3) \to \sin(k r_3 + \delta)$$
$$\Phi(0) = 0 \tag{49}$$

and letting αt be the variational approximation to the scattering length,

$$\lim_{k \to 0} \Phi(r_3) \to r_3 - \alpha_t \tag{50}$$

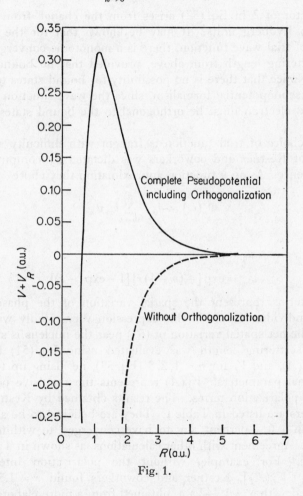

Fig. 1.

The stationary value of the scattering length, here denoted μ, then satisfies the condition

$$\mu \leqslant \alpha_t + \int_0^\infty \Phi(r_3)\, L\Phi(r_3)\, \mathrm{d}^3 r_3 \tag{51}$$

with α_t determined by actually taking the limit (49), and where

$$L = -\left(\frac{d^2}{dr^2}\right) + 2V_T \tag{52}$$

The factor of 2 in Eq. (52) arises from the change from atomic units to Rydberg units. It may be shown that for the correct choice of trial wave function, there is a monotonic convergence to the scattering length from above, provided that no bound states exist. Notice that there is no possibility for bound states to occur in the pseudopotential formalism, since the wave function for the scattered electron must be orthogonal to the bound states of the atom.

The choice of trial function is fraught with difficulty and the choice of Kestner and coworkers was dictated by computational convenience. As an n function approximation they chose

$$\Phi_n(r_3) = u_0 + \sum_{m=1}^{n} b_m u_m \tag{53}$$

where

$$u_0 = r$$

$$u_m = \exp\left[-(m-1)r\right]\left[1 - \exp\left(-r\right)\right] \tag{54}$$

These terms represent the spatial variation of the phase shift. While individual terms in this expression vary rapidly with distance, the net spatial variation of (53) near the nucleus is small.

The scattering length was evaluated using Eq. (51) for V_A, $V_A + V_R^S(A)$, and V_T for $n = 1, 2, 3$ (Eq. 53), i.e. using up to three variational parameters. $V_R^S(A)$ represents the repulsive potential without polarization terms. The results obtained by Kestner and coworkers are listed in Table I. They are believed to be accurate to within a few percent, i.e. to have converged to within a few percent. Agreement with other calculations, as shown in Table II, is good. For example, without the polarization interaction $[V = V_A + V_R^S(A)]$, Kestner and coworkers found $\alpha = 1.502$ a.u. compared with $\alpha = 1.442$ a.u. obtained from a more elaborate and more precise calculation by Moiseiwitsch.[15]

It should now be clear that the use of the pseudopotential representation has the advantage of providing a simple physical picture of the electron–helium atom scattering process. Even in this

TABLE I. Variational Calculations of Scattering Lengths for Electron–Helium Atom Scattering Using the Kohn–Ohmura Procedure

Potential	$\mu(0)$ (a.u.)	$\mu(1)$ (a.u.)	$\mu(2)$ (a.u.)	$\mu(3)$ (a.u.)
1. No pseudopotential or polarization: V_A	-4.887	Undefined		
2. No polarization effects: $V_A + V_R(A)$	3.0425	1.713	1.5056	1.5016
3. Total potential	2.7412	1.473	1.278	1.193

TABLE II. Comparison of Theoretical and Experimental Zero-energy Scattering Lengths

Electron–helium atom scattering

Theory

Moiseiwitsch[a]	1.442	(Static exchange approximation)
LaBahn and Callaway[b]	1.097	
Bauer and Browne[c]	1.21	(Adiabatic exchange approximation)
Williamson and McDowell[d]	1.146	
Kestner, Jortner, Cohen and Rice[e]	1.193	
LaBahn and Callaway[f]	1.186	(Dynamic exchange approximation)

Experiment

Ramsauer and Kollath[g]	1.19
Golden and Bandel[h]	1.15
Frost and Phelps[i]	1.18
Crompton and Jory[j]	1.18

[a] Reference 15.
[b] LaBahn, R. W., and Callaway, J., *Phys. Rev.* **135**, A1539 (1964).
[c] Bauer, E. G., and Browne, H. N., in *Atomic Collision Processes*, North Holland Publ. Co., Amsterdam, 1964, p. 16.
[d] Williamson, J. H., and McDowell, M. R. C., *Proc. Phys. Soc. (London)* **85**, 719 (1965).
[e] Reference 10.
[f] LaBahn, R. W., and Callaway, J., *Phys. Rev.* **147**, 28 (1966).
[g] Ramsauer, C., and Kollath, R., *Ann. Physik.* **3**, 536 (1929); **12**, 529 (1932).
[h] Golden, D. E., and Bandel, H. W., *Phys. Rev.* **138**, A14 (1965).
[i] Frost, L. S., and Phelps, A. V., *Phys. Rev.* **136**, A1538 (1964).
[j] Crompton, R. W., and Jory, R. L., in *IVth International Conference on Physics of Electronic and Atomic Collisions*, Science Bookcrafters, Hastings-on-Hudson, New York, 1965, p. 118.

extreme case, where a combination of exchange, orthogonality and polarization effects is encountered, the calculations of Kestner and coworkers show that the concept of a potential can be used, if care is taken to define the situation in which this effective potential is to be used. Knowledge of the potential also enables us to understand better the qualitative features of the scattering results. It appears that most properties of the potential are given within 10% by this useful approximation.

We feel confident that the methods described can be applied with reasonable accuracy to other electron–rare gas and inert-molecule scattering problems without some of the conversions to parameters often used for the exchange, orthogonalization and polarization effects.

B. Atomic Rydberg States in the One-electron Approximation

As a second application of the one-electron pseudopotential theory of Sections 2 and 3, we consider the calculation of the energies of atomic Rydberg states. We remind the reader that in the simplest one-electron orbital description of the electronic structure of atoms and molecules the electrons are assigned to atomic or molecular orbitals. Those states in which one electron is excited to an orbital large in size relative to the singly charged core, and for which the term values can be expressed in the form

$$T = \mathrm{Ry}\, Z_c^2/(n-\delta)^2 \qquad (55)$$

are identified as Rydberg states.[16] In Eq. (55) Ry is the Rydberg constant, Z_c the charge on the core, n the principal quantum number, and δ the quantum defect. The form of Eq. (55) suggests that the Rydberg states are essentially hydrogenic in character, which in turn implies that the core acts primarily as a unit monopole. Of course, the very existence of the finite quantum defect δ testifies to the existence of deviations from the simple hydrogenic picture. Deviations arise from exchange between the outer electron and the core, from correlation between the instantaneous position of the outer electron and the core electrons, etc. In the case of molecules the field deviates from that characteristic of a monopole, and both the symmetry of the charge distribution and the magnitude of the charges on the several atomic centres are influential in determining the magnitude of δ. Although the electrostatic effects of the core may be approximated by the first few terms in a multipole expansion, it is clear that the Pauli exclusion principle imposes severe constraints on the wave function of the system. Indeed, in the one-electron approximation, it is the orthogonality between the outer orbital and the core orbitals which prevents the collapse of any variational wave function for the outer electron. Just as in the problem described in Section 4A, it is the orthogonality constraint which leads us to seek, in place of the classical methods of analysis, a pseudopotential representation of the Rydberg states of an atom.

Hazi and Rice[17] have developed a description of Rydberg states using an effective Hamiltonian in the one-electron approximation. Because of the conceptual advantages of working with an effective Hamiltonian they introduce a pseudopotential which represents the interaction between the outer electron and the core. They show that the pseudopotential has two parts: the Hartree–Fock potential of the core and the potential which prevents the collapse of a variational wave function for the outer electron. An advantage of a pseudopotential formalism is that it suggests new approximations applicable to polyatomic molecules where a treatment of all electrons is at present impractical.

Hazi and Rice's formalism is contained within the general theory discussed in Section 3. Essentially they consider a variational solution of (20) where $\mathscr{H}$ is the (open-shell) Hartree–Fock operator and the core orbitals are fixed to be those of the ion in the absence of the Rydberg electron. The interested reader is referred to their paper for details of the analysis,[17] which is expressed in somewhat less transparent form that that of Section 3.

The results of Hazi and Rice's calculations on the Rydberg states of helium and beryllium are compared with experiment in Table III. Although they made no attempt to extract the most refined possible numerical answer (e.g. by optimizing orbital exponents) their results compare favourably with other calculations of similar complexity and with experimental observations.

Since we later consider a number of approximations in the theory developed in Section 3 and apply these to the calculation of Rydberg states, it should perhaps be mentioned that, formally speaking, the theory is exact. Hazi and Rice showed by explicit calculations on the Rydberg states of helium and beryllium that the solution of a pseudopotential equation does indeed give the same numerical results as does the original constrained variation problem, provided the exact form of the pseudopotential is used.

TABLE III.   Comparison of Calculated and Experimental Term Values for the ^{3}S States of Helium and Beryllium

He	1s2s	1s3s	1s4s	1s5s	1s6s
Expt.	4.7675[a]	1.8688	0.9934	0.6154	0.4184
Calc.	4.7411	1.8635	0.9914	0.6144	0.4178

Be	1s²2s3s	1s²2s4s	1s²2s5s	1s²2s6s
Expt.	2.8651	1.3247	0.7665	0.4997
Calc.	2.7285	1.2853	0.7496	0.4911

[a] Energy in electron volts (eV).

For basis sets used in variational calculation see reference 17.

Hazi and Rice conclude:

(a) Despite its complex appearance, the role of the pseudo-potential can be simply and accurately described as altering the potential near the ion cores so as to simultaneously account for the deep Coulomb well, the increased kinetic energy of the electron in that well, and the requirements imposed by the Pauli principle.

(b) In the absence of suitable functions, or for the construction of simple analyses, it is often sufficiently accurate to make guesses at the form of the pseudopotential which will reproduce the major features of the complete non-local potential. Condition (a) is, of course, just a restatement of the philosophy leading to the intro-duction of the pseudopotential analysis. However, in so far as it leads to (b), the analysis may be used in new ways.

Consider, for example, the Rydberg states of complex molecules. Now it seems unlikely that useful SCF functions for benzene or larger molecules will become available in the near future. We might, however, ask whether or not a guess at the effective potential of the core might not provide a pseudo-Hamiltonian of sufficient accuracy that the Rydberg states of complex molecules could be consistently classified and ordered. One might assume, for example, that the effective potential of the positive ion core of benzene is adequately approximated by the sum of the pseudopotentials of the carbon (and hydrogen ?) atoms in their appropriate valence state (s.) The pseudo-Hamiltonian then has the same symmetry as the multipole field and simple calculations are possible. The results of such considerations are reported in Section 6 following a survey of the formal implications of the use of model potentials.

5. GENERAL ONE-ELECTRON MODEL POTENTIALS

As is clear from the derivation of Section 3, the use of the exact Phillips–Kleinman pseudopotential, Eq. (21), in a variational solution of Eq. (20) is equivalent to solving the original problem, Eq. (10), with constraints defined by Eq. (11). One does exactly the same work and will get the same rate of convergence. One must also remember that, formally, the projection operators must be defined with exact eigenfunctions if we are to have an upper bound to the lowest valence eigenvalue. With these points in mind, it seems to us that the advantage of a pseudopotential formalism lies not in the formal exact solution but in the physical insights it gives and the models it suggests. It is rather difficult to think of a simple model potential for the wave function of a valence-electron of an atom because of the orthogonality nodes it must have. It is easier to think of a model potential for the pseudowave function where, for example, the arbitrary part is chosen to remove the inner nodes of the valence orbital. The pseudopotential equation gives a formal representation of the kind of equation a particular pseudowave function must satisfy. It is then feasible to consider replacement of the exact pseudopotential by some kind of model potential suggested from the general formalism.

From a general point of view, the model potential formalism as developed by Heine and coworkers[18, 19] lies outside the pseudopotential framework. Abarenkov and Heine[18] define a model potential over a specified energy range of interest as one giving the same eigenvalues or phase shifts, modulo π, as does the real potential. Thus a model potential equation, like a pseudopotential equation, is designed to give the eigenvalues correctly. The model potential is more general than the pseudopotential because the form of the model potential has not yet been specified and there is no necessary requirement that the model wave function take the form of a pseudowave function given in (8). An essential feature of the use of a pseudopotential or model potential is the replacement of $\mathscr{H}$, which can have low-lying core solutions, by $(\mathscr{H} + V_R)$ or by some model Hamiltonian $\mathscr{H}_M$ whose lowest eigenvalue is equal to the energy of the lowest valence state one wishes to consider. By construction, then, one is safe from variational collapse past the desired energy.

In this Section we examine the criteria with which it is possible to define a model potential for an atom or molecule and describe several applications of the formalism to the calculation of molecular Rydberg states. In molecules the quantum defect (see Eq. 55) is determined mainly by three factors: the symmetry of the charge distribution in the core, the penetration of the Rydberg electron into the core, and the restrictions imposed by the Pauli exclusion principle on the wave function of the Rydberg electron. Clearly, if the model potential is to yield the correct eigenvalues it must properly account for the major factors governing the magnitude of δ. Accordingly, the model potential should have the symmetry defined by the molecular core and in some way should account for the penetration and exclusion effects inside the core. We now have the following guidelines for the selection of suitable potentials:

(a) asymptotic behaviour at large r: $-Z_c/r$,

(b) correct symmetry,

(c) penetration and exclusion effects inside the core.

It is relatively easy to find potentials satisfying criteria (a) and (b), but criterion (c) needs a more careful examination.

The reader should realize that the model potential V_M is not an accurate approximation to the true potential $V + V_R$ everywhere in space. In particular, the true potential is extremely complicated inside the core, and it is hard to find a simple, analytic representation of $V + V_R$ in this region. Instead, we look for a potential V_M which is a good approximation to $V + V_R$ outside the core, but which is a fictitious potential inside the core. The model potential eigenvalue equation, nevertheless, gives the correct eigenvalues provided the potential V_M imposes the proper boundary conditions on the wave function. Criterion (a) already ensures the correct asymptotic behaviour of the wave function at infinity, so that the fictitious potential inside the core need only impose the proper boundary condition at the imaginary surface separating the two regions under consideration. In other words, we can think of the potential V_M as a potential which has no physical significance inside the core, but which to a good approximation imposes the same boundary conditions as does the true potential.[20] Similarly, the resulting wavefunction χ is a good approximation to the Rydberg orbital only outside the core and one should not attach any physical significance to the behaviour of χ in the core region.

We can now proceed to consider criterion (c) more explicitly. It is well known[16] that a Rydberg orbital with several precursors (occupied core orbitals of the same symmetry as the Rydberg orbital) is more penetrating in general than one with fewer or no precursors. This behaviour is related to the fact that the inner loops of a particular Rydberg orbital almost completely recapitulate the loops of its precursors, whereas this same orbital is orthogonal to all non-precursors due to the symmetry of the states. Accordingly, the model potential and the boundary conditions it must impose strongly depend on the symmetry of the Rydberg series under consideration. Furthermore, as a result of the exchange terms in the potential $V + V_R$, Rydberg states of the same spatial symmetry with different spin angular momentum have different energies. This behaviour implies that the potential V_M must depend not only on the spatial symmetry, but also on the spin state of the Rydberg state. All of the above conclusions can be summarized by saying that inside the core the model potential is quite non-local and can be written in the general form[18]

$$V_s(r) = \sum_\Omega V_\Omega(r) P_\Omega \qquad (56)$$

In Eq. (56) the subscript s labels the potential applicable to a particular spin state and the summation runs over all the possible symmetries Ω of the Rydberg orbital. The operator P_Ω projects out from an arbitrary function the component with spatial symmetry Ω. The explicit form of $V_\Omega(r)$ is still not specified, but the previous investigation by Abarenkov and Heine of model potentials for atomic ions[18] has shown that the functional form of $V_\Omega(r)$ is not too important as long as $V_\Omega(r)$ contains one or more adjustable parameters. Indeed, Abarenkov and Heine obtained the best results with a potential $V_\Omega(r) =$ constant inside the spherically symmetrical core. Thus they suggest a model Hamiltonian for an atomic ion of the form

$$H_M = -\tfrac{1}{2}\nabla^2 + A_\Omega; \quad r < R_\Omega$$
$$H_M = -\tfrac{1}{2}\nabla^2 - Z/r; \quad r \geqslant R_\Omega \tag{57}$$

where R_Ω and A_Ω are adjustable constants for the particular symmetry Ω.

Let us consider now the Rydberg states of a homonuclear diatomic molecule, for simplicity H_2. We assume that outside the core region the Rydberg electron moves in the potential

$$V_{\text{out}} = \tfrac{1}{2}(-1/r_A - 1/r_B),$$

where $r_A(r_B)$ is the distance of the electron from nucleus $A(B)$. Clearly, V_{out} satisfies criteria (a) and (b), i.e. V_{out} behaves as $-1/r$ for large r and it has the correct cylindrical symmetry. Next we must choose the boundary defining the core region. An examination of the electron density map of H_2^+ for $R_{AB} = 2.0$ a.u. shows that the constant density surfaces are very nearly spherical for $r \geqslant 2.0$ a.u. Accordingly, we define the core region by a sphere centred on the molecular midpoint with a radius of 2.0–2.5 a.u. In view of the arguments given above, we take the potential inside the core to be constant. We can write potential V_1 explicitly in the form

$$V_1 = \frac{1}{2}\left(-\frac{1}{r_A} - \frac{1}{r_B}\right); \quad r \geqslant R_0 \tag{58a}$$

$$V_1 = A_\Omega^s; \qquad\qquad r < R_0 \tag{58b}$$

where r is the distance from the molecular midpoint, R_0 is the radius of the spherical core, and the constant A_Ω^s depends on the spin state and spatial symmetry of the Rydberg state under consideration. The magnitude of A_Ω^s is determined using the prescription of Abarenkov and Heine:[18] for a given R_0, A^s is selected such that the binding energy computed with V_1 for a particular member of the Rydberg series agrees with the experimental binding energy. Now, if the computed energies of all the other states of this same series agree with the experimental values, the model potential is considered to provide a good representation of the boundary conditions applicable to this series.

One possible objection to V_1 is that the boundary defining the core region was selected in view of the nearly spherical electron density surfaces of H_2^+. One can argue that H_2 is a special case in this respect, and the constant electron density surfaces of most complex molecular ions are far from spherical. Thus, although V_1 may give a good description of the Rydberg states of H_2, its applicability to other diatomic and polyatomic molecules is questionable.

To overcome the objection raised above we now propose a second model potential, which, in the case of H_2, is identical to V_1 outside the core, but which defines a more complicated core region. To facilitate the application of the potential to complex molecules we assume that the molecular potential is a sum of "atomic" contributions, i.e.

$$V_2 = \sum_i^{\text{atoms}} \mathscr{V}_i \qquad (59)$$

By an "atomic" contribution $\mathscr{V}_i$ we mean a potential centred on one of the constituent atoms, and which depends on both the partial charge[21] and the valence state of this atom in the particular molecular environment. We assume that each atom i has its own core region, which is taken to be a sphere of radius R_i^0 centred on the ith nucleus. We can write $\mathscr{V}_i$ explicitly in the form

$$\mathscr{V}_i = -\delta Z_i/r_i; \quad r_i \geqslant R_i^0 \qquad (60a)$$

$$\mathscr{V}_i = A_i; \qquad r_i < R_i^0 \qquad (60b)$$

where δZ_i[21] is the partial charge associated with atom i and r_i is the distance from the ith nucleus. The constants A_i are again to be determined empirically. The reader should note that if the molecular core has net charge Z_c then

$$Z_c = \sum_i^{\text{atoms}} \delta Z_i \qquad (61)$$

It is clear that the model potential defined by Eqs. (59) and (60) satisfies criteria (a) and (b) discussed above. Equations (60a) and (61) imply that V_2 behaves as $-Z_c/r$ for large r. Equation (59) and the proper choice of the Z_i ensure that V_2 has the correct spatial symmetry. As far as criterion (c) is concerned, we can say only that both the boundary defining the total molecular core region and V_2 inside this boundary are rather complicated, though certainly adjustable through the proper choice of the parameters R_i^0 and A_i. Finally, the reader should note that unlike V_1, V_2 is not constant everywhere inside the molecular core.[22]

6. CALCULATION OF THE ENERGIES OF MOLECULAR RYDBERG STATES USING MODEL POTENTIALS

As one of several applications of the molecular model potential formalism, Hazi and Rice[23] calculated the term values of the $(1\sigma_g, n\pi_u)\,^3\Pi_u$ states of H_2 at an internuclear separation of 2.0 a.u. Potential V_1 was explicitly selected for the core in the Rydberg states of H_2 and is given by Eq. (58). Hazi and Rice also tested potential V_2. In the case of a neutral homonuclear diatomic molecule the core has a single positive charge, and each of the atoms in the core carries a partial charge of $\delta Z = \frac{1}{2}$. Accordingly, V_2 can be written in the form

$$\mathscr{V}_2 = \mathscr{V}_A + \mathscr{V}_B \tag{62a}$$

$$\mathscr{V}_A = -1/2r_A; \quad r_A \geqslant R^0_A \tag{62b}$$

$$\mathscr{V}_A = A_A; \qquad r_A < R^0_A \tag{62c}$$

The equation defining $\mathscr{V}_B$ can be obtained from Eqs. (62b) and (62c) by exchanging subscripts A and B, and, of course, $R^0_A = R^0_B$. The potentials V_1 and V_2 were inserted into the effective one-electron Hamiltonian and the resulting eigenvalue equation was solved by both a variational procedure and by using first order perturbation theory. Details of the calculation may be found in the paper of Hazi and Rice.[23]

To further test the model potential form (62), Hazi and Rice studied the Rydberg states of nitrogen and of benzene. The term values of the $(^2\Sigma_g, np\sigma_u)\,^1\Sigma_u$ states of nitrogen were computed for an internuclear separation of 2.113 a.u. The explicit form of V_2 applicable to the $^1\Sigma_u$ states of nitrogen is given by Eq. (62). The effective one-electron eigenvalue equation containing V_2 was again solved using a variational procedure. First order perturbation theory (with the hydrogen atom as the zero order problem) is not applicable here, because the quantum defect of the $^1\Sigma_u$ states of nitrogen is too large ($\simeq 0.7$).

In the calculation of the term values of the R and R' Rydberg states of benzene[24] using model potential V_2, Hazi and Rice neglected the effect of the hydrogen atoms, i.e. they assumed that the benzene molecule-ion consists of six carbon atoms, each of which occupies a corner of a hexagon with sides equal to 2.636 a.u. and carries a partial charge of $\delta Z = \frac{1}{6}$. The potential applicable to the

R and R' states can be written in the form

$$V_2 = \sum_{i=1}^{6} (\mathscr{V}_i^{(R)} P_R + \mathscr{V}_i^{(R')} P_{R'}) \tag{63a}$$

$$\mathscr{V}_i^{(R)} = -1/6r_i; \quad r_i \geqslant R_0^{(R)} \tag{63b}$$

$$\mathscr{V}_i^{(R)} = A_i^{(R)}; \quad r_i < R_0^{(R)} \tag{63c}$$

where the carbon atoms are labelled 1 to 6 and r_i is the distance from the ith carbon atom. The equations defining $\mathscr{V}_i^{(R')}$ can be obtained from Eqs. (63b) and (63c) by exchanging superscripts R and R'. In Eq. (63a) P_R and $P_{R'}$ are the projection operators defined in Eq. (56). The constants $A_i^{(R)}$ and $R_0^{(R)}$ are the adjustable parameters, which, in general, are different for the two sets of states.

In the orbital approximation one can form five different sets of Rydberg states (symmetries E_{1g}, E_{1u}, A_{1u}, A_{2u} and E_{2u} in D_{6h}) by exciting an electron from the highest filled π orbital (e_{1g}) into an s or a p atomic-like Rydberg orbital. The p_0 and $p\pm$ Rydberg orbitals give rise to states with symmetries E_{1u} and A_{2u}, respectively, to which electric dipole transitions from the ground state $(^1A_{1g})$ are allowed. Electric dipole transitions to the other two states (A_{1u} and E_{2u}) arising from the (core, $p\pm$) configuration and to the E_{1g} (core, s) states are forbidden. Because the experimentally observed transitions to the R and R' states appear to be both spin and symmetry allowed,[25] Hazi and Rice assumed that these two states can be assigned as either $^1E_{1u}$ or $^1A_{2u}$. Of course, there are two possible assignments if the R and R' states have different symmetries, and they tried to determine the more consistent of these two assignments.

An examination of the results presented in Tables IV, V, and VI indicates that, in general, one can reproduce the experimental term values of molecular Rydberg states using simple model potentials with adjustable parameters as the effective one-electron core potential. Furthermore, to a good approximation the adjustable parameters depend only on the spin state and spatial symmetry, but not on the principal quantum number, n, of the states under consideration. The numerical results obtained provide quantitative confirmation of the qualitative arguments used to derive appropriate model potentials. In other words, these arguments seem essentially correct.

TABLE IV. Term Values of $^3\Pi_u$ States of H_2, $R_{AB} = 2.0$ a.u.

	Experi-(ment[a])	Model potential 1 $R_0 = 2.0$ (a.u.) $A = -0.85$ (a.u.)		Model potential 2 $R_A^0 = 1.5$ (a.u.) $A_A = -0.85$ (a.u.)	
		Perturbation	Variational[b]	Perturbation	Variational[b]
$2p\pi$[c]	3.689[d]	3.555	3.589	3.520	3.566
$3p\pi$	1.588	1.566	1.573	1.558	1.571
$4p\pi$	0.885	0.874	0.877	0.871	0.877
$5p\pi$	0.565	0.556	0.558	0.555	0.558
$6p\pi$	0.388	0.385	0.386	0.384	0.387
$4f\pi$	—	0.852	0.852	0.852	0.852

[a] Matsen, F. A., and Browne, J. C., in *Molecular Orbitals in Chemistry, Physics, and Biology*, P. O. Löwdin, and B. Pullman, Eds. Academic Press, New York, 1964, p. 151.

[b] For basis sets see reference 23.

[c] UAO symbols designating various members of the Rydberg series.

[d] Term values in electron volts (eV).

TABLE V. Term values of $^1\Sigma_u$ States[a] of N_2, $R_{AB} = 2.113$ a.u.

	Model potential parameters $R_A^0 = 2.5$ a.u., $A_A = 0.125$ a.u.	
	Experiment[b]	Theory
$3p\sigma$[c]	2.648[d]	2.567[e]
$4p\sigma$	1.253	1.223
$5p\sigma$	0.735	0.720
$6p\sigma$	0.481	0.475

[a] Configuration: $(1\sigma_g)^2 (1\sigma_u)^2 (2\sigma_g)^2 (2\sigma_u)^2 (1\pi_u)^4 (3\sigma_g) (n\sigma_u)$.

[b] Ogawa, M., and Tanaka, Y., *Can. J. Phys.* **40**, 1593 (1962).

[c] UAO symbols designating various members of the Rydberg series.

[d] Term values in electron volts (eV).

[e] For basis set used see reference 23.

TABLE VI. Term values of the R and R' States[a] of Benzene Computed with Model Potential 2

	Parameters			States[a]			
Assumed symmetry	R_0 (a.u.)	A (a.u.)	$n = 2$	$n = 3$	$n = 4$	$n = 5$
R series							
$^1E_{1u}$	2.5	0.45	2.154[b, c]	1.086	0.657	0.441
$^1A_{2u}$	2.0	0.40	2.198	1.090	0.657	0.440
Experiment[d]			2.319	1.100	0.660	0.442
R' series							
$^1E_{1u}$	4.0	0.85	1.660	0.884	0.556	0.383
$^1A_{2u}$	4.0	0.29	1.672	0.885	0.555	0.382
Experiment[d]			1.834	0.875	0.562	0.393

[a] Designation used in reference 25.
[b] Term values in electron volts (eV).
[c] For basis sets used see reference 23.
[d] Reference 25.

In particular, the good agreement between the theory and experiment obtained in the case of the $^3\Pi_u$ states of H_2 using either V_1 or V_2 shows that the computed term values are not sensitive to the exact form of the potential, provided it has the correct asymptotic behaviour and in some way imposes the proper boundary conditions on the model wave function χ. The inclusion of adjustable parameters in the model potential allows one to meet this latter condition. The good agreement between theory and experiment obtained with V_2 establishes the utility of the procedure in which the molecular potential is built from a sum of atomic contributions. Such a potential has the advantage that it not only gives as accurate term values as does a potential specifically selected for a particular molecule (e.g. potential V_1), but it can also be easily applied to almost any polyatomic molecule.

It is clear that once the appropriate potential is defined the solution of the resulting one-electron eigenvalue equation is not difficult. At least in the case of the potentials considered by Hazi and Rice, adequate approximate solutions can be found by expanding the model function in basis sets of Slater-type orbitals and minimizing the energy with respect to the expansion coefficients. The good agreement between the term values of the $^3\Pi_u$ states of H_2 obtained variationally and those computed using first order perturbation theory indicates that reasonable sets of Slater-type orbitals form adequate bases for the model functions associated with potentials V_1 and V_2. Furthermore, for a given potential the computed term values are found to be insensitive to small variations in the orbital exponents of the basis Slater-type orbitals.

The numerical results obtained for the Rydberg states of benzene (Table VI) dramatically show the limitations of model potentials containing adjustable parameters. In this case the assignment of the symmetries of the experimentally observed states has not been definitely established. Although Hazi and Rice were able to fit the experimental term values quite accurately using model potential V_2, they could not determine the symmetry assignment of the states on *a priori* grounds. As Table VI shows, they obtained equally good results, assuming two different assignments and then determining the appropriate potential for each by adjusting the parameters in the potential. General experience indicates that the use of model potentials with adjustable parameters is limited to problems where the adjustable parameters can be determined using experimental quantities *independent* of the quantities to be calculated with the resulting potential. We discuss in later Sections results of such a use of model potentials.

An alternative approach is the use of model potentials that contain *no* adjustable parameters. Now the question immediately arises, how does one determine the potential applicable to a particular problem, e.g. the Rydberg states of a polyatomic molecule? Intuitively, one would like to retain the idea of the "sum of atomic potentials" because, at least at the level of the Hartree–Fock approximation, accurate wave functions of atoms, but not of large polyatomic molecules, are at present available. It is suggested that in certain problems the effective one-electron molecular potential may be approximated by a sum of atomic potentials, i.e. the molecular Hartree–Fock charge density may be adequately approximated by the superposition of undistorted atomic Hartree–Fock charge distributions. Hazi and Rice have made some preliminary calculations of the term values of the $^3\Pi_u$ Rydberg states of H_2 assuming that the electron density of H_2^+, $|\psi_{core}|^2$, can be approximated by $\frac{1}{2}(|\phi_A|^2+|\phi_B|^2)$, where ϕ_A and ϕ_B are 1s hydrogen atom wave functions centred on nuclei A and B. The accuracy obtained with this approximation, using no adjustable parameters, is approximately the same as that obtained with V_1 and V_2, although the discrepancy between the computed and experimental values has opposite signs in the two cases. For example, in the case of 3p state $E_{exp} = -1.588$ eV; $E(V_1) = -1.573$ eV, $\Delta = -0.015$ eV; $E(V_{atomic}) = -1.608$ eV, $\Delta = +0.020$ eV.

7. MODEL PSEUDOPOTENTIALS: MODEL POTENTIALS SUGGESTED FROM PSEUDOPOTENTIAL THEORY

In Section 5 we discussed model potentials from a general point of view and stressed the fact that the model potential is not restricted to be some local approximation to the effective potential $(V+V_R)$. The particular form the model potential may take and the justification for its use lie outside pseudopotential theory.

Of course any local approximation of the exact pseudopotential is a possible model potential. A commonly used local form arises by choosing some smooth (nodeless) χ^s and forming

$$V_L = \frac{(V_R\chi^s)}{\chi^s} \tag{64}$$

However, as Abarenkov and Heine[18] point out, there is a complementarity effect here for bound state problems—the use of a smooth χ^s produces a V_L which has several oscillations in it.

This can readily be seen when one recalls that χ^s may be written in the form

$$\chi^s = \psi_v + \sum_c a_c^s \phi_c \tag{8}$$

where the coefficients a_c^s are chosen to cancel out the oscillations in ψ_v. Then $\sum_c a_c^s \phi_c$, the core part of χ^s, must itself be oscillating since it is chosen to smooth out the oscillations in ψ_v. However, using the Phillips–Kleinman pseudopotential, Eq. (7), one sees that for this χ^s:

$$(V_R\chi^s) = \sum_c (E_v - E_c) a_c^s \phi_c \tag{65}$$

which is essentially the oscillating core part of χ. Thus, choosing the denominator χ in (64) to be smooth produces an oscillating numerator, and vice versa. These oscillations arise solely from the local approximation to the true V_R one has taken, and have no real physical meaning.

The oscillation effect cited can be seen very clearly in the recent results of Szasz and McGinn.[27] They give "self-consistent" solutions of the Hartree–Fock equations with a Phillips–Kleinman pseudopotential for a number of ions. The pseudowave functions they find are fairly smooth but the local pseudopotentials of the form (64) they give have numerous rapidly varying oscillations. One should also recall that the use of the exact Phillips–Kleinman pseudopotential permits an arbitrary core part in the pseudowave function. Thus the functions and potentials to which the Szasz and McGinn numerical procedure converged are not unique, but instead are part of a large class of mathematically acceptable solutions. We feel, following Cohen and Heine,[7] that one should let physical insight, rather than the arbitrary choice of the computer, determine which solution one considers.

The point we wish to stress here is that simple and physically appealing model potentials can reproduce many of the advantages of the exact pseudopotential formalism, while allowing a great simplification in computational effort. The formal pseudopotential equations (20) or (6) are probably more difficult to solve exactly than the original unmodified equations. The use of model potentials simplifies computations considerably and the freedom to use experimental data to calibrate such potentials allows one to take into account, in an approximate way, effects not considered in the formal theory.[18]

Although model potentials can be justified from an independent point of view, it is useful to consider more carefully the possible relation between the model potential and the pseudopotential. Can one use pseudopotential theory to suggest the form of a model potential and to gain additional physical insight into the nature of the solution of a particular model? In this Section we discuss model potentials from this point of view.

We consider the special case of a single valence-electron outside a closed atomic core. Let the self-consistent field equation for the wave function of the single valence-electron moving outside the closed N electron core be

$$H^{\text{ion}}\psi_v = E_v\psi_v \tag{66}$$

$\psi_v \equiv \psi_{N+1}$ is orthogonal to the first N "core-like" solutions of H^{ion}; this is true even in Roothaan's open shell Hartree–Fock formalism.[38] If we neglect polarization of the core by the valence-electron then both the Hartree–Fock core orbitals ϕ_c and ψ_v are eigenfunctions of the same Hamiltonian H^{ion}, and hence are orthogonal to each other. From the general results cited earlier we may replace the orthogonality constraints on ψ_v with the pseudopotential equation

$$(H^{\text{ion}} + V_R^{\text{ion}})\chi = E_v\chi \tag{67a}$$

or

$$[(1-P)H^{\text{ion}}(1-P) + E_v P]\chi = E_v\chi \tag{67b}$$

where P is the projection operator defined over "core-like" solutions or, ignoring polarization, core Hartree–Fock orbitals. The pseudowave function χ is of the form

$$\chi = \psi_v + \sum_c a_c\phi_c \tag{68}$$

for arbitrary a_c since $P\psi_v = 0$ and $P\phi_c = \phi_c$.

We now make the physically suggestive approximation of "core-valence non-penetrability". With this we assume that there exists a radius R_c such that all the core solutions ϕ_c are essentially zero outside this radius and the valence function ψ_v is essentially zero inside. This is simply the shell model, which we adopt as exact. Naturally the assumption that the valence function does not penetrate the core is the weak point here—indeed, it is this very penetration showing up in orthogonality oscillations that leads us to desire a nodeless pseudowave function. Nonetheless, this approximation contains the essential physics of the situation. As we will mention later, there exist for beryllium and magnesium radii such that the core overlap outside (i.e. integrated from R_c to ∞) is less than 0.01 and the valence overlap is about 0.9, so our approximation is good to about 90%.

Making these assumptions, then,

$$\phi_c = 0, \quad r > R_c; \quad \psi_v = 0, \quad r < R_c \tag{69}$$

and assuming completeness of the eigenfunctions of H^{ion}, we see that the projection operator P, which divides function space into two regions, also divides real space into a core region (0 to R_c) and a valence region (R_c to ∞).

Consider now the pseudopotential equation (67) where χ is of the form (68) under these assumptions. For $r < R_c$ we have only the arbitrary core part of χ non-zero by assumption and for this the effective Hamiltonian is just the constant E_v [see Eq. (67)]. For $R > R_c$ the effective operator on χ is just H^{ion}; here we have only valence functions and these are untouched by $(1 - P)$. For simplicity, since we consider H^{ion} only outside R_c we take its asymptotic form

$$H^{ion} \simeq -\tfrac{1}{2}\nabla^2 - \frac{Z}{r}; \quad Z = M - N \qquad (70)$$

where Z is the net charge of the nucleus surrounded by the N electron core.

These arguments suggest to us a model Hamiltonian, H_m, in which $(H^{ion} + V_R)$ is replaced by

$$\begin{aligned} H_m &= E_v; & r < R_c \\ H_m &= -\tfrac{1}{2}\nabla^2 - Z/r; & r \geqslant R_c \end{aligned} \qquad (71)$$

An amusing feature of this model is the existence of degenerate "core" solutions, just as in the actual Phillips–Kleinman pseudo-potential. Functions which have essentially all their density inside the core clearly satisfy

$$\langle \chi | H_m | \chi \rangle = E_v \qquad (72)$$

identically and represent the "core" solutions of Eqs. (71). To remove this degeneracy and obtain a well-defined model potential we consider the "smoothest" function, i.e. that combination of the core and valence functions which minimizes the kinetic energy. For this function we can extend the kinetic energy integration into the core with little effect on Eq. (72). Thus we obtain a "smoothest function" model potential where

$$\begin{aligned} H_m^s &= -\tfrac{1}{2}\nabla^2 + E_v; & r < R_c \\ H_m^s &= -\tfrac{1}{2}\nabla^2 - Z/r; & r \geqslant R_c \end{aligned} \qquad (73)$$

We have at last alighted on the Abarenkov and Heine model potential [Eq. (57)], and these considerations suggest that the depth of the well should be about equal to the lowest valence-state energy

E_v, and that R_c should be chosen to be near the core radius, say the region within which 95% of the core electron density is confined. In addition to suggesting reasonable values for the parameters, the pseudopotential formalism gives hope that the solutions χ_m of this equation might resemble a "smooth" pseudowave function, i.e. that to a good approximation

$$\chi_m \simeq \psi_v + \sum_c a_c \phi_c \tag{74}$$

for some choice of the core coefficents a_c.

It is clear that since this model potential has two adjustable parameters (R_c and the depth A) there is a wide range of variation of each over which the same lowest eigenvalue E_v can be found. The above considerations provide a motive coupled with physical intuition for the choice of these parameters.

Of course the exact solution of any model potential can be approximated in the form (74), where we look on the right-hand side as the expansion of χ in the first members of the complete set of eigenfunctions of H^{ion}. The accuracy of (74) depends on the particular form of the model potential chosen. We suggest that there are smooth, physically appealing model potentials which have solutions quite accurately approximated by (74). These might be called model pseudopotentials. To demand that a model have solutions exactly in the form (74) then requires large, non-physical oscillations in the model potential such as occur in the local approximation to a pseudopotential (64). Since the use of any model potential inevitably involves some approximation, we feel it is best to relax slightly (74) and thus gain a simple model which may be visualized physically. "Exact" pseudopotentials expressed in the form (64) are themselves local *approximations* to the correct non-local pseudopotential. Their form, and thus their physical interpretation, depends explicitly on the particular pseudowave function chosen. The large oscillations produced by the local approximation (64) tend to destroy the physical picture involved and may well produce numerical difficulties. We feel that formal pseudopotential theory is primarily a conceptual device—useful for gaining physical insight into a problem and a deeper understanding of various models proposed—rather than a practical computational method.

8. RYDBERG STATES OF MODEL PSEUDOPOTENTIALS FOR Be⁺ AND Mg⁺

Weeks and Rice[9] obtained approximate variational solutions of the model equation (57) for Be^+ and Mg^+ by expansion of a trial function in a finite basis set. For the potentials for s states they set the depth of the well parameter A_s equal to the lowest valence energy level E_{vs}, in accordance with the physical ideas discussed above, and varied R_c until the lowest eigenvalue of the secular equation agreed with the experimental energy E_{vs}. Several different basis sets were used to try to find the limiting value of R_c of the exact solution of Eq. (57). It is, of course, important to fit the parameters in a model potential independent of the particular basis set used to approximate the solution. This has not always been carried through with model potentials proposed in the literature.[28] We feel that these variationally determined parameters are about as accurate as those that would be found by an exact solution, following Abarenkov and Heine,[18] because of the interpolations in the existing tables of Coulomb wave functions their method requires. Having thus fixed the parameters of the model potential, one may obtain upper bounds to higher eigenvalues of the model equation by taking the higher eigenvalues of the secular equation.

The p model potential parameters were fitted in a fashion similar to that described. Following Abarenkov and Heine's[18] suggestion, Weeks and Rice kept R_c fixed at the value determined from the s potential and varied the well depth A_p till the experimental E_{vp} value was obtained as the lowest eigenvalue of the secular equation.

The parameters A and R_c determined in this way are given in Table VII for Be^+ and Mg^+ for the s and p model potentials. The R_c's determined in this way have the property of achieving good valence–core separateness, consistent with the physical assumptions discussed above, as will be shown later in Table IX.

A further test of the model potentials is in the Rydberg spectrum they predict. Table VIII gives the predicted s Rydberg series for Be^+ and Mg^+. Details of this calculation are given in the paper by Weeks and Rice.[9] Clearly, the calculations give a good representation of the Rydberg series. It is, of course, not surprising that with a model potential having the correct asymptotic form one can predict a Rydberg series. By adjusting the parameters in the

TABLE VII. Model Potential Parameters for Be$^+$ and Mg$^+$

	Symmetry	Model radius R_0	Depth A_l
Be$^+$	s	1.305[a]	−0.66928[b]
	p	1.305	−2.28[c]
Mg$^+$	s	1.73	−0.55255[b]
	p	1.73	−0.75[c]

[a] All parameters are in atomic units.

[b] Experimental valence energy.

[c] Note the much deeper well for the Be$^+$ p model potential than for the Mg$^+$ model potential. This can be understood when one remembers that Be$^+$ has no p core functions while Mg$^+$ does. Thus there is no positive pseudopotential for Be$^+$ to cancel out the true negative potential.

TABLE VIII. Comparison of Calculated and Experimental Term Values for the Rydberg States of Be$^+$ and Mg$^+$

Be$^+$	1s^{2}2s	1s^{2}3s	1s^{2}4s	1s^{2}5s	1s^{2}6s
Expt.	18.21[a]	7.27	3.90	2.42	1.66
Calc.[b]	18.09	7.32	3.93	2.45	1.66
Mg$^+$	KL3s	KL4s	KL5s	KL6s	KL7s
Expt.	15.03	6.38	3.53	2.24	1.55
Calc.	14.97[c]	6.51	3.61	2.28	1.51

[a] Term values in electron volts (eV).

[b] Model potential parameters given in Table VII. See reference 9 for basis sets used in variational calculation.

[c] The one-parameter quantum defect model (δ fitted to the Mg$^+$ 3s state energy) is also a less accurate representation of this experimental series for Mg$^+$ than it is for Be$^+$. More parameters are required to represent this series with high accuracy.

potential one essentially is fitting the quantum defect. The upper states are not very sensitive to the precise form of the model potential inside R_c, as the results of Hazi and Rice on benzene (Section 6) show. It is the use of the model potentials in other problems, such as the two-electron calculations discussed in Sections 9 and 10, that provides one with a more difficult and meaningful test of the usefulness of such potentials.

The results reported in Table VIII suggest that one should be able to fit the first two valence levels of an E_{nl} series by proper choice of the parameters, R_c and A_l (although fitting the second level exactly may not be possible because even it is relatively insensitive to variations of the parameters) and thus obtain a very accurate fit for the entire Rydberg series. However, since our primary purpose here is not the construction of very accurate one-electron model potentials for a Rydberg series, we have effectively eliminated one of the parameters by taking the well depth for s states to be exactly that given by the experimental valence level E_{vs} in order to test the reasonableness and consistency of our assumptions.

9. TWO-ELECTRON PROBLEMS AND THE USE OF MODEL POTENTIALS

We now apply the ideas of the previous Sections to many-electron problems where explicit notice is taken of the correlation between valence-electrons. We consider, as an example, the case of two valence-electrons outside a closed core of N electrons.

Fock, Wesselov and Petrashen[29] showed that if one approximates the wave function of the $N+2$ particle system as

$$\psi = (N+2)^{-\frac{1}{2}} \tilde{a}_p [\hat{\psi}(N+1, N+2) \det \phi_1(1) \phi_2(2) \dots \phi_N(N)] \quad (75)$$

where $\tilde{a}_p$ is the partial antisymmetrizer between the two groups of electrons, and $\hat{\psi}(N+1, N+2)$ is a general correlated wave function for the valence-electrons obeying the orthogonality constraints

$$\int \hat{\psi}(1, 2) \phi_c(1) \, d\tau_1 = \int \hat{\psi}(1, 2) \phi_c(2) \, d\tau_2 = 0, \quad c = 1, \dots, N \quad (76)$$

then the total energy of the system can be written

$$E_A \leqslant \frac{\int \psi H \psi}{\int \psi \psi} = \int \hat{\psi}(1, 2) [H_1^{\text{core}} + H_2^{\text{core}} + 1/r_{12}] \hat{\psi}(1, 2) \, d\tau_1 \, d\tau_2 + E_c$$

$$\equiv E_v + E_c \quad (77)$$

In Eq. (77) E_A is the total energy of the atom, E_c is the energy of the core calculated using the set of core orbitals ϕ_c, and H_1^{core} is the Hartree–Fock Hamiltonian defined with these core orbitals for electron 1. We now hold the ϕ_c fixed in some convenient and physically meaningful set and vary our valence function $\hat{\psi}(1,2)$ subject to the orthogonality constraints Eq. (76). Note that we will have an upper bound to the total energy of the atom regardless of the choice of the ϕ_c. This procedure of constrained variation lies within the scope of the general theory discussed in earlier sections. Define the one-electron projection operator over the the core space of electron 1 as

$$P_1 = \sum_{c=1}^{N} |\phi_c(1)\rangle \langle \phi_c(1)| \tag{78}$$

Then a function

$$\hat{\psi}(1,2) = (1 - P_1)(1 - P_2)\Phi(1,2) \tag{79}$$

will satisfy the orthogonality constraints [Eq. (76)] for arbitrary variations of $\Phi(1,2)$. The effective equation for the non-orthogonal $\Phi(1,2)$ is given by Eq. (19) or Eq. (23) in the form

$$(1 - P_1)(1 - P_2)[H_1^{core} + H_2^{core} + 1/r_{12} - E_v](1 - P_1)(1 - P_2)\Phi(1,2) = 0 \tag{80}$$

This is a formal pseudopotential type of equation for motion of the valence electrons. Before we show how the one-electron model potentials found in Section 8 can be used to simplify this equation, we wish to show the relation of Eq. (80), as an exact equation, to those of Sinanoğlu and Szasz.

Equation (80) reduces immediately to that of Szasz[30] when one makes the assumptions he makes in his derivation. Szasz assumed that the ϕ_c are the Hartree–Fock orbitals of the N-electron core system, that is, that $H_1^{core}\phi_c(1) = \varepsilon_c \phi_c(1)$. In that case H_F^{core} and $(1 - P)$ commute. Using Eq. (81) and the idempotent property Eq. (13), Eq. (80) becomes

$$[H_1^{core} + H_2^{core} + (1 - P_1)(1 - P_2)1/r_{12}]\hat{\psi}(1,2) = E_v\hat{\psi}(1,2) \tag{81}$$

which is just the result derived by Szasz. One should note that Eq. (81) is not Hermitian and that in a variational calculation one must orthogonalize the trial function to the ϕ_c as given in Eq. (76) or else one can collapse below the lowest eigenvalue E_v. (The proof is similar to that for the Austin, Heine and Sham pseudopotential.[9])

Sinanoğlu's "exact pair" equation[4] is also contained within Eq. (80) in this simple case. Sinanoğlu considers the effective equation for the "correlation part" of $\hat{\psi}$, writing

$$\hat{\psi}(1,2) = \beta(1,2) + \hat{u} \qquad (82)$$

where $\beta(1,2)$ is the Hartree–Fock approximation to $\hat{\psi}$ (anti-symmetrized product of spin orbitals) and $\hat{u}$ is the correlation part which, under his choice of normalization, is orthogonal to all Hartree–Fock orbitals of the $N+2$-electron system, including those given by $\beta(1,2)$. He fixes the ϕ_c as the N core solutions of the Hartree–Fock equations for the $N+2$-electron system. Under these assumptions Sinanoğlu notes that the "exact pair" equation [Eq. (20) or (26) of reference 4] can be written in the form

$$\mathcal{H}_{KL}\hat{\psi}_{KL} = E_{KL}\hat{\psi}_{KL} \qquad (83)$$

[Eq. (27) of reference 4], where

$$\mathcal{H}_{KL} = H_K^{core} + H_L^{core} + 1/r_{12}$$

[Eq. (11) of reference 4]. Formally, Eq. (83) should be multiplied from the left by $(1-P_K)(1-P_L)$ since $\mathcal{H}_{KL}$ operating on $\hat{\psi}_{KL}$ can have components in the core space and these must be eliminated for the right- and left-hand sides of Eq. (83) to agree. When this operation is carried through, and Eq. (79) recalled, Eq. (80) is obtained.

Thus Eq. (80) is equivalent to either Szasz's equation (81) or to Sinanoğlu's "exact pair" equation (83) depending on whether we fix the core orbitals ϕ_c as the Hartree–Fock orbitals of the N-electron core system or the total $N+2$-electron system. For the case we consider we expect the differences between the two possible assumptions to be small; indeed we will later use this fact to simplify Eq. (80). The main difference between our formulation and those of Szasz and Sinanoğlu is the conceptual shift we have made, characteristic of pseudopotential formulations, of considering the equation for the unconstrained Φ rather than the orthogonal $\hat{\psi}$. Equations of the Sinanoğlu type [Eq. (83)] are difficult to solve. Approximate solutions have been obtained for beryllium by several authors,[31-33] and Nesbet has succeeded in solving very similar equations (what he calls "second-order Bethe–Goldstone equations") with high accuracy for beryllium and neon.[34] These calculations are all rather lengthy and complex. We shall show that for the special case we are considering the theory can be simplified by making approximations suggested from pseudopotential theory, i.e. by using the model (pseudo)potentials for the one-electron ion to simplify the exact equation (80).

With this approach in mind, let us formally write the total energy of the valence-electron pair as

$$E_{12} = E_1^{ion} + E_2^{ion} + E_{12}^{int} \tag{84}$$

where E_1^{ion} is the energy of the valence-electron in the $N+1$-electron ion (i.e. when the other valence-electron is absent) and E_{12}^{int} is the interaction energy between the two valence-electrons, including all effects of adding the second electron (core polarization, etc.). Noting that $(1-P_2)$ and H_1 commute since they refer to different electrons, we rewrite Eq. (80) using Eq. (84) as

$$(\{(1-P_1)[H_1^{core} - E_1^{ion}](1-P_1)\}(1-P_2)$$

$$+ \{(1-P_2)[H_2^{core} - E_2^{ion}](1-P_2)\}(1-P_1)$$

$$+ (1-P_1)(1-P_2)[1/r_{12} - E_{12}^{int}](1-P_1)(1-P_2))\,\Phi(1,2) = 0 \tag{85}$$

Equation (85) has an obvious physical interpretation in analogy with the Phillips–Kleinman one-electron model potential discussed in Section 8. We recall that the essential effect of the operator $(1-P_1)(1-P_2)$ is to remove the electron density from the core, i.e. keep the electrons out of the core. Thus we interpret Eq. (85) as saying, essentially, that the two electrons move under the influence of the ionic Hamiltonians while interacting directly through the $1/r_{12}$ term *while outside the core*.

Equation (85) is still exact and we simplify it further. We pick the N core orbitals ϕ_c, as Sinanoğlu does, to be the Hartree–Fock core orbitals of the $N+2$-electron system, using these in the projection operators and H^{core}. We now assume the core orbitals for the $N+1$-electron ion to be very similar to these, which is a reasonable and frequently made approximation—one expects the addition of a second valence-electron to have little effect on the closed core orbitals. In this case, the terms in brackets { } can be represented in terms of the one-electron pseudopotential equation of the positive

12

ion, using Eqs. (22), (23) and (67), and thus (85) becomes

$$\{[H_1^{ion} + V_R^{ion} - E_1^{ion}](1 - P_2) + [H_2^{ion} + V_R^{ion} - E_2^{ion}](1 - P_1)$$

$$+ (1 - P_1)(1 - P_2)[1/r_{12} - E_{12}^{int}](1 - P_1)(1 - P_2)\} \Phi(1, 2) = 0 \quad (86)$$

We may interpret the first term of Eq. (86) as follows: the "electron 2 part" of our pseudowave function $\Phi(1, 2)$ is kept out of the core [due to the $(1 - P)$ operator] while the "electron 1 part" moves under the influence of the ionic Hamiltonian plus ionic pseudo-potential. The third term remains unchanged. Little core penetration is allowed here: the electrons "see" each other while moving outside the core. To simplify further we use the model potentials developed in Section 8 as approximations to the one-electron pseudopotentials. We then have a two-electron model equation:

$$\{[H_1^m - E^{ion}](1 - P_2) + [H_2^m - E^{ion}](1 - P_1)$$

$$+ (1 - P_1)(1 - P_2)[1/r_{12} - E_{12}^{int}](1 - P_1)(1 - P_2)\} \Phi(1, 2) = 0 \quad (87)$$

Note that no new parameters are introduced. Equation (87) may be solved self-consistently for E_{12}^{int}, starting with some approximate value and iterating. E^{ion} is taken from experimental results and the parameters in H_m^{ion} have been fixed from the previous one-electron ion calculations.

The model equation (87) has a different physical interpretation from that given for the exact equation (80). In the latter we interpreted the effect of the operator $(1 - P_1)(1 - P_2)$ as essentially keeping both valence-electrons out of the core, i.e. presenting a "hard core" to the valence-electrons. The first two terms of the model equation (87) allow for a "softening of the core". For example, the first term of (87) may be thought of as follows: The "electron 2 part" of the wave function feels a hard core while the "electron 1 part" moves under the influence of the model Hamiltonian, which allows for core penetration without variational collapse. In both equations the electron–electron interaction is given by the valence part of Φ—the part outside the core.

The main advantage Eq. (87) has over the exact equation (80) is in the replacement of the Hartree–Fock Hamiltonian in the latter by the much simpler model Hamiltonian, which does not require knowledge of the core orbitals and which greatly reduces the number and difficulty of the integrals involved. However, in order to treat the projection operators exactly, knowledge of the core orbitals is required.

10. APPLICATION OF THE TWO-ELECTRON MODEL POTENTIAL TO THE STUDY OF THE ELECTRONIC ENERGY OF BERYLLIUM AND MAGNESIUM

A. General Considerations

Weeks and Rice[9] have obtained variational approximations to solutions of Eq. (87) using the method of configuration interaction for the cases of beryllium and magnesium. One would expect the configuration interaction series to converge fairly rapidly to a good approximation to the energy of the outer valence-electrons because a large part of the correlation energy of the $2s^2$ pair in beryllium arises from s–p near degeneracy: configuration interaction is the appropriate method for treating this type of effect. Furthermore, the "correlation hole" effects, expressed in configuration interaction language as a slowly converging expansion of the interelectronic r_{12} singularity, will be less important for this "loose pair".

Equation (87) is an effective two-electron equation for the electron pair in a ^1S state outside the closed core. One can form a trial function leading to a ^1S state for the valence-pair exactly like that of the helium atom:[39]

$$\Phi(1,2) = \sum_l F_l(r_1, r_2) \frac{\sqrt{2l+1}}{4\pi} P_l(\cos\gamma) \frac{1}{\sqrt{2}} [\alpha(1)\beta(2) - \beta(1)\alpha(2)] \quad (88)$$

where the sum is over all l values of angular momentum, $P_l(\cos\gamma)$ is the lth Legendre function of γ, the angle between the radius vectors of the two electrons. F_l is expanded in configurations ω which are symmetrical products of orthonormal one-electron radial functions U_i^l having the analytic form

$$U_i = \sum_j c_{ij} S_j(n_j, z_j) \quad (89)$$

The normalized (Slater) basis functions S_j are of the form

$$S_j(n_j, z_j) = A_j r^{n_j-1} \exp(-z_j r) \quad (90)$$

with the normalization constant A_j given by

$$A_j = (2z_j)^{n_j+\frac{1}{2}} [(2n_j)!]^{-\frac{1}{2}} \quad (91)$$

B. Beryllium Results

For the $2s^2$ pair of the beryllium atom, Weeks and Rice chose for the basis functions, $S_j^{(s)}$, the set used by Roothaan, Sachs and Weiss[35] in their Hartree–Fock solution for beryllium. For the p basis set they chose a set very similar to the p set of Watson,[36] leaving out one function that contributed very little to the outer electronic correlation energy. Watson's results for beryllium show that d and higher l configurations make a negligible contribution to the $2s^2$ energy (< 0.0004 a.u.), so these were neglected.

The first two of the one-electron functions, u_i of s symmetry, Weeks and Rice chose as the Roothaan Hartree–Fock solutions for the 1s and 2s orbitals of beryllium.[35] Two other functions of s symmetry were chosen and Schmidt orthogonalized to the Hartree–Fock results. These were the most effective in lowering the energy of several sets tried.

TABLE IX. Valence–core Separability in Beryllium and Magnesium: The Overlap Integrals of Hartree–Fock Functions from the Model Radius to Infinity in Beryllium and Magnesium

	Model radius R_c[a] (a.u.)	$\langle 1s \mid 1s \rangle_{r>R_c}$[b]	$\langle 2s \mid 2s \rangle_{r>R_c}$	$\langle 3s \mid 3s \rangle_{r>R_c}$
Be	1.305	0.005[c]	0.915[d]	—
Mg	1.73	5×10^{-8} [c]	0.006[c]	0.912[d]

[a] See Table VII.

[b] For notation see Eq. (94).

[c] Should be 0 for approximation to be exact.

[d] Should be 1 for approximation to be exact.

Table IX gives the overlap integrals of the Hartree–Fock 1s and 2s functions integrated from outside the model beryllium radius, $R_m = 1.305$ a.u. to infinity. Note that there is a well-defined separation into "core" and "valence" regions for the Hartree–Fock 1s and 2s functions. The other two functions tend to "span" this region, and it is to be expected that functions of this type would be of most importance in augmenting the 1s and 2s set.

The first function of p symmetry chosen was, essentially, the p_I function which Watson found to give almost all of the p correlation energy of the $2s^2$ pair. Two other functions of p symmetry were chosen by trial and error. These produced little energy lowering.

One may now form configurations $\omega(1, 2)$ as symmetrical products of the U_i. The results of Watson show that the wave function of the outer electron pair in beryllium is closely approximated by a linear combination of the Hartree–Fock $2s^2$ configuration and a $2p^2$ configuration. As already noted, he found that configurations of higher l (d, f, ...) had very little effect on the energy of the outer pair (< 0.0004 a.u.) and that additional s and p configurations had even less effect on the energy of the outer pair. From general pseudopotential theory one knows the solution of a pseudopotential equation is given by a combination of core and valence functions, and we have suggested that the solutions of the model potential equations are close to this form. Thus one would expect that the most important configurations for the model equation (87) for the outer electron pair of beryllium are the Hartree–Fock $2s^2$ configuration, the Watson $2p^2$ configuration and the Hartree–Fock 1s, 2s configuration. The latter configuration is permitted by the removal of orthogonality constraints allowed in use of the model equation (87). One can look on it as allowing for a "smoother" wave function than that given by the $2s^2$ configuration alone.

In Table X we report the results of calculations using different choices of configurations. Essentially all the energy lowering is given by the three configurations mentioned above, the Hartree–Fock $2s^2$, the Hartree–Fock 1s, 2s and the Watson $2p^2$. A 12 configuration calculation with 9 configurations of s symmetry and 3 of

TABLE X. Calculated Results for Beryllium Valence Pair Energy

Number of configurations	Brief description	Pair energy (eV)
3[a]	$2s^2$; $2p^2$; 2s, 1s	27.18
12	9 configurations of s symmetry 3 p configurations	27.58

[a] Represents a pseudowave function.

p symmetry lowered the energy only about 0.01 a.u. This energy lowering does not represent correlation energy for the most part. Watson's results show that within our accuracy (~ 0.005 a.u.) correlation energy is negligible once the s–p degeneracy is taken care of. The 0.01 a.u. energy lowering represents, rather, the degree to which the solution of the model potential equations cannot be expressed in the form of a pseudowave function, that is, in the form of some linear combination of valence plus core solutions. [One should note that in this two-electron problem any function annihilated by $(1 - P_1)(1 - P_2)$ satisfies the exact pseudo-potential equation (80) identically and can be thought of as a core solution.] The fact that relatively little energy lowering occurs on the addition of other electronic configurations implies that, to a good approximation, the model potential equations have a solution close to a pseudowave function. This is particularly clear in the case of the $2p^2$ configuration. There is no p "core" function in beryllium so one would hope that the p component of the model equation wave function would be the same as that corresponding to the exact equation. The very small mixing of other configurations shows this to be true to a good approximation.

TABLE XI. Comparison of Beryllium Results with Other Methods and Experiment

	Model calculation	Hartree–Fock	"Theoretical" pair energy	Experimental pair energy
Energy of Be $2s^2$ electron pair (eV)	27.58[a]	26.10	27.40[b]	27.526 eV[c]

[a] Best result from Table X.

[b] Hartree–Fock energy plus $2s^2$ correlation energy estimated by Geller, Taylor, and Levine.[32]

[c] Sum of ionization potentials of two valence-electrons. Model results are best compared with this figure, since experimental results from Be+ were used to calibrate model potential.

In Table XI the results of Weeks and Rice's calculation are compared with the calculations of Watson and others solving equations of the same type as Eq. (80) without approximations. The agreement between this calculation and all others is very good. Furthermore, since Weeks and Rice calibrated the model potential using experimental data for the ion, one expects their calculated pair energy to approximate the experimental pair energy computed by adding the ionization potentials for the two electrons. The results of Table XI show good agreement there also.

C. Magnesium Results

Weeks and Rice obtained approximate solutions of the model equation (87) for the outer electron pair of magnesium using the Mg^+ model potential discussed in Section 7 in a manner similar to that described for beryllium. Table IX gives the radial overlap of the magnesium Hartree–Fock 1s, 2s and 3s functions, from the model Mg^+ radius of 1.73 a.u. to infinity. Again there is good valence–core separability. The reader is referred to their paper[9] for details of the calculation.

Weeks and Rice found that the solution to the model equation is quite accurately represented in the form of a pseudowave function, as in Eq. (74). A very good approximation to the model equation solution is given by a linear combination of the Hartree–Fock $3s^2$, 3s, 2s and 3s, 1s configurations, the latter two representing s core mixing in the model wave function, together with p configurations to account for the s–p degeneracy. As in beryllium, a good representation of the p function which interacts strongly because of s–p degeneracy can be found by taking the lowest (3p-like) solution of the ionic Mg^+ p model potential. Since magnesium has p core functions, this 3p-like function must be orthogonalized to the magnesium Hartree–Fock 2p core function. This orthogonalization produces little change in the 3p function, since it turns out that it is already very nearly orthogonal to the magnesium 2p function. The p configurations that intereact strongly are then the $3p^2$ and the core configuration 3p, 2p. It is quite important to include core configurations when using the model potentials. Although the mixing coefficients are small, these core configurations contribute strongly to lowering the energy because of the existence of very large off-diagonal matrix elements.

Using the five configurations mentioned above, Weeks and Rice obtained a pair energy of 22.63 eV. Adding 4 other s configurations and 3 other p configurations, formed with other U_i, lowered the energy only slightly to 22.66 eV. These calculated energies are to be compared with the experimental energy 22.674 eV. A rough estimate of the s–p degeneracy can be found by assuming, as in beryllium, that it is the cause of essentially all the energy lowering from the p configurations. The result obtained is about 0.8 eV which is the same order of magnitude as that found by Watson in beryllium (1.14 eV).

As would be expected, the model wave function for magnesium is a better approximation to a pseudowave function, i.e. Eq. (74) holds better, than is the beryllium model wave function, since there are more core functions in magnesium with which to expand the model function. In general, the larger the system the better is the approximation that a given model potential leads to a solution which has the form of a pseudowave function, Eq. (74).

11. FURTHER APPROXIMATIONS TO TWO-ELECTRON MODEL POTENTIAL EQUATIONS

In the previous calculations, because of the appearance of projection operators, we have had to know the Hartree–Fock core orbitals. Thus our two-electron model equation (89) is not as simple as is the one-electron model equation (75), where no knowledge of core orbitals is required. That core orbitals need not be known is certainly a desirable feature of Eq. (73) and we might hope to find some further approximation to (87) which would also have this property. The basic problem we face is that one cannot remove all orthogonality constraints on a two-electron wave function using only a one-electron model potential or pseudo-potential. The model equation (87) still contains projection operators, the use of which requires knowledge of the core functions. We can make additional approximations in (87) to eliminate the troublesome projection operators, but these approximations will not be as good as were those leading to Eq. (87).

Let us recall the valence–core non-penetrability approximation discussed in Section 7. We assume that the normalized pseudo-wave function (or model wave function) can be written in the form

$$\chi_m = \alpha_m \psi_m^v + \beta_m \phi_m^c \tag{92}$$

with

$$\langle\phi^c|\phi^c\rangle = \langle\psi^v|\psi^v\rangle = 1; \quad \langle\psi^v|\phi^c\rangle = 0; \quad \alpha_m^2 + \beta_m^2 = 1 \qquad (93)$$

ψ^v is the true valence wave function and ϕ^c is the core part. (It may be the sum of several core orbitals, etc.) The valence–core non-penetrability approximation then suggests that

$$\langle\psi^v|\psi^v\rangle_{r>R_c} \simeq 1; \quad \langle\phi^c|\phi^c\rangle_{r>R_c} \simeq 0 \qquad (94)$$

where we describe by the notation $\langle\ \rangle_{r>R_c}$ a radial integration taken from outside the core (model) radius to infinity, instead of over all space as in Eq. (93). Using Eq. (94) we may approximate an integral over all space involving a projection operator as

$$\langle\chi_m(1-P)\chi_n\rangle \simeq \langle\chi_m\chi_n\rangle_{r>R_c} \qquad (95)$$

assuming that all functions inside R_c can be expanded in terms of core orbitals and thereby are eliminated by the operator $(1-P)$. We may make this approximation in the terms $[H_m^{\text{ion}}-E](1-P)$ in Eq. (87) and thus effectively eliminate the projection operators from that equation. Note that the net effect is that in replacing the projection operator we multiply instead by the valence "fraction" outside the core. Finally we consider integrals involving the last term in Eq. (87):

$$I \equiv \langle\chi_m(1)\chi_p(2)|(1-P_1)(1-P_2)[1/r_{12}-E_{12}^{\text{int}}]$$
$$\times (1-P_1)(1-P_2)|\chi_n(1)\chi_q(2)\rangle \qquad (96)$$

The following approximation appears sensible: ignore the projection operators and instead perform the radial integration outside the core from R_c to ∞, taking as exact the physical picture of two electrons interacting with each other outside the core. This approximation will almost certainly underestimate the magnitudes of these integrals—the true valence function does penetrate the core to an extent of about 10%, and this core penetration will be more important for integrals of the type (98) because $1/r_{12}$ is larger inside the core, where the electrons are forced to be close to one another. However, most of the repulsion energy comes from the part of the pseudowave function outside the core. In view of these considerations we choose to approximate Eq. (96) as follows:

$$\langle\chi_m(1)\chi_p(2)|(1-P_1)(1-P_2)[1/r_{12}-E_{12}^{\text{int}}](1-P_1)(1-P_2)|\chi_n(1)\chi_q(2)\rangle$$
$$\simeq \langle\chi_m(1)\chi_p(2)|1/r_{12}-E_{12}^{\text{int}}|\chi_n(1)\chi_q(2)\rangle_{r>R_c} \qquad (97$$

The approximations (95) and (97) eliminate the need to know the core orbitals when solving the model equation (87). They are, however, less accurate than the approximations leading to (87).

Weeks and Rice[9] performed calculations for beryllium and magnesium using the sets of approximations (95) and (97) with the same basis sets as in the exact calculations reported above. The results, given in Table XII, show that one still obtains a rather good total energy, but an energy which is lower than that given by the exact calculations (and slightly lower than that observed) because of the underestimation of interelectronic repulsion inherent in the approximation (97). The good agreement between the exact and the approximate calculations gives us encouragement that our physically reasoned approximations contain the essential features of the exact calculations.

Another approximate scheme was suggested intuitively by Szasz and McGinn[28] and also by Hazi and Rice.[17] Szasz and McGinn[37] later gave a derivation of the approximation, which

TABLE XII. Calculated Results for Beryllium and Magnesium Valence Pair Energy using Various Approximations to Eliminate Projection Operators in Two-electron Model Equation (89)

	Number of configura- tions	Result using valence–core non-penetrability approximation	Result ignoring projection operators	Exact[a] result	Experi- mental[a] result
Be	3[c]	27.30[b]	26.29	27.18	27.526
	12	27.63[d]	27.08[e]	27.58	
Mg	5[c]	22.80	21.11	22.63	22.674
	12	22.85[d]	21.77[e]	22.66	

[a] See Section 10.

[b] All energies in electron volts eV.

[c] Represents a pseudowave function (see Section 10).

[d] Under valence–core non-penetrability approximation, additional configurations produce about the same energy lowering as in the exact configuration.

[e] Additional configurations mix in much more than in the exact calculation. The model wave function is not so accurately a pseudowave function.

simply ignores all projection operators in (87). This leads to the simple physical picture of two electrons moving under the influence of the interelectronic repulsion plus the sum of the one-electron ionic pseudopotentials. One solves the equation

$$\{[H_1^m - E^{\text{ion}}] + [H_2^m - E^{\text{ion}}] + [1/r_{12} - E^{\text{int}}]\}\,\Phi(1,2) = 0 \qquad (98)$$

or, using (84) and (86),

$$\{[H_1 + V_R] + [H_2 + V_R] + 1/r_{12}\}\,\Phi(1,2) = E_{12}\Phi(1,2) \qquad (99)$$

In their derivation of (99) Szasz and McGinn[37] start from the same expression of Fock, Wesselov and Petrashen (Eq. 77) that we do. However, they fail to require that the effective two-electron function be kept orthogonal to the core functions during variation. This orthogonality constraint is necessary if the energy is to be written in the form given by Fock, Wesselov and Petrashen, and it shows up in our equation (Eq. 87) by the presence of projection operators even after the introduction of the one-electron pseudo-potentials. As mentioned before, a one-electron pseudopotential cannot remove all orthogonality requirements on a two-electron wave function.

This latter fact is disturbing since the remaining projection operators require knowledge of the core orbitals. The approximate treatment discussed above eliminates this necessity, but introduces errors whose precise effects are difficult to determine. We feel that the errors introduced in this approximation are greater than those arising from the introduction of the one-electron model potentials. We consider Eq. (98) or Eq. (99) as yet another approximate treatment of the projection operators.

Let us compare Eq. (98) with the more exact equation (87) to determine the effects of ignoring the projection operators. In solving (98) rather than (87) variationally one in effect adds to (87) the terms

$$\langle \Phi(1,2)|\,[H_1^m - E^{\text{ion}}]\,P_2 + [H_2^m - E^{\text{ion}}]\,P_1\,|\Phi(1,2)\rangle$$
$$+\langle \Phi(1,2)|\,1/r_{12} - E^{\text{int}}\,|\Phi(1,2)\rangle$$
$$-\langle \Phi(1,2)|\,(1-P_1)(1-P_2)\,[1/r_{12} - E^{\text{int}}]\,(1-P_1)(1-P_2)\,|\Phi(1,2)\rangle \qquad (100)$$

The first term can be shown to be positive definite.[9] This comes about because the operator $(H^m - E^{\text{ion}})$ is positive definite by choice of the parameters in the model potential.

Next we have the difference between the expectation value of the electron repulsion term for the complete pseudowave function Φ and that when the core part of Φ is removed by the projection operators. This difference is expected to be positive: we remove core electron density in the second integral by the projection operators and this will reduce the value of the integral by removing electron density in the region where $1/r_{12}$ is largest. Let us recall that the functions minimizing the integrals over the terms involving the one-electron model potentials have a non-negligible core penetration. The exact equation (87) requires that in computing electron–electron repulsion the core density of this function be removed by projection operators. If one solves (98), ignoring the projection operators, one will then overestimate the correct inter-electronic repulsion by adding in that of the core part of the pseudowave function also.

One might hope, because of its attractive simplicity, that Eq. (98) could prove useful in calculations by giving upper bounds to the pair energy of Eq. (87). The variation principle could perhaps make the additional positive terms of Eq. (98) small enough that its pair energy would be close to that found by Eq. (87).

Weeks and Rice calculated the valence pair energies of beryllium and magnesium using (98) with the same set of configurations as in the calculations described above. The results of this calculation, given in Table XII, show that the energy found by the use of (98) is somewhat higher than the experimental result. One should also note that the solutions under this approximate method do not take on the form of a pseudowave function as accurately as does the exact solution or the solution corresponding to the first approximate scheme. The large energy lowering which occurs when additional configurations are added to those representing a pseudowave function shows that the configuration interaction series is converging rather slowly from this starting point. The results given in Table XII thus may not represent the lowest energy that one could find using (98).

Since they obtained slow convergence of the configuration interaction series when the orthogonal functions U_i were chosen on the assumption that the model wave function would be close to a pseudowave function, Weeks and Rice abandoned this assumption and made a new choice of orthogonal functions U_i. They took the lowest solutions of the ionic (Be^+ and Mg^+) model potential equations and used these directly as the new set of orthogonal functions, rather than orthogonalizing them to the Hartree–Fock functions as they did before. Convergence of the configuration interaction series was faster with the new functions and much better results were obtained. A calculation of the valence pair energy of beryllium with 6 configurations of s symmetry and 3 configurations of p symmetry using the new set of functions U_i gave a valence pair energy of 27.48 eV. Good results were also obtained for magnesium where a 12 configuration calculation with the new functions gave a valence pair energy of 22.51 eV. Thus good results can be obtained using (98) although the resulting model wave function is not so well represented in the form of a pseudowave function.

12. FINAL REMARKS

A few general remarks about the applications of model potential and pseudopotential theory discussed in this review are in order. First, we should stress that the model potential we have used [see Eq. (57)] is only one of many possible forms. Other models may be more convenient for different purposes. Abarenkov and Antonova[40] have suggested a model potential of the form

$$V_l = \frac{1}{r}[Z + B_l \exp(-k_l r)] \tag{101}$$

where B_l and k_l are adjustable parameters, which they find gives a model wave function quite close to a pseudowave function. This model may prove useful in molecular calculations where the spherical symmetry of Eq. (57) is no longer an advantage.

The two approximate treatments of the remaining projection operators in (87) which we have discussed in Section 9 bracket the results of an exact treatment for the cases of beryllium and magnesium, the first giving an energy below that given by (87) and the second an energy above that of (87). The rather close agreement between the two approximate schemes with their very different treatments of the projection operators shows that the effects of the remaining projection operators in (87) are relatively unimportant.

By replacing the Hartree–Fock Hamiltonians in (80) by the model Hamiltonians as in (87) one has avoided essentially all danger of variational collapse when the orthogonality constraints are removed. The remaining projection operators lead to small perturbations, and can be treated by whatever approximate method is most appealing. Great simplifications are then possible in the treatment of valence electrons in atoms and molecules.

In the previous Sections we have reviewed some applications of pseudopotential theory in several problems dealing with forces ranging from the unbound interactions in electron–helium atom scattering, to the loosely bound Rydberg electron, to the tightly bound valence-electrons in two-electron atoms. The pseudopotential formalism has provided the conceptual framework, and the use of model potentials has provided the mathematical simplification that together make simple calculations with a physical motive possible. There are a number of other possible applications of pseudopotential theory, since a wide variety of problems can be formulated in terms of a constrained variational principle. The method of pseudopotentials with its stress on "core" and "valence" electrons fits in well with chemical intuition and provides a link between these concepts and the more formal theories based on mathematically constrained functions. We hope it will prove a useful and practical tool in Quantum Chemistry.

ACKNOWLEDGEMENTS

This research was supported by the Directorate of Chemical Sciences, United States Air Force Office of Scientific Research. We have also benefited from the use of facilities provided by the Advanced Research Projects Agency for Materials Research at the University of Chicago.

APPENDIX

Since the completion of this manuscript (February 1968) several relevant papers have appeared. We can here mention only briefly a few of these.

Schneider, Weinberg, Tully and Berry[41] have considered the use of model potentials in the description of inelastic scattering processes in atoms. Their formalism is applied to the photodetachment of electrons from O^-, where it gives good results. This work helps to tie together the connection between the pseudo-potential scattering methods discussed here in Section 4A and the model potential methods discussed in later sections. In a later paper Schneider and Berry[42] apply their method to a number of molecules with good results.

A theoretical approach similar to that described in Sections 3 and 9 has been taken by Logatchov[43] and applied to a study of the luminescence centres of the potassium chloride (T1) type. In this regard one should also see the papers by Szasz, McGinn and Schroeder,[44] Kutzelnigg, Koch and Bingel,[45] and Szasz.[46]

References

1. See, for example: Hellmann, H., *Acta Physicochim. USSR* **1**, 913 (1935); *J. Chem. Phys.* **3**, 61 (1935); Gombas, P., *Z. Physik.* **94**, 473 (1935).
2. Phillips, J. C., and Kleinman, L., *Phys. Rev.* **116**, 287 (1959).
3. Harrison, H., *Pseudopotentials in the Theory of Metals*, W. A. Benjamin, Inc., New York, 1966. For a different view of the pseudopotential based on statistical theories see Gombas, P., *Pseudopotentiale*, Springer-Verlag, Wien, New York, 1967.
4. Sinanoğlu, O., *J. Chem. Phys.* **36**, 706 (1962); *J. Chem. Phys.* **36**, 3198 (1962).
5. Szasz, L., *Phys. Rev.* **126**, 169 (1962); *Phys. Rev.* **132**, 936 (1963).
6. Löwdin, P. O., *Phys. Rev.* **139**, A357 (1965).
7. Cohen, M. H., and Heine, V., *Phys. Rev.* **122**, 1821 (1961).
8. Austin, B. J., Heine, V., and Sham, L. J., *Phys. Rev.* **127**, 276 (1962).
9. Weeks, J. D., and Rice, S. A., *J. Chem. Phys.* **49**, 2741 (1968).
10. Kestner, N. R., Jortner, J., Cohen, M. H., and Rice, S. A., *Phys. Rev.* **140**, A56 (1965).
11. Temkin, A., and Lamkin, J. C., *Phys. Rev.* **121**, 788 (1961).
12. Temkin, A., *Phys. Rev.* **107**, 1004 (1957); **116**, 358 (1959).
13. Dalgarno, A., *Advan. Phys.* **11**, 281 (1962).
14. Since scattering calculations are less familiar to many, we will report more details here than in the bound state problems considered later on.
15. Moiseiwitsch, B. L., *Proc. Roy. Soc.* (*London*) **77**, 721 (1960).
16. Mulliken, R. S., *J. Am. Chem. Soc.* **86**, 3183 (1964).
17. Hazi, A. U., and Rice, S. A., *J. Chem. Phys.* **45**, 3004 (1966).
18. Abarenkov, I. V., and Heine, V., *Phil. Mag.* **12**, 529 (1965).
19. Animalu, A. O. E., and Heine, V., *Phil. Mag.* **12**, 1249 (1965).
20. Johnson, K., *J. Chem. Phys.* **45**, 3085 (1966).
21. For convenience, we assign a fictitious partial charge δZ_i to each atom i in the molecular ion. In case of highly symmetrical molecules this procedure is straightforward. For example, we assign a charge of $\frac{1}{2}$ to each of the H atoms in H_2^+.

342   JOHN D. WEEKS, ANDREW HAZI AND STUART A. RICE

22. Let us consider a diatomic molecule AB. What is the potential V_2 at a point with coordinate r_A (or $R_{AB}+r_B$), inside the molecular core ? A point is inside the molecular core if either $r_A \leqslant R_A^0$ or $r_B \leqslant R_B^0$. Clearly, three possibilities exist: (i) if $r_A < R_A^0$ and $r_B < R_B^0$, which is possible provided $R_A^0 + R_B^0 \geqslant R_{AB}$, then $V_2 = A_A + A_B$; (ii) if $r_A < R_A^0$ and $r_B > R_B^0$, then $V_2 = A_A - \delta Z_B/r_B$; (iii) if $r_A > R_A^0$ and $r_B < R_B^0$, then $V_2 = A_B - \delta Z_A/r_A$.

23. Hazi, A. U., and Rice, S. A., *J. Chem. Phys.*, **48**, 495 (1968).
24. Wilkinson (reference 25) has observed four Rydberg series of benzene (denoted R, R', R'', and R''') converging to the same ionization limit. Previously, Price and Walsh (reference 26) have also identified two Rydberg series; these correspond to the R and R' series of Wilkinson.
25. Wilkinson, P. G., *Can. J. Phys.* **34**, 596 (1956).
26. Price, W. C., and Walsh, A. D., *Proc. Roy. Soc. (London)* **191A**, 22 (1947).
27. Szasz, L., and McGinn, G., *J. Chem. Phys.* **47**, 3495 (1967).
28. Szasz, L., and McGinn, G., *J. Chem. Phys.* **42**, 2363 (1965).
29. Fock, V., Wesselov, M., and Petrashen, M., *Zh. Eksperim i Theor. Fiz.* **10**, 723 (1940); see also reference 30.
30. Szasz, L., *Z. Naturforsch.* **14a**, 1014 (1959).
31. Tuan, D. F., and Sinanoğlu, O., *J. Chem. Phys.* **41**, 2677 (1964).
32. Geller, M., Taylor, H. S., and Levine, H. B., *J. Chem. Phys.* **43**, 1727 (1965).
33. Szasz, L., and Byren, J., *Phys. Rev.* **158**, 34 (1967).
34. Nesbet, R. K., *Phys. Rev.* **155**, 51 (1967); *Phys. Rev.* **155**, 56 (1967).
35. Roothaan, C. C. J., Sachs, L. M., and Weiss, A. W., *Rev. Mod. Phys.* **32**, 186 (1960).
36. Watson, R. E., *Phys. Rev.* **119**, 170 (1960).
37. Szasz, L., and McGinn, G., *J. Chem. Phys.* **45**, 2898 (1966).
38. Roothaan, C. C. J., *Rev. Mod. Phys.* **32**, 179 (1960).
39. Slater, J. C., *Quantum Theory of Atomic Structure*, McGraw-Hill, New York, 1960.
40. Abarenkov, I. V., and Antonova, I. M., *Phys. Stat. Sol.* **20**, 643 (1967). See also references 1 and 28.
41. Schneider, B., Weinberg, M., Tully, J., and Berry, R. S., "1. General Method and Photodetachment of O^-", in *A Pseudopotential Method for Low Energy Electron Scattering* (to be published).
42. Schneider, B., and Berry, R. S., "2. Photoionization of N_2", in *A Pseudopotential Method for Inelastic Processes in Atoms and Molecules* (to be published).
43. Logatchov, Yu. A., *Phys. Stat. Sol.* **25**, 763 (1968).
44. Szasz, L., McGinn, G., and Schroeder, J., *Z. Naturforschg.* **22a**, 2109 (1967).
45. Kutzelnigg, W., Koch, R. J., and Bingel, W. A., *Chem. Phys. Letters*, **2**, 197 (1968).
46. Szasz, L., *J. Chem. Phys.* **49**, 679 (1968).

PHASE TRANSITIONS IN VAN DER WAAL'S LATTICES

LOTHAR MEYER, *Department of Chemistry and James Franck Institute,*
University of Chicago, Chicago, Illinois 60637

CONTENTS

1. INTRODUCTION

Crystals of substances with a low melting point and low boiling point, such as the solids of the rare gases, are held together essentially by the same forces which produce the deviations from the ideal gas law at higher temperatures. They are often referred to as van der Waals' lattices. The forces include not only the van der Waal's or London forces proper, due to induced fluctuating dipoles, but also interactions due to permanent electric dipoles and/or higher electrical poles, especially quadrupole and octupole moments. The state of our knowledge of the solid rare gases has been extensively reviewed by Pollack[1], Boato,[2] Hollis Hallet[3] and Hingsammer and Lüscher[108] recently, and earlier by Dobbs and Jones.[4] Many properties of the solid rare gases, especially the lattice energy (heat of sublimation), can be quite well represented by summing over a two-body potential consisting of a short range repulsion (either exponential or decaying with a high power of the distance r, e.g. r^{-12} is often used) and an attractive potential falling off with the sixth power of the distance as suggested by the London potential. By adjusting the parameters of a potential such as $V(r) = 4\varepsilon[(\sigma/r)^{12} - (\sigma/r)^6]$ often called the Mie–Lennard-Jones potential, where ε is the depth of the potential well, and σ the distance r at which attraction and repulsion just cancel each other, and by eventually introducing an additional third term (Kihara,[5] Horton and Leach[6]) rather good agreement with experimental data can be achieved without too much difficulty. All these potentials are spherically symmetrical around any atom or molecule in the lattice and are a function of distance only. However, one serious problem remains. Any reasonable potential of the Mie or Lennard-Jones type leads to the result that of the two close-packed lattices, face-centred cubic (space group $Fm3m$) with the stacking of the close-packed planes $ABCABC$ (compare reference 7, page 228) has a higher potential than the hexagonal close packed lattice (space group $P6_3/mmc$) with the stacking $ABAB$ (reference 7, page 225). Errors in the stacking of planes—for example, $ABABCACBA$ instead of perfect ordering $ABAB$ or $ABCABC$—are called stacking faults (reference 7, page 451). The nearest neighbour distances and the next nearest neighbour distances in the same plane are the same for both lattices. There is a difference in distance to the next nearest neighbour in the plane following the neighbouring plane. In a hexagonal close packed lattice, the distance from an atom in an A plane to the next nearest neighbour in the next A plane—these two planes are separated by a plane in the B position—is somewhat shorter than the distance from an atom in an A plane to the next nearest neighbour in a C plane in a face-centred cubic lattice. Only quite special potentials increasing the repulsive contribution at the distance of next nearest neighbours can make the face-centred cubic more stable than the hexagonal close packed lattice (compare reference 109). All rare gases crystallize in the face-centred cubic lattice with the exception of solid helium, where the zero-point energy plays a dominating role.

Many attempts have been made to overcome this difficulty. Jansen and coworkers[8, 9] analysed three-body forces due to electron exchange between a central atom and its nearest neighbours. This contribution—assuming a Gaussian charge distribution for the electron—favours the face-centred cubic structure found in nature. However, the two-body interaction has to be artificially reduced (see also Lucas,[10] Götze and Schmidt[11]). Recently Gillis, Werthamer and Koehler[12] suggested that the stability of the face-centred cubic structure can be derived from a self-consistent harmonic approximation.

A recent series of experimental investigations by Professor C. S. Barrett and the author of the rare gas solids, especially argon and its alloys, showed clearly that the supposedly "simple" van der Waal's lattices present in reality quite complicated many-body problems.

The experimental study of van der Waal's lattices, especially by X-ray diffraction, was quite intense in the 1920's and early 1930's with Vegard's pioneering work,[13] the work of the Leiden group, Keesom and coworkers[14] and Simon's group,[15] but apparently went out of fashion later. Simultaneously with our recent investigations, the high precision work on rare gas single crystals by Simmons and his coworkers became available with its important implications for lattice dynamics, vacancy content, expansion coefficient and compressibility, all within the stability range of the face-centred cubic structure.[16] This review is mainly concerned with structural changes and new phases especially in some two-component systems of van der Waal's lattices. Since practically no X-ray diffraction studies on two-component systems existed, and much of the older work was found unreliable, this review is unfortunately somewhat biased towards our own work. The many unusual features we found were completely unexpected; there is no rigorous theoretical explanation and we are still quite unable to predict what will happen in similar systems. All these facts make a really systematic survey impossible and the presentation could not be completely freed from the personal element of historical, heuristic development.

Our investigation was initiated by the observation that in freezing argon from the liquid, hexagonal close packed crystals were observed by X-ray diffraction from time to time in addition to the stable face centred cubic crystals.[17] A similar result was obtained for neon (reference 102). Both types of crystals showed diffraction lines which were essentially instrumentally sharp. The metastable hexagonal close packed crystals survived in contact with face-centred cubic crystals as long as the refrigeration was maintained, for many hours and even days. Introducing stacking faults into the crystallites by mechanical deformation and then annealing reduced the amount of metastable hexagonal close packed phase relative to the stable face-centred cubic phase. However, annealing reproduced hexagonal close packed crystals with instrumentally sharp diffraction lines; the sharpness implies that the crystallites contained only small perturbations of the long range order.

The energy or free energy difference between face-centred cubic and hexagonal close packed is not exactly known. However all estimates lead to values of, at best, 0.1% of the lattice energy, the latter being approximately the heat of sublimation.[8, 9] Using Trouton's rule that the heat of vaporization is of the order of 10 RT,

TABLE I. Thermodynamic and Structural Data for the Atoms

	B.P. (°K)	M.P. (°K)	Transition temperature (°K)	Space group	Crystal structure	Nearest neighbour distance (Å)
Argon	87.3	83.81		$Fm3m$	Face-centred cubic	3.7477
			86	$P6_3/mmc$	Hexagonal close packed	3.748
Neon	27.1	24.66		$Fm3m$	Face-centred cubic	3.1564
Krypton	119.8	115.775		$Fm3m$	Face-centred cubic	3.9841
Xenon	165.04	161.4		$Fm3m$	Face-centred cubic	4.3368
Nitrogen	77.33	63.13		$P6_3/mmc$	Hexagonal	(4.0)
			35.61	$P2_13$ or Pa_3	Cubic	—
Carbon monoxide	81.61	68.09		$P6_3 mmc$	Hexagonal	(3.98)
			61.57	$P2_13$ or Pa_3	Cubic	—
Oxygen	90.13	54.39		$Pm3n$	Cubic	3.4
			43.76	$R3m$	Rhombohedral	—
			23.66	$C2/m$	Monoclinic	—
Fluorine	85.02	53.54		$Pm3n$	Cubic	—
			45.55	$C2/m$	Monoclinic	—
Hydrogen	20.39	13.96		$(P6_3/mmc)$	Hexagonal	3.6
			~2.0		Cubic	—
Deuterium	23.57	18.72		$(P6_3/mmc)$	Hexagonal	3.8
			~2.5		Cubic	—

we find that the energy difference between hexagonal close packed and face-centred cubic near the melting point is at best of the order of 1% of the thermal energy. In spite of this fact the freezing process produces the coexistence of practically perfect crystals of both forms and *not* random stacking. Calculations of the stacking-fault energy[18] using three-body interactions lead, according to reference 8, to a strong stacking-fault energy but to lower values according to reference 9b. Experimental evidence from electron microscope observations of fault lines[19] showed that the actual stacking fault energy is quite small, at least 15 times smaller than the value calculated in reference 18 ($\gamma < 0.7$ erg/cm²). This fact makes the absence of random stacking in crystals frozen from the melt still more puzzling. Similar effects have been observed by Jagodzinski[111a-c] in silicon carbide around 2000°C where a martensitic type transformation between close packed structures produces mainly crystals with long range order and very few

and Molecules being Components of the Systems Investigated

V_m (cm³)	0.002 electron density contours		Electric moment		ΔH_m ΔH_{trans} (Cal/mole)		ΔS_m ΔS_{trans} (Entropy units)	
	Length	Width	Dipole (Debye)	Quadrupole (10^{-26} e.s.u.)				
22.415	3.78	3.78	0	0	284.5	—	3.39	—
—	—	—	—	—	—	Very small	—	—
13.310	3.18	3.18	0	0	79.20	—	3.21	—
26.932	—	—	0	0	392.0	—	3.39	—
34.73	—	—	0	0	548.2	—	3.40	—
—	4.34	3.39	0	−1.52	172.3	—	2.73	—
—	—	—	—	—	—	54.71	—	1.54
—	5.39	3.67	0.117	−2.5	199.7	—	2.93	—
—	—	—	—	—	—	151.3	—	2.46
23.0–23.6	4.18	3.18	0	−0.39	106.3	—	1.96	4.06
20.9–21.8	—	—	—	—	—	177.6	—	0.95
20.90	—	—	—	—	—	22.42	—	—
20.9	4.18	2.86	0	+0.88	121.98	—	2.28	—
19.3–19.8	—	—	—	—	—	173.90	—	3.82
22.42	—	—	0	+0.66	28.0	—	2.01	—
—	—	—	—	—	—	Very small	—	—
19.97	—	—	0	+0.65	47.0	—	2.51	—
—	—	—	—	—	—	Very small	—	—

stacking faults. The argon results immediately raised the question whether or not the stability of the hexagonal close packed phase was due to an impurity. Our preliminary experiments demonstrated that the phase diagrams published in the literature for two-component systems containing argon based on thermal analysis were unreliable. This led to a systematic study of the phase diagrams of argon with nitrogen, carbon monoxide, oxygen and fluorine, which all have similar vapour pressures and molecular size.

Table I contains boiling points, melting points, temperatures of phase transitions, structures of the different phases and the size and shape of the molecules for those substances which were components of the binary systems investigated. The data of Table I are collected from the following sources: electric dipole and quadrupole moments from reference 20, dimensions of molecules based on self-consistent Hartree–Fock electron density calculations from references 21, 22, 23; the values of the 0.002 electron density distance for neon and argon were kindly supplied by Professor J. A. Hinze from the exact basis set of Bagus of the Laboratory for Molecular Structure and Spectra at University of Chicago (private communication).

Substances	Thermodynamic data	Structure data
Ar	Reference 24 (contains a detailed discussion of earlier work)	References 17, 16a, 25
Ne	Reference 26	References 16a, 16b
Kr	References 27, 28	See reference 1
Xe	References 28, 29	See reference 30
N_2	Reference 31	References 32, 33, 34
CO	References 35, 36	References 13a, 13b. No recent determination of the structure of α-CO and β-CO is available. Vegard concluded that the structures of β-CO and β-N_2 are identical, just as those of α-CO and α-N_2, a result confirmed by the diffraction patterns obtained in references 25 and 37. The corresponding nitrogen space groups are listed for α-CO and β-CO
O_2	Reference 38	γ-O_2. References 39, 13f β-O_2. References 40, 41 α-O_2. References 42, 43
F_2	Reference 44	β-F_2. Reference 45 α-F_2. Reference 46
H_2	References 47, 48, 49, 50	References 14a, 51, 52. 53 contains detailed list of older literature
D_2	References 54, 55	References 51, 52, 53

2. EXPERIMENTAL TECHNIQUE

The experimental technique is described in detail in reference 56. The samples were prepared from gases of the highest commercially available purity by condensing the mixtures of pure gases to liquid and then freezing. Liquid helium, liquid hydrogen and liquid nitrogen were used as refrigerants in the cryostat described in reference 57. The cell in which the condensing and freezing took place contained a copper grid on which the sample froze and which ensured good thermal contact. The front of the cell was covered in the early experiments with a 0.002 in. "Mylar" window. Later a 0.002 in. "Teflon" window (Teflon sheet made without plasticiser by Dupont) was used. The Teflon window produces with copper Kα-radiation only a liquid-like diffraction peak at 2θ of about 15° which is usually outside the range of interest. The structures were determined by X-ray diffraction powder method. In the few cases where recrystallization was so fast that it was impossible to maintain a fine-grained sample for the time of the scanning (about $\frac{1}{2}$ to 1 hour), a pseudo-single crystal technique was used as described in reference 56. The sample was rotated around its axis and those reflections were used which were not too far from the Bragg–Brentano parafocusing condition.

A major feature of the equipment was a Nylon or Teflon tipped chisel-like tool, mounted via O-rings and bellows for wide mobility, which could be used to deform the sample or crush bigger crystals at the temperature of the experiment. This tool not only allowed making fine-grained samples out of big grains by crushing the grains but also enabled us to test whether or not a phase was thermodynamically stable. Very often the observed diffraction pattern contained the diffraction lines of more phases than the phase rule permitted for equilibrium conditions. Deformation of the sample with the help of the tool and annealing usually yielded a clear distinction between stable and metastable phases. The intensity of the diffraction lines of the stable phase increased at the expense of the intensities of the lines of metastable phases, if the latter did not disappear altogether. This method proved especially valuable in studying the equilibrium lines between the hexagonal close packed and face-centred cubic lattices.

All previous investigations of phase diagrams in van der Waal's lattices used thermal analysis. We found that many of the solid–solid transformations in these lattices were very sluggish and could easily be supercooled, often indefinitely, and that the heat effects, especially of the hexagonal close packed–face-centred cubic transformations, were quite small even compared with the specific heat. Furthermore, thermal equilibrium in a fine-grained heterogeneous sample was very slow. All these reasons make thermal analysis a method of extremely dubious value for the study of van der Waal's lattices. The results of such investigations are reviewed in references 3 and 105.

3. RESULTS FOR TWO-COMPONENT SYSTEMS

The phase diagrams of argon with nitrogen and oxygen had been investigated by thermal analysis many times,[105] all agreeing that solid argon accepts 50–60% nitrogen or oxygen in its lattice as solid solution. When the question arose whether or not the presence of hexagonal close packed crystals in addition to the known stable face-centred cubic phase in solid argon frozen from the melt was due to impurities, nitrogen and oxygen were naturally the most likely candidates. Adding at first 1–2% of air and later small amounts of pure nitrogen or oxygen to the argon produced, surprisingly, exclusively the hexagonal close packed phase. Since none of the earlier investigations even hinted at this possibility, it became imperative to redetermine the complete phase diagrams.

The following two-component phase diagrams, most of them having argon as one component, were completed: (1) argon–nitrogen (reference 25), (2) argon–carbon monoxide (reference 37), (3) argon–oxygen (reference 58), (4) argon–fluorine (reference 59), (5) nitrogen–carbon monoxide (reference 60), (6) neon–hydrogen (reference 52), (7) oxygen–nitrogen (reference 61). This study of the phase diagrams revealed interesting and previously not recognized features of some of the pure components and led to a more detailed study of the solid phases of oxygen[41-43], fluorine[46] and hydrogen.[52]

The first two binary systems were chosen because nitrogen and carbon monoxide are isoelectronic and have, as Table I shows, very similar sizes, boiling points and melting points.

The only real difference between the properties of nitrogen and carbon monoxide is in the temperature of the α–β transition in the solid; 35.6°K for nitrogen and 61.6°K for carbon monoxide. Also, their quadrupole moments differ appreciably (Table I). Carbon monoxide has a weak dipole moment (0.12 Debye) which, according to standard theoretical estimates, should not be of any great influence on the lattice structure or the lattice energy.[101] However, disordering of the dipoles seems to contribute almost $R \ln 2$ to the entropy of carbon monoxide.[35, 36]

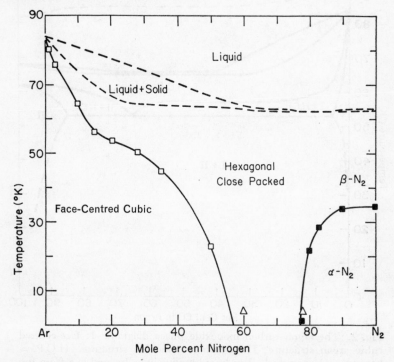

Fig. 1. The argon–nitrogen phase diagram. Dashed lines: liquidus and solidus line.

The phase diagrams for the systems argon–nitrogen and argon–carbon monoxide are shown in Figs. 1 and 2. There is a certain similarity at the argon-rich end of the diagram at higher temperatures. Small amounts of both nitrogen and carbon monoxide produce the hexagonal close packed phase. Hexagonal close packed is essentially the phase in equilibrium with the melt and the face-centred cubic phase becomes stable only at lower temperatures for

higher concentrations of the second component. Extrapolating the
equilibrium line between hexagonal close packed and face-centred
cubic to pure argon, leads in both cases, to a temperature 3–5° above
the melting point of argon. Since oxygen in argon shows a very

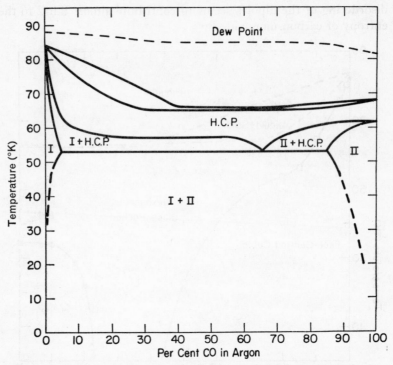

Fig. 2. The argon–carbon monoxide phase diagram. I: face-centred
cubic argon structure; II: α-carbon monoxide structure. H.C.P. =
hexagonal close packed. Top line: liquidus line. Second line from top
line: solidus line.

similar behaviour, this evidence indicates that in pure argon there
exists a hypothetical equilibrium temperature, face-centred cubic–
hexagonal close packed, just a few degrees above the melting point.
Therefore, at the melting point, the free energy difference between
face-centred cubic and hexagonal close packed is still quite small
and it is not surprising that from time to time metastable hexagonal
close packed crystals are formed when freezing liquid argon. Since

pure argon is face-centred cubic, a peritectic reaction must lie between 1% of the second component and pure argon around $82 \pm 1°K$. This area has not been investigated in detail in any of the diagrams because recrystallization and grain growth in this temperature range are too fast to allow the use of the powder method. The pseudo-single crystal technique is rather unreliable for searching for a new phase because absence of a characteristic line is not necessarily due to the absence of that phase, but can be caused by the fact that none of the crystals is near parafocusing conditions.

Transformations from face-centred cubic to hexagonal close packed take place by rearrangement of the stacking of close-packed planes, mostly through slip. Transitions of this kind are often called martensitic* (reference 7, pages 487 and 521). An important characteristic of these transformations is that they can easily be supercooled and are very sensitive to strain which can enhance and retard the transformation. If they start spontaneously, they usually stop before the transformation is complete. Metallurgists distinguish between: (1) M_s, spontaneous martensitic, transformations which usually occur far below the temperature of equilibrium, and thus involve appreciable supercooling before the reaction starts, and (2) M_d, martensitic by deformation, which is usually not far from the temperature of equal free energy.

The fact that these transformations, especially in van der Waal's lattices, are rarely complete even after repeated deformation and annealing, can easily lead to misinterpretation of experimental evidence. For instance, the equilibrium lines between hexagonal close packed and face-centred cubic could only be established by plastic deformation and annealing. Only in a few cases did the metastable phase disappear completely. Mostly, the intensity of the diffraction lines of the metastable phase decreased, whereas the intensities of the diffraction lines of the stable phase increased, indicating a tendency to transform in one direction.

The temperature at which no change in relative intensities of the diffraction patterns of the two phases could be detected after deformation and annealing for a given concentration was considered

* We are using here the expression "martensitic transformation" somewhat loosely for all transformations whose main character is that major macroscopic groups rearrange themselves with or without very little rearrangement inside the groups themselves.

to be the equilibrium temperature. Deformation and annealing weakened the intensity of the diffraction lines of one phase at higher temperatures, but enhanced the same lines at lower temperatures. This method is unfortunately limited to temperatures where annealing still takes place within reasonable time—above about 30°K for the lattices of oxygen, nitrogen, carbon monoxide and fluorine. At lower temperatures plastic deformation introduces stacking faults which cause severe broadening of many diffraction lines. Keeping the sample for a day at these low temperatures did not produce annealing and consequently did not improve the diffraction pattern.

Another characteristic of the martensitic-type transformation is the ease with which a metastable phase can be supercooled and frozen in without the slightest indication that the path of the thermal treatment had crossed phase boundaries. For instance any argon–nitrogen mixture between 10–50% which was cooled relatively fast after freezing to 21° in about 10 to 20 minutes, is pure hexagonal close packed, at which temperature this phase is actually unstable.

This is probably the main reason that calorimetric techniques, especially thermal analysis, completely missed this type of transformation, as in the argon–nitrogen and argon–oxygen phase diagrams. This type of transformation is also often misinterpreted in the literature as second order, or as still more complicated transformations, for instance, the case of the α–β oxygen transformation[43] or the hexagonal–cubic transformation in ortho-rich solid hydrogen.[52, 53]

Any similarity between the argon–nitrogen and the argon–carbon monoxide diagrams is limited to the argon-rich side and the transformation of hexagonal close packed to face-centred cubic. In spite of the fact that the physical properties and the sizes of nitrogen and carbon monoxide are almost identical (see Table I), the phase diagrams with argon differ radically. The argon–nitrogen phase diagram shows essentially solid solubility in spite of the transition from hexagonal close packed to face-centred cubic and the field of α-nitrogen. The nearest neighbour distances in the two close-packed structures are identical within the error of the experiment, at the same temperature and concentration. This is also valid for pure argon.[56] Even the formation of the α-nitrogen structure at constant temperature as a function of nitrogen concentration does not cause a discontinuity in the nearest neighbour distance. Only the slope of the curve of this distance against concentration changes (see Fig. 3). At 21°K cubic α-nitrogen accepts about one argon atom per unit cell, then it transforms into the hexagonal β-nitrogen-like structure.*

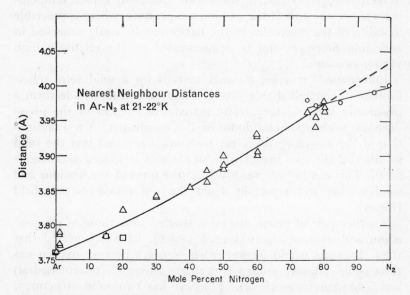

Fig. 3. Nearest neighbour distance in the argon–nitrogen phase diagram at 21–22°K as a function of concentration.

No indication of unmixing was observed down to 21°K. The third law of thermodynamics requires that any two-component system is either completely ordered or unmixed for temperatures approaching 0°K. The critical temperature for either forming the superstructure or unmixing is probably so low in the argon–nitrogen system that the mobility of the molecules is already negligibly small and the process unobservable.

---

* The intensities of the different diffraction lines were evaluated in order to determine whether or not the argon-containing hexagonal structure allows the nitrogen molecule the same type of limited rotational freedom as suggested for pure β-nitrogen (Jordan and coworkers[34]).

In contrast to the solubility of nitrogen in argon, the argon–carbon monoxide phase diagram shows a miscibility gap from 5–85% carbon monoxide at a temperature of 55°K. In many binary systems with van der Waal's lattices, molecules of quite similar dimensions either form special, ordered structures, as the delta phase in the argon–oxygen system and the x-phase in the oxygen–nitrogen system, or show wide miscibility gaps at temperatures above one-half of the melting temperature, where appreciable mobility of the molecules in the lattice can be easily observed in annealing processes and is demonstrated by the relatively high vapour pressure.

Ruhemann,[62] trusting thermal analysis for a solid–solid transformation, concluded that nitrogen and carbon monoxide form a practically ideal mixture. The extreme differences of the phase diagrams with argon cast doubt on this conclusion. A re-examination of the diagram[60] by X-ray techniques revealed that the solid solution of the two isoelectronic molecules is indeed quite nonideal. This can be seen readily from the form of the liquidus and solidus lines and especially from the α–β transformation field (Fig. 4).

Another pair of phase diagrams studied were those of oxygen–argon and fluorine–argon (Figs. 5 and 6). It was expected that these diagrams would be somewhat complex because oxygen has three solid phases: γ-oxygen (cubic), β-oxygen (rhombohedral) and α-oxygen (monoclinic). Fluorine has two solid structures: β-fluorine (cubic) and α-fluorine (monoclinic) (Table I). The argon-rich end of the argon–oxygen diagram shows quite a similarity with the argon–nitrogen and argon–carbon monoxide systems. Adding the second component again changes the structure of argon from face-centred cubic to hexagonal close packed and again the equilibrium line between the two close-packed phases extrapolates for pure argon to a temperature a few degrees above the melting point of argon. The case of oxygen is especially striking since a $\frac{1}{2}\%$ of oxygen is sufficient to transform face-centred cubic argon into a purely hexagonal close packed structure, whereas pure oxygen itself crystallizes in a cubic structure at these temperatures. At this concentration the oxygen molecules in the argon matrix are still separated by several argon atoms. The face-centred cubic field of the oxygen–argon diagram is quite narrow even at 20°K.

The oxygen-rich side shows the expected complexity with one new and unexpected feature: all three solid forms of oxygen accept about 10% argon in the lattice and even the high temperature phase γ-oxygen does not accept much more. This is surprising since γ-oxygen is soft and recrystallizes within minutes after plastic

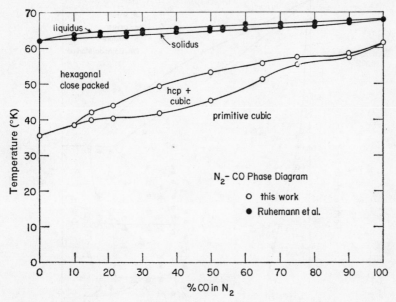

Fig. 4. The nitrogen–carbon monoxide phase diagram. Data from references 62 and 60.

deformation, even with 10% argon dissolved in it, indicating a structure which easily permits rearrangement. Gamma-oxygen and β-fluorine could be probably classified as plastic crystals (see L. A. K. Staveley, reference 63). Visual observation of the specimen during the transformations revealed that the γ–β transformation of oxygen and the β–α transformation of fluorine are quite spectacular, an observation which led to a detailed study of the low temperature phases of oxygen and fluorine.[41–43, 46] In both cases and at almost the same temperature the soft, transparent, high temperature phase transforms with an unusually high volume change and a heat of transformation of about one and a half times the heat of melting (see Table I) into an opaque, hard and brittle substance visually resembling grey tin, α-fluorine having a yellowish tint. This quite spectacular transformation proceeds spontaneously

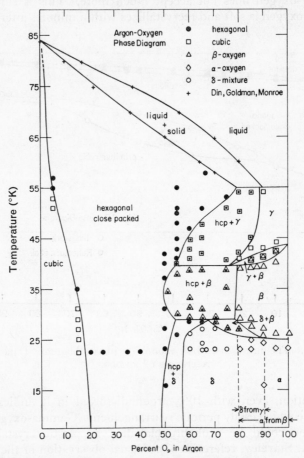

Fig. 5. The argon–oxygen phase diagram. The liquidus and solidus lines are from reference 107.

in pure oxygen or fluorine after a slight supercooling and the rate seems to be controlled by the speed with which the high heat of transformation can be removed. However, γ-oxygen containing 10% argon needs more than 10 hours to transform to β-oxygen.

If this transformation was allowed to run to completion by leaving the sample at a temperature around 40°K overnight, then further cooling to 21°K led to the formation of the α-oxygen structure near 23°K instantaneously. Cooling γ-oxygen with 10–20% argon to

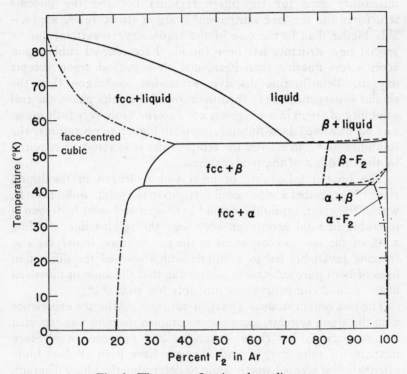

Fig. 6. The argon–fluorine phase diagram.

21°K within 10 to 20 minutes produced an entirely new structure, δ-oxygen, which is cubic with a unit cell length of 13.23 Å and contains 64 lattice sites. Delta-argon–oxygen is apparently a super-structure to γ-oxygen where the argon atoms are not randomly distributed, but occupy well-defined positions.[110]

The contrast between the argon–oxygen and the argon–fluorine diagrams is appreciable but not quite as great as between the argon–nitrogen and argon–carbon monoxide systems. In spite of the similarity in size of the molecules and the crystal structures of γ-oxygen and β-fluorine, fluorine is the only substance investigated with a boiling point near the argon boiling point which does not change the face-centred cubic structure of argon to hexagonal close packed. Near the solidus line 35% fluorine can be dissolved in argon and no hexagonal close packed phase is formed. The miscibility gaps (or two-phase regions) between the fluorine structures and the face-centred cubic argon structure are appreciably bigger than in the case of the argon–oxygen system, but no special new structure has been found. Face-centred cubic argon accepts less fluorine than hexagonal close packed argon accepts oxygen. Beta-fluorine dissolves somewhat less argon than the similar γ-oxygen. Due to the formation of the delta phase, the real solubility of argon in α-oxygen is not known. Only very little argon can be dissolved in α-fluorine, even at higher temperatures, the solubility going to zero as the temperature is lowered, as required by the third law of thermodynamics.

The limited miscibility of neon and hydrogen in the liquid state[64-66] suggested a wide miscibility gap in the solid. Solid neon is, within the error, insoluble in solid hydrogen and solid hydrogen is insoluble in solid neon even very near the solidus line. Adding a $\frac{1}{2}$% of the second component to the gas mixture, liquefying and freezing invariably led to a sample which showed the diffraction lines of both pure substances, indicating that the amount dissolved of the second component was probably less than 0·1%.

The last system studied is oxygen–nitrogen. After the experience with the argon systems, a somewhat complex diagram was expected because oxygen has three structures and nitrogen two. Since mixtures of solid oxygen and nitrogen have been studied quite often[62, 67-70] it seemed worth while to determine the phase diagram. The diagram (Fig. 7) shows an even higher complexity than expected in spite of the fact that oxygen and nitrogen form a nearly ideal mixture in the liquid.[71] It demonstrates again that similarities in size and chemical bonding of molecules is not enough to produce mutual solid solubility even near the melting line. The oxygen–nitrogen system demonstrates both wide miscibility gaps, even near the solidus line, and an ordered structure (x-phase). The x-phase is orthorhombic, containing probably 12 molecules per unit cell with the axes $a = 9.27$ Å, $b = 3.77$ Å, $c = 12.77$ Å. Furthermore, the x-phase shows unusual metastability; once it is formed, it is almost impossible to make it disappear again, at least at temperatures below 40°K. It is interesting to note that Trepp[67] in his study of the hardness of solid oxygen–nitrogen mixtures found it impossible to get hardness data in the concentration range of the x-phase because his tool, a loaded cone, penetrated the sample in a completely irregular manner.

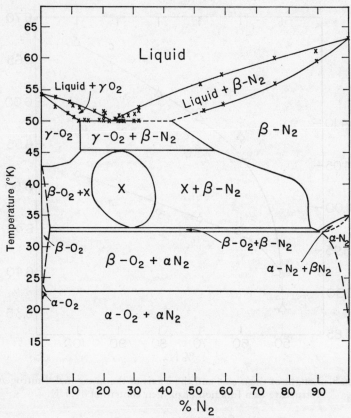

Fig. 7. The oxygen–nitrogen phase diagram.

The persistent metastability of x required long annealing, up to 24 hours. Very often the diffraction lines of x did not disappear in regions where the phase rule clearly indicated that x was only metastable. However, long annealing showed a diminishing of the intensities of the x diffraction lines, and an increase in the intensities of the diffraction lines of the stable phases.

Another important feature of the oxygen–nitrogen diagram is the range of solubility of oxygen in β-nitrogen. Figure 8 shows the lattice parameters of the hexagonal β-nitrogen structure as a function of oxygen content at 47°K. Both a and c parameters decrease almost linearly between 100% nitrogen and 60% nitrogen,

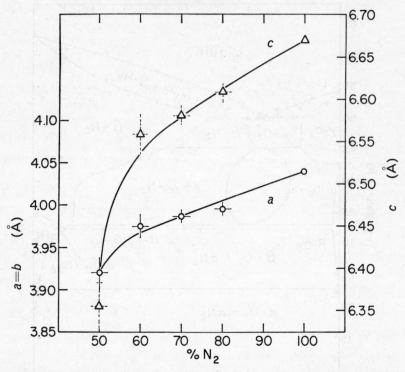

Fig. 8. Length of the a axis and the c axis of the hexagonal β-nitrogen structure as a function of oxygen content at 47.5°K.

but between 60% and 50% the steep drop of the curve is possibly related to the impending change to the two-phase region (γ-oxygen + β-nitrogen). In total, the volume of the unit cell of β-nitrogen decreases by as much as 10% on the addition of 50% oxygen. It is also interesting that the solubility of oxygen in β-nitrogen is about 53% at 50°K, but only 15% at 40°K—a decrease by two-thirds in a temperature range of only 10°. However, it is in just this temperature range where pure oxygen transforms from γ-oxygen to β-oxygen, both having extreme values of expansion coefficients.[41]

The oxygen–nitrogen diagram strikingly demonstrates that the phase diagrams of simple, chemically saturated molecules of similar size can be as complex as any metallurgical system where the changes in structure are usually related to changes in the number of conduction electrons, in band structures and Brillouin zones.

4. THE CRYSTAL STRUCTURES OF PURE ELEMENTS

A. Introduction

The study of the phase diagrams offered a possibility to study the transformations in the pure components, especially oxygen, fluorine, hydrogen and carbon monoxide. The structures of α-oxygen and α-fluorine were unknown. There is an extensive literature on the transformation of hydrogen and deuterium around 2°K from supposedly face-centred cubic to hexagonal close packed as a function of ortho- and para-concentration.[52, 53, 112, 115, 116] Carbon monoxide is of interest because the ordering of the molecules in the lattice seems to be incomplete, leading to the much discussed zero-point entropy.[35, 36] In the course of the study of the phase diagrams quite good diffraction patterns of α-oxygen and α-fluorine were obtained. However, powder patterns are not very favourable for determining low symmetry structures. Furthermore, the intensities of the diffraction lines varied badly from run to run because of local preferred orientation. The low temperature phases of oxygen and fluorine form out of γ-oxygen and β-fluorine respectively, which recrystallize within minutes. The rather violent transformations around 45° break up the big grains of the high temperature phases very effectively. However, a degree of preferred orientation remains, making the evaluation of diffraction intensities in connection with structure analysis doubtful. Determining a structure from the X-ray diffraction of a single crystal is far more straightforward and unambiguous. However, growing good single crystals of these low temperature structures seems to be still a formidable task. In spite of all the rather unpleasant limitations of the powder method and the less than ideal quality of the samples, the structures of α-oxygen and α-fluorine could be determined and rather striking properties of γ-oxygen and β-oxygen elucidated.

B. Solid Oxygen and Fluorine

Pure solid oxygen has been studied extensively because the ground state of the molecule is a triplet state, making oxygen paramagnetic.[72] The magnetic properties have been measured quite often with qualitative agreement, indicating that γ-oxygen and β-oxygen are paramagnetic and α-oxygen antiferromagnetic.[68, 73, 74] However, the values found for the susceptibility in the γ- and the β-phase varied from one investigation to the other, far beyond the error normally encountered in measuring magnetic susceptibilities.

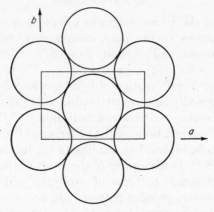

Fig. 9. Basal plane of monoclinic α-oxygen. The size of the molecules is drawn to scale according to the 0.002 electron density contour of Fig. 11.

In a discussion remark at a regional meeting in Prague, 1963,[75] Alikhanov suggested that from neutron scattering data α-oxygen is monoclinic, space group $C\,2/m$. The diffraction patterns obtained for α-oxygen in the phase diagram study fitted this structure, and the observed intensities allowed the determination of the atom positions. The unit cell is shown in Figs. 9 and 10.[42, 43] The dimensions are $a = 5.403$ Å, $b = 3.429$ Å, $c = 5.086$ Å, $\beta = 132.53°$.

A real understanding of the structure can be achieved by using the electron density contours calculated by Bader, Henneker and Cade[21] from self-consistent Hartree–Fock ground-state wave functions for oxygen (Fig. 11). The values of the contours are in atomic units (one atomic unit $= e/a_0^3 = 67.49\ e/\text{Å}^3$). From a careful

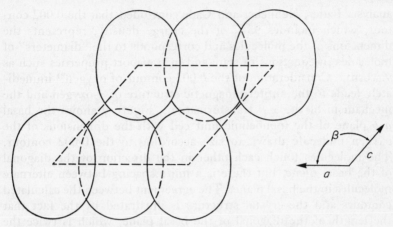

Fig. 10. *a–c* plane of monoclinic α-oxygen. The molecules are drawn to scale according to the 0.002 electron density contour of Fig. 11. The dotted lines indicate the molecule in the centre of the *a–b* plane.

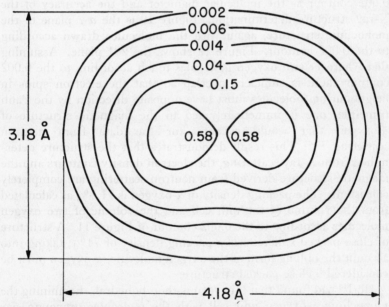

Fig. 11. Electron density contours of the oxygen molecule from self-consistent Hartree–Fock calculations (reference 21). The values of the contours are in atomic units: $e/a_0^3 = 67.49e/\text{Å}^3$.

analysis Bader, Henneker and Cade concluded that the 0.002 contour, which includes 95% of the charge density, represents the
dimensions of the molecules and corresponds to the "diameters" of
molecules in equations of state and in transport properties such as
viscosity. Consideration of the 0.002 contour of oxygen[21] immediately leads to the antiferromagnetic structure of α-oxygen and the
mechanism of the α–β transformation. Figure 9 shows the basal
a–b plane of the monoclinic unit cell with the dimensions of the
oxygen molecule drawn to size according to the 0·002 contour.
The molecules touch each other in the direction of the diagonal
of the basal plane, but there is a finite spacing between alternate
molecules in the basal plane. The agreement between the calculated
contours and the crystal structure is illustrated by the fact that
the length of the diagonal of the basal plane, which is twice the
diameter of a molecule, is 6·40 Å, yielding 3.20 Å for the molecule
diameter, whereas the 0.002 contour requires 3.18 Å. This
discrepancy is completely within the error of interpreting the
0.002 contour as the molecular diameter and the accuracy of the
X-ray structure determination. Figure 10 is the a–c plane of the
monoclinic structure, again with the molecules drawn according
to the 0·002 contour. Figure 12 shows the $\bar{2}01$ plane. Assuming
that wherever two oxygen molecules touch according to the 0.002
contour there is sufficient overlap so that the electron spins in
neighbouring molecules must have opposite direction by the Pauli
principle, one is immediately led to the magnetic structure of
α-oxygen that would account for the data from neutron
scattering.[75-77] This result demonstrates that the structure determined from X-ray scattering, the electron density contours and the
magnetic structure derived from neutron scattering are completely
consistent. The packing density in α-oxygen is 71.4% as calculated
from the volume of the unit cell and the volume of two oxygen
molecules according to the 0.002 contour of Figure 11. A structure
of close packed spheres has a packing density of 74%; taking into
account the oblong form of the oxygen molecule, α-oxygen must be
considered a close packed structure.

Hörl[40] had found that β-oxygen is rhombohedral. Examining the
monoclinic α-oxygen unit cell with the molecules drawn to size
(or a corresponding model) reveals the mechanism of the α–β
transformation. Two steps are necessary in order to produce the

rhombohedral β-oxygen structure from the monoclinic α-oxygen (Fig. 13): (a) to give the molecules in the basal plane equal spacing and hexagonal symmetry, (b) to shift consecutive layers by about

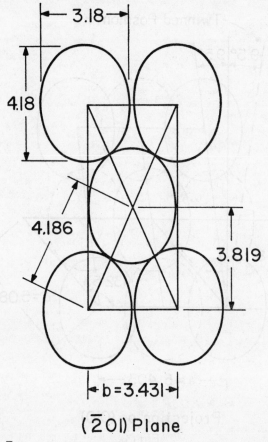

(2̄01) Plane

Fig. 12. 2̄01 plane of monoclinic α-oxygen with the molecules drawn according to the 0.002 contour of Fig. 11, showing the rows of oxygen molecules having direct contact.

9.5° to rhombohedral stacking sequence. The α–β-oxygen transformation has often been discussed in the literature[69, 78, 79] as being second order because specific heat data and susceptibility data seemed to indicate that the transition is smeared out over a temperature range of more than 1°. However, the transformation must be first order by a theorem due to Landau and Lifshitz,[106] even if the volume change is quite small; a second order transformation is only possible if the number of symmetry elements changes in specific

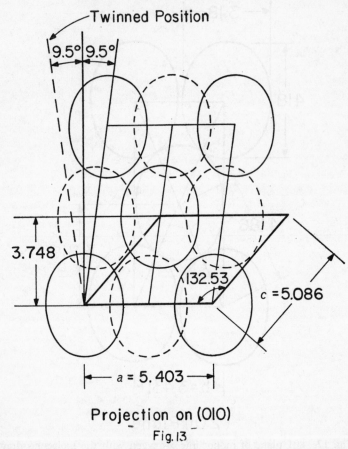

Projection on (010)

Fig. 13

Fig. 13. View of the monoclinic α-oxygen unit cell projected on the *a–c* plane, demonstrating the shift required to transform α-oxygen to rhombohedral β-oxygen.

ways which always include a factor of 2, but never a factor of 3. Landau and Lifshitz explicitly state that a transformation in which the number of symmetry elements changes by a factor of 3 cannot be second order. Alpha-oxygen, space group $C2/m$, has 4 symmetry elements; β-oxygen, space group $R\bar{3}m$, has 12 symmetry elements. This clearly shows that the α–β transformation cannot be second order.

However, it is easy to understand why the transition is simulating a second order transformation and is not occurring at a fixed temperature. The transformation requires slipping of lattice planes as an important part of the mechanism. Such slip is extremely sensitive to strain, which can enhance and inhibit the transformation. It is again a martensitic type transformation where the spontaneous transformation is not necessarily related to the temperature of equal free energy of the two different phases, but is influenced by kinetic conditions and depends not just on temperature but also on strain, which can vary considerably over the sample.

Jahnke[80] observed that the X-ray density of α-oxygen differed from the density calculated from the measured liquid density, the volume changes in melting and in the γ–β and β–α transformations by more than 10%—far outside the normal experimental errors. The reason for this discrepancy is the unusual behaviour of the expansion coefficients of γ-oxygen and β-oxygen.[41] The expansion coefficient of γ-oxygen is very high, $\Delta l/l \simeq 800 \times 10^{-6}$ per degree, independent of the temperature, causing a volume change by about 2.5% in the 10° stability range of γ-oxygen. After a 5% contraction in the γ–β transformation the expansion coefficient in the a direction of the rhombohedral β-structure increases to a value of about 1200×10^{-6} per degree. $\Delta l/l$ for argon near the melting point at 84°K is only about 600×10^{-6}, diminishing rapidly with decreasing temperature. The high expansion coefficient in the a direction of β-oxygen is essentially temperature independent over the 20° stability range of this phase, in spite of the fact that it is far below the melting point. The expansion coefficient in the c direction of the rhombohedral structure is negative, its measured value depending on the grain size in the sample. This is not surprising since temperature changes must produce strong local strain in fine-grained material of such an extremely anisotropic substance. The value of the expansion coefficient in the c direction is expected to be still higher for a single crystal, where the change in length with the change in temperature is not hindered by neighbouring grains of different orientation. A negative expansion coefficient in the c direction, which means an expansion with decreasing temperature, can be rationalized by assuming that the barrel-shaped molecules (at the 0.002 electron density contour, see Fig. 11) standing slightly separated but essentially parallel in the a–b plane are pulled together so strongly with decreasing temperature (as evidenced by the unusually high expansion coefficient in this direction) that consecutive planes are forced apart. The total volume change in the stability range of β-oxygen is again unusually high, about 4.6%. Taking these unexpected volume changes in the stability ranges of γ- and β-oxygen into account, the discrepancy between the X-ray density and the density derived from directly measured values disappears. Figure 14 shows the molar volume of oxygen as a function of temperature.

Solid fluorine has a transition from a soft, transparent, high-temperature phase, β-fluorine, to an opaque, hard and brittle substance, α-fluorine, almost at the same temperature as has oxygen and also with a heat of transformation appreciably exceeding the heat of melting (compare Table I). The α-fluorine structure is not an analogue of the orthorhombic structure of the other halogens, but is similar to α-oxygen. It is monoclinic, the unit cell dimensions being $a = 5.50$ Å, $b = 3.28$ Å, $c = 10.01$ Å, $\beta = 134.66°$.[46] The main difference between α-oxygen and α-fluorine is the doubling of the c axis of the unit cell of α-fluorine because the molecules are tilted in alternating directions by about 11° against the normal to the basal plane. The most probable unit cell is $C\,2/m$, just as for α-oxygen. Figure 15 shows the basal planes of α-fluorine, Fig. 16 the a–c planes in $C\,2/m$. However, α-fluorine has 4 molecules per unit cell which admits space group $C\,2/c$ as a possiblity. In $C\,2/m$ the tilt of the molecules has to be in the a–c plane. Its value can assume any arbitrary value in each of the two layers of the unit cell. In contrast, the tilt of the molecules in $C\,2/c$ has to be out of the a–c plane; the a–c plane being a glide plane, the tilt has to alternate in opposite directions by equal amounts in consecutive layers. The size of the molecules, as given in reference 21 (see Fig. 17) and the spacing of the c planes restricts the tilts to about 11°. In principle, it should be possible to derive the exact amount of tilt and its direction from the intensities of the X-ray diffraction patterns. Unfortunately, the fluctuations in line intensities due to incomplete randomness of the orientation of the crystallites is too great to allow a definitive decision between $C\,2/m$ and $C\,2/c$, although a better agreement between calculated and observed intensities is obtained for $C\,2/m$. The packing density of α-fluorine is slightly higher than 70% as calculated from the volume of the

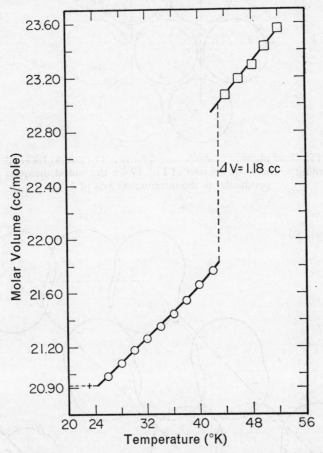

Fig. 14. Molar volume of solid oxygen in cm³/mole as a function of temperature.

unit cell and the volume of four fluorine molecules according to the 0.002 contour of Fig. 17, taking however as the smaller axis not the value of 2.86 Å at the "waist" but the maximum dimension perpendicular to the internuclear axis: 3.0 Å. Alpha-fluorine must be considered a close packed structure just as α-oxygen.

Fig. 15. Basal plane of monoclinic α-fluorine. The molecules are drawn according to the 0.002 contour of Fig. 17 for the widest diameter perpendicular to the internuclear axis of 3 Å.

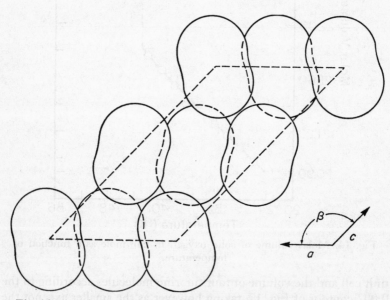

Fig. 16. *a–c* plane of monoclinic α-fluorine in space group *C* 2/*m*. The molecules are drawn according to the 0.002 electron density contour of Fig. 17, the tilt of the molecules is slightly exaggerated. Dotted lines indicate the molecules at the centres of the *a–b* planes.

It should be emphasized that there is excellent agreement between the molecular contours of reference 21 and the structures of α-oxygen and α-fluorine. The 0.002 contours of argon and neon based on self-consistent Hartree–Fock calculations, are quite consistent with the nearest neighbour distance in solid argon and neon

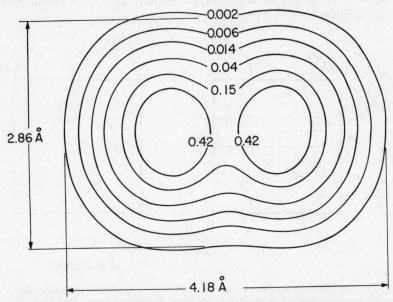

Fig. 17. Electron density contours of the fluorine molecules from self-consistent Hartree–Fock calculations (reference 21). The values of the contours are in atomic units: $67.49e/Å^3$.

at 4°K (see Table I). The size of the fluorine molecules according to these contours was an important element in elucidating the structure from the available powder patterns. The contours could possibly explain why fluorine has a high temperature structure almost identical with γ-oxygen but transforms immediately into α-fluorine, which is very similar to α-oxygen. Beta-oxygen is the phase which has the extreme expansion coefficient in the *a* direction, indicating very strong forces are pulling the molecules together with decreasing temperature. These forces act in the direction of the smaller axis of the molecules. A glance at Figs. 11 and 17 immediately reveals that the oxygen and fluorine contours at the

0.002 level even differ qualitatively just at the ends of the smaller axes. The oxygen 0.002 contour is concave, the fluorine 0.002 contour is convex. It is highly probable that fluorine does not form a rhombohedral structure analogous to β-oxygen because of this difference in electron density contours. The energy difference between the β-oxygen and α-oxygen structures is quite small, only 22 cal/mole.

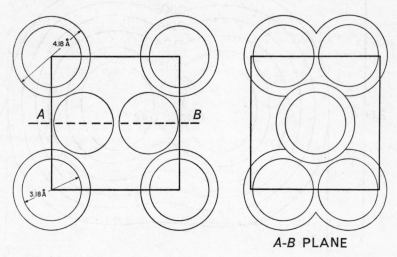

A-B PLANE

Fig. 18. Side-view of the cubic unit cell of γ-oxygen with the molecules drawn according to the 0.002 electron density contour of Fig. 11.

The agreement of the low temperature structures of α-oxygen and α-fluorine even with subtle details of the electron density contours is so excellent that conclusions drawn on the basis of older determinations of the size of these molecules or their atoms in covalent bonding[39, 45] may have to be re-examined. For instance, it is assumed that the molecules at the centre and the corner of the cubic unit cell of γ-oxygen and β-fluorine are free to rotate in all directions, whereas those molecules which are located in the faces of the cube rotate in one direction only. Figure 18 shows the cubic γ-oxygen structure. The circles around each lattice point are drawn according to the length of the two axes of the oxygen molecule, 4.18 Å and 3.18 Å, respectively. Free rotation requires a spherical space with a diameter of 4.18 Å. The space requirement for rotation in one direction only is a disc 4.18 Å wide and 3.18 Å high. The figure shows that the molecules in the corners of the cube can perform free rotation in all directions if the molecules in the face of the cube rotate only around an axis perpendicular to the centre plane (A–B) of the drawing. However, a cut through the cubic unit cell along the A–B plane shows that such a rotation interferes with the free rotation of the molecules in the centre of the cube, which are, however, equivalent to the molecules at the corners of the cubic unit cell. The postulated rotational motion is only consistent with the 0.002 electron density contours of reference 21 if all molecules move essentially in a synchronized way like a meshed gear. This is not very probable. It is possible that the rotation is not really free and that the molecules show, on average, random orientation with regard to a rotational degree of freedom. A molecule can easily perform partial rotational motion whenever the rotational position of the neighbouring molecules creates the necessary vacant space. Such a process is different from what is usually called "hindered rotation" where an energy barrier separates positions of lower energy, but is a rotational motion analogous to Brownian motion or diffusion, where the degree of motion of one molecule is determined by the fluctuations in the positions of its neighbours. In such a way the results of the intensities of the X-ray scattering from single crystals could be reconciled with the space requirements of the molecules given in reference 21.

C. Solid Nitrogen and Carbon Monoxide

Both nitrogen and carbon monoxide have two structures: hexagonal high temperature phases β-nitrogen and β-carbon monoxide with 2 molecules per unit cell, and low temperature phases, α-nitrogen and α-carbon monoxide (Table I) which are essentially face-centred cubic arrangements of the molecular centres where the molecules are oriented in the directions of the body diagonals, in this way reducing the symmetry to a primitive cubic space group $P2_13$ or $Pa3$. From a discussion of the intensities of the X-ray diffractions of a single crystal of α-nitrogen, Jordan and coworkers[34] concluded that the centres of the molecule are slightly displaced from position at the corners and the centres of faces along the body diagonal. However, the vibrational modes made infrared active by this asymmetry have not been found in infrared spectroscopy.[81, 82] The question of whether α-nitrogen and α-carbon monoxide are space group $P2_13$ or $Pa3$ is, therefore, still open.

Alpha-carbon monoxide is of special interest because of the orientational disorder of the direction of the carbon monoxide dipole moment. This seems to be frozen in and leads to a residual entropy at $T = 0°\text{K}$ (compare references 35 and 36). The residual entropy found in calorimetric studies is always lower than $R \log 2 \simeq 1.4$ entropy units, indicating that some ordering takes place in cooling carbon monoxide through the α–β transformation at $61°\text{K}$ and then to liquid helium temperatures. Such a partial ordering would also explain why different studies lead to different values for the residual entropy due to orientational disorder.[35, 36] Some preliminary unpublished results and the experience with the transformations in oxygen and fluorine suggest that cycling through the α–β transformation in carbon monoxide might improve the degree of orientational ordering.* Ordering of the carbon monoxide molecule should produce superstructure lines in the diffraction patterns and can therefore be followed by X-ray diffraction methods.

D. Solid Hydrogen and Deuterium

Both solid hydrogen rich in ortho-hydrogen and also deuterium rich in para-deuterium show a phase transition from hexagonal, supposedly hexagonal close packed, to cubic, supposedly face-centred cubic. As discussed in the case of the phase diagrams of argon–nitrogen, argon–carbon monoxide and argon–oxygen, such a transformation involves a change in stacking order from $ABAB$ stacking to $ABCABC$ stacking. It was pointed out[52] that this transformation shows all the signs of the martensitic-type transformation, taking place via a slip mechanism and being therefore extremely sensitive to strain. Plastic deformation could induce this transformation in solid hydrogen at a temperature near $3°\text{K}$; a spontaneous transformation is usually observed only below $2°\text{K}$, demonstrating clearly the well-known difference between the M_d point (martensitic by deformation) and the M_s point (spontaneous martensitic) (reference 7, page 531). Since martensitic-type

* A similar result was obtained by A. Ron (private communication).

transformations are mostly incomplete even after plastic deformation, all the unusual features of these transformations found in NMR work[83-86] and in specific heat measurements,[47-50] which could not be fitted to any simple theory, can probably be explained as being due to the martensitic nature of the transformation.

Schuch and coworkers[53] recently published a detailed X-ray diffraction study of the hydrogen transformations. They cycled both hydrogen and deuterium many times through the transformation. They observed the intensity of the (002) diffraction line for the cubic phase and that of the (10.1) diffraction line for the hexagonal phase as an indication of which phase was present and in what amounts. They found that after repeated cycling the hexagonal form does not reappear at all, or only incompletely. Heating to $9°K$ is necessary to regain the hexagonal structure in a reasonably short time. They conclude that the cycling produces an intermediate structure. The diffraction lines Schuch and coworkers used in their study of the transformations are quite sensitive to stacking faults (see reference 7, page 460). The fact that after cycling through the transformation the (10.1) line does not reappear can easily be explained by assuming that the sample retained a sufficient number of stacking faults to broaden the (10.1) line so much that it disappears in the background. Warming to $9°K$ simply allows the stacking faults to anneal out. The same behaviour can be found for the face-centred cubic–hexagonal close packed transformation in the argon-rich systems with nitrogen, for instance, if the transformation and/or deformation takes place below about $45°K$. It is not surprising that strain-induced slip at first is accompanied by many stacking faults which, at these low temperatures $(T \simeq 0.1\ T_{melting})$, require appreciable time to anneal out. The fact that supercooling of the anisotropic hexagonal phase below the M_s point led to an apparently more complete transformation than the warming of the isotropic cubic phase is quite consistent with the kinetics of a martensitic-type transformation. The cooling of the anisotropic phase produces more strain in the sample than the warming of an isotropic phase.

Extreme sensitivity of the hydrogen system to small strain is evidenced by the observation[52] that cube-texture gold, as background, stabilized the cubic phase up to $4°K$ and even changed the lattice parameters outside the error of the experiment.

X-ray diffraction patterns from hydrogen and deuterium are relatively poor, mainly for two reasons. The scattered intensity is weak so that only few of the strongest lines are seen. Furthermore, X-rays penetrate deeply into the weakly absorbing and scattering sample, so that the diffracted lines come from an unknown depth, are correspondingly broad and the peak position much more uncertain, than in cases where the beam penetrates the sample by less than 1 micron. For these reasons it is quite possible that the structures of solid hydrogen and deuterium do not really possess the high symmetry usually assumed. The weak diffraction lines indicating lower symmetry just escaped observation.

Schnepp and coworkers[112] have recently shown by far-infrared spectroscopy that the symmetry of the cubic phases of solid hydrogen and solid deuterium is indeed lower than face centred cubic and probably similar to α-nitrogen. Neutron diffraction patterns of solid deuterium [115, 116] confirm these results.

5. GENERAL CONCLUSIONS

The amazing complexity of the two-component phase diagrams of these relatively simple molecules and the astonishing properties of the low temperature structures of pure substances, such as oxygen and fluorine, demonstrate clearly that the existing theories for van der Waal's lattices are inadequate to predict any structure, or even to explain why a certain structure is formed. This is in contrast to the theory of metals where in some cases even the structure seems to be predictable from simple general principles like pseudopotential calculations (Heine and coworkers).[89] Experiments *and* theory for metals have been carried to an accuracy of $10^{-2}\%$ for band structures and Fermi surfaces. In the field of van der Waal's lattices both experiment and theory are almost a factor 100 from this level of precision. A theory is considered quite satisfactory if the value of the lattice energy, for instance, can be derived to a few percent accuracy.[6, 8, 91-92a] The difference in energy between the different structures is usually below this value. We know from experimentally determined phase diagrams the lines of equal free energy for the different phases, but we do not know how the free energy depends on concentration and temperature in detail. It would be of great interest to study the changes in free energy between stable and metastable phases by measuring the vapour pressures over these phases. This could be done by mass spectroscopic techniques. For the time being, probably the simplest and fastest way to gain more knowledge of these lattices is to continue careful experimental studies of the simpler binary systems.

Many of the low temperature transformations in van der Waal's lattices are of the martensitic type (involving slip, not being spontaneous and often incomplete) such as the face-centred cubic–hexagonal close packed transformation in argon and its alloys, the hydrogens, and the α–β-oxygen transformation. This result raises the question of whether or not other transformations, described in the literature as second order, are really of higher order or are instead a first order transformation of the martensitic type simulating a higher order transformation. One of these cases is the transformations in methane and similar molecules, which from thermal data seem to be of higher order. Early theories[93] tried to relate the apparent higher order transformations to the onset of rotational motion in the crystal. Recent evidence from neutron scattering seems to indicate that the methane molecule is not freely rotating even in the liquid state.[94-96] A careful crystallographic study of such cases is urgently needed. The phase diagram argon–methane has been determined recently by Greer and coworkers.[113] It shows a miscibility gap with a critical point at 63°K and 35% methane.

The crystal structures of α- and β-carbon monoxide should be checked, especially with regard to the ordering of the carbon monoxide dipoles eventually with $^{12}C^{18}O$ to increase the difference in scattering.

The system oxygen–fluorine should be studied. The structures of α-oxygen and α-fluorine are very similar; the structures of γ-oxygen and β-fluorine almost identical. It is of great interest whether or not oxygen and fluorine accept the other component in their structure, what the solubility of fluorine in β-oxygen would be and what the influence of dissolved fluorine would be on the expansion coefficients of β-oxygen.

Some, admittedly superficial, conclusions can be drawn from more general arguments suggested by the recent experimental evidence, to indicate where refinement and improvement of the theoretical approach are needed.

The van der Waal's or London forces in the form of Lennard-Jones or Kihara potentials are discussed as function of distance only from a molecule chosen as the origin. This is probably a good approximation for two gas molecules in the calculation of the second virial coefficient or even for nearest neighbours in a close packed structure. The London forces are fluctuating dipole forces. Argon, for instance, has a refractive index of $\simeq 1.5$. It is quite clear that next nearest neighbours, or molecules separated by several other molecules, will experience a force which is not only a function of distance but also of the medium between the two molecules (reference 10). That is, calculations of lattice energies from London forces have not taken into account the effect of the medium as have calculations for ion lattices,[98] which include dielectric screening. It is quite obvious that any such screening effect destroys the spherical symmetry of the van der Waal's forces and creates appreciable directional differences, depending on the electron density encountered in a certain direction.

It has been suggested that the forces due to quadrupole moments cause some of the low temperature transformations, especially the α–β transformations in nitrogen and carbon monoxide.[99-101] The contribution of the quadrupole moments to the lattice energy is definitely quite important. However, it is not certain that any structural changes are caused by quadrupole moment interaction. Kohin[101] has already pointed out that the quadrupole–quadrupole interactions in nitrogen and carbon monoxide are much greater than the heats of transformation of the α–β transformations. The quadrupole moments as compiled by Stogryn and Stogryn[20] increase the discrepancy, giving even higher values for the nitrogen and carbon monoxide quadrupole moments than the values used by Kohin. The quadrupole–quadrupole interaction is probably already an optimum both in the α- and the β-structures of nitrogen and carbon monoxide (the small dipole moment of carbon monoxide is of only minor importance[101]) because nearest neighbours are oriented almost perpendicular to each other in both structures. According to reference 34, the molecules are tilted by about $56°$ against the crystal axis in the β-phase and they are oriented in the direction of the body diagonal of the face-centred cube in the α-structure. The quadrupole–quadrupole interaction is actually so strong that such an orientation of nearest neighbours should persist to a certain degree in the liquid.

A further, somewhat speculative, conclusion is suggested by the low symmetry of α-oxygen and α-fluorine and the perplexing expansion coefficients of solid oxygen and fluorine, namely, that a weak sort of chemical "bonding" due to overlap and exchange energy (delocalized electrons) determine the structure and cause the exceptional contraction.[102] Such an effect can be highly selective and anisotropic. The molecules arrange themselves in such a fashion that overlap produces the maximum "bonding". For instance, such "bonding" would take place at the end of the smaller axis of the oxygen molecule (Fig. 11) in β-oxygen but in a direction halfway between the two main axes in α-oxygen and α-fluorine. The fluorine molecule has—at the 0.002 level—a convex electron density contour at the end of the smaller axis (Fig. 17) which is obviously less favourable for this type of overlap bonding and this may be the reason that fluorine does not form a rhombohedral structure analogous to β-oxygen. It is an extremely subtle problem whether the dominant effect of overlap is repulsion or attraction due to exchange energy and rigorous calculations will not be available in the near future. The unexpected complexity found in these simple two-component systems of molecules of quite similar size can be due to just this problem and could be the reason for the stability of superstructures like the delta phase in the argon–oxygen system and the x-phase in the oxygen–nitrogen system.

If one admits the possibility that overlap of the molecular wave functions can lead to weak "bonding", some aspects of the behaviour of pure argon can be brought into a consistent picture. Any Lennard-Jones type interaction between pairs predicts that hexagonal close packed argon should be more stable than face-centred cubic, and none of the explanations for the observed stability of face-centred cubic has been sufficiently rigorous to be completely convincing. The possibility that the cubic form of argon is in essence the consequence of overlap "bonding" of delocalized electrons cannot be ruled out. Since the effect of overlap depends very strongly on the distance between the atoms, overlap effect would be sharply weakened by thermal expansion. We would expect under these conditions that the high temperature phase, where due to thermal expansion overlap effects are reduced, would show the structure predicted by Lennard-Jones type potentials, namely hexagonal close packed. The investigation of the alloys of argon shows just this result: the hexagonal close packed phase is the high temperature phase of argon.

Both argon and neon[16b, 16c, 56] show that the Grüneisen constant γ^{104} decreases sharply at low temperatures, indicating that the expansion coefficient goes to zero faster than the specific heat as $T \to 0$. Such a behaviour requires a degree of anharmonicity beyond a Lennard-Jones potential.[92] A model involving overlap bonding would explain this extreme anharmonicity quite naturally.

London forces between molecules are usually calculated in an approximation using the ground state polarizability and the ionization potential. The influence of quite low lying excited electronic states, such as the singlet state of the oxygen molecule, on the intermolecular forces especially at short distances is not well known (compare reference 114). It is not impossible that these low lying states are at least in part responsible for the unusual expansion coefficients observed for the different phases of solid oxygen.

ACKNOWLEDGEMENTS

The results presented here are due to the collaboration with Professor C. S. Barrett over the last years. Discussions with Leo Falicov helped clarify many points. The research was supported by NSF Grant GP5859 and in part by the facilities provided by Advanced Research Projects Agency at the University of Chicago.

References

1. Pollak, G. L., *Rev. Mod. Phys.* **36**, 748–791 (1964).
2. Boato, G., *Cryogenics* **4**, 65–75 (1964).
3. Hollis Hallet, A. C., in *Argon, Helium and the Rare Gases*, G. A. Cook, Ed., Interscience, New York, 1961.
4. Dobbs, E. R., and Jones, G. O., *Rep. Progr. Phys.* **20**, 516 (1957).
5. Kihara, T., in *Advances in Chemical Physics*, **5**, I. Prigogine, Ed., John Wiley and Sons, New York, 1963, p. 186.
6. Horton, G. K., and Leach, J. W., *Proc. Phys. Soc. London* **82**, 816 (1963).
7. Barrett, C. S., and Massalski, T. B., *Structure of Metals*, 3rd Edition, McGraw-Hill, New York, 1966.
8. Jansen, L., *Phys. Letters* **4**, 91 (1963); *Phys. Rev.* **125**, 1798 (1962); *Phys. Rev.* **135**, A1292 (1964).
9a. Jansen, L., and Zimring, S., *Phys. Letters* **4**, 95 (1963).
9b. Lombardi, E., and Jansen, L., *Phys. Rev.* **167**, 822 (1968).
10. Lucas, A., *Physica* **35**, 353 (1967).

11. Götze, W., and Schmidt, H., *Z. Physik* **192**, 409 (1966).
12. Gillis, N. S., Werthamer, N. R., and Koehler, T. R., *Bull. Am. Phys. Soc.* **12**, 899 (1967).
13a. Vegard, L., *Z. Physik* **61**, 185 (1930).
13b. Vegard, L., *Z. Physik* **88**, 235 (1934).
13c. Vegard, L., *Z. Physik* **98**, 1 (1935).
13d. Vegard, L., *Z. Physik* **58**, 497 (1929).
13e. Vegard, L., *Z. Physik* **79**, 471 (1932).
13f. Vegard, L., *Nature, Lond.* **136**, 720 (1935).
14a. Smedt, J. de, and Keeson, W. H., Communication 53a, Kamerlingh Onnes Lab., University of Leiden; *Proc. Amsterdam Acad.* **25**, 118 (1922); **27**, 846 (1924).
14b. Keeson, W. H., Smedt, J. de, and Mooy, H. H., Communication 209d, Kamerlingh Onnes Lab., University of Leiden; *Proc. Amsterdam Acad.* **33**, 814 (1930).
14c. Keeson, W. H., and Mooy, H. H., Communication 209b, Kamerlingh Onnes Lab., University of Leiden; *Proc. Amsterdam Acad.* **33**, 447 (1930).
14d. Mooy, H. H., Communication 213d, Kamerlingh Onnes Lab., University of Leiden; *Proc. Amsterdam Acad.* **34**, 550 (1931).
14e. Mooy, H. H., Communication 216a, Kamerlingh Onnes Lab., University of Leiden; *Proc. Amsterdam Acad.* **34**, 660 (1931).
14f. Smedt, J. de, Keeson, W. H., and Mooy, H. H., Communication 202a, Kamerlingh Onnes Lab., University of Leiden; *Proc. Amsterdam Acad.* **32**, 745 (1929).
14g. Keesom, W. H., and Taconis, K. W., *Physica* **2**, 463 (1935).
14h. Keesom, W. H., and Taconis, K. W., *Physica* **3**, 141 (1936).
14i. Keesom, W. H., and Taconis, K. W., *Physica* **3**, 327 (1936).
14j. Keesom, W. H., and Taconis, K. W., *Physica* **5**, 161 (1938).
14k. Keesom, W. H., Communication Suppl. 64c, Kamerlingh Onnes Lab., University of Leiden, 1928.
15a. Simon, F., and Simson, C. V., *Z. Physik* **21**, 168 (1924); **25**, 160 (1924).
15b. Ruhemann, B., and Simon, F., *Z. Phys. Chem.* **B 15**, 389 (1932).
16a. Batchelder, D. N., and Simmons, R. O., *J. Appl. Phys.* **36**, 2864 (1965).
16b. Batchelder, D. N., Losee, D. L., and Simmons, R. O., *Phys. Rev.* **162**, 767 (1967).
16c. Peterson, O. G., Batchelder, D. N., and Simmons, R. O., *Phys. Rev.* **150**, 703 (1966).
16d. Urvas, A. O., Losee, D. L., and Simmons, R. O., *J. Phys. Chem. Solids* **28**, 2269 (1967).
16e. Peterson, O. G., Batchelder, D. N., and Simmons, R. O., *Phil. Mag.* **12**, 1193 (1965).
16f. Peterson, O. G., Batchelder, D. N., and Simmons, R. O., *J. Appl. Phys.* **36**, 2682 (1965).
16g. Losee, D. L., and Simmons, R. O., *Phys. Rev. Letters* **18**, 451 (1967).
17. Meyer, L., Barrett, C. S., and Haasen, P., *J. Chem. Phys.* **40**, 2744 (1964).

18. Plishkin, Yu. M., and Greenberg, B. A., *Phys. Letters* **19**, 375 (1965).
19. Bullough, R., Clyde, H. R., and Venables, J. A., *Phys. Rev. Letters* **17**, 249 (1966).
20. Stogryn, D. E., and Stogryn, A. P., *Mol. Physics* **11**, 371 (1966)
21. Bader, R. F., Henneker, W. H., and Cade, P. E., *J. Chem. Phys.* **46**, 3341 (1967).
22. Huo, W. M., *J. Chem. Phys.* **43**, 624 (1965).
23. Wahl, A. C., *Science 1966*, p. 961.
24. Flubacher, P., Leadbetter, A. J., and Morrison, J. A., *Proc. Phys. Soc. London* **78**, 1449 (1961).
25. Barrett, C. S., and Meyer, L., *J. Chem. Phys.* **42**, 107 (1965).
26. Clusius, K., Flubacher, P., Piesbergen, U., Schleich, K., and Sperandio, A., *Z. Naturforschung* **15A**, 1 (1960).
27. Beaumont, R. H., Chihara, H., and Morrison, J. A., *Proc. Roy. Soc. London*, **78**, 1462 (1961).
28. Lovejoy, D. R., *Nature, London* **197**, 353 (1963).
29. Clusius, K., and Riccoboni, L., *Z. Phys. Chem.* **B38**, 81 (1937).
30. Sears, D. R., and Klug, H. P., *J. Chem. Phys.* **37**, 3002 (1962).
31. Giauque, W. F., and Clayton, J. O., *J. Am. Chem. Soc.* **55**, 4875 (1933).
32. Bolz, L. H., Boyd, M. E., Mauer, F. A., and Peiser, H. S., *Acta Cryst.* **12**, 247 (1958).
33. Streib, W. E., Jordan, T. H., and Lipscomb, W. H., *J. Chem. Phys.* **37**, 2962 (1962).
34. Jordan, T. H., Smith, H. W., Streib, W. E., and Lipscomb, W. N., *J. Chem. Phys.* **41**, 756 (1964).
35. Clayton, J. O., and Giauque, W. F., *J. Am. Chem. Soc.* **54**, 2610 (1932).
36. Gillard, E. K., and Morrison, J. A., *J. Chem. Phys.* **45**, 1585 (1966).
37. Barrett, C. S., and Meyer, L., *J. Chem. Phys.* **43**, 3502 (1965).
38. Giauque, W. F., and Johnston, H. C., *J. Am. Chem. Soc.* **51**, 2300 (1929).
39. Jordan, T. H., Streib, W. E., Smith, H. W., and Lipscomb, W. N., *Acta Cryst.* **17**, 777 (1964).
40. Hörl, E. M., *Acta Cryst.* **15**, 845 (1962).
41. Barrett, C. S., Meyer, L., and Wasserman, J., *Phys. Rev.* **163**, 851 (1967).
42. Barrett, C. S., Meyer, L., and Wasserman, J., *J. Chem. Phys.* **47**, 592 (1967).
43. Barrett, C. S., and Meyer, L., *Phys. Rev.* **160**, 694 (1967).
44. Wu, J. H., White, D., and Johnston, H. L., *J. Am. Chem. Soc.* **75**, 5642 (1953).
45. Jordan, T. H., Streib, W. E., and Lipscomb, W. N., *J. Chem. Phys.* **41**, 760 (1964).
46. Meyer, L., Barrett, C. S., and Greer, S. C., *J. Chem. Phys.* **49**, 1902 (1968).
47. Simon, F., Mendelssohn, K., and Ruhemann, M., *Z. Phys. Chem.* **B15**, 121 (1937).
48. Hill, R. W., and Ricketson, B. W. A., *Phil. Mag.* **45**, 277 (1954).

49. Orttung, W. H., *J. Chem. Phys.* **36**, 652 (1962).
50. Ahlers, G., and Orttung, W. H., *Phys. Rev.* **133**, A1642 (1964).
51. Bostanjoglo, O., and Kleinschmidt, R., *J. Chem. Phys.* **46**, 2004 (1967).
52. Barrett, C. S., Meyer, L., and Wasserman, J., *J. Chem. Phys.* **45**, 834 (1966).
53. Schuch, A. F., Mills, R. L., and Depattie, D. A., *Phys. Rev.* **165**, 1032 (1967).
54. Gonzalez, O. P., White, D., and Johnston, H. C., *J. Phys. Chem.* **61**, 773 (1957).
55. Grenier, G., and White, D., *J. Chem. Phys.* **37**, 1563 (1962).
56. Barrett, C. S., and Meyer, L., *J. Chem. Phys.* **41**, 1078 (1964).
57. Barrett, C. S., *Acta Cryst.* **9**, 621 (1956); *Trans. Am. Soc. Metals* **49**, 53 (1957).
58. Barrett, C. S., Meyer, L., and Wasserman, J., *J. Chem. Phys.* **44**, 998 (1966).
59. Barrett, C. S., Meyer, L., and Wasserman, J., *J. Chem. Phys.* **47**, 740 (1967).
60. Angwin, M. J., and Wasserman, J., *J. Chem. Phys.* **44**, 417 (1966).
61. Barrett, C. S., Meyer, L., Greer, S. C., and Wasserman, J., *J. Chem. Phys.* **48**, 2670 (1968).
62. Ruhemann, M., Lichter, A., and Komarow, P., *Physik Z. Sowjetunion* **8**, 326 (1935).
63. Staveley, L. A. K., in *Annual Review Phys. Chem.* 1962.
64. Simon, M., *Phys. Letters* **5**, 319 (1963).
65. Streeb, W. B., and Jones, C. H., *J. Chem. Phys.* **42**, 3989 (1965).
66. Brouwer, J. P., Hermans, L. F. J., Knaap, H. F. P., and Beenakker, J. J. M., *Physica* **30**, 1409 (1964).
67. Trepp, C., *Schweizer Archiv Angew. Wiss u. Tech.* **24**, 191 (1958).
68. Jamieson, H. C., M.A. Thesis, Toronto, 1966, unpublished.
69. Fagerstrom, C. H., and Hollis Hallett, A. C., *Ann. Acad. Sci. Fenn. A* **VI 210**, 210 (1966).
70. Prikhoto, A., *Acta Physico Chimica U.R.S.S.* **X**, 913 (1939).
71. Knobler, C. M., van Heijningen, R. J. J., and Beenakker, J. J. M., *Physica* **27**, 296 (1961).
72. Herzberg, G., *Molecular Spectra and Molecular Structure*, D. van Nostrand, Princeton, New Jersey, 1950, p. 446.
73. Kanda, E., Haseda, T., and Otsubo, A., *Physica* **20**, 131 (1954).
74. Perrier, A., and Kamerlingh Onnes, H., *Comm. Leiden* 139d (1914).
75. Alikhanov, R. A., in "Physics and Techniques of Low Temperatures", *Proc. Regional Conf. 3rd Prague*, 1963, p. 127.
76. Collins, M. F., *Proc. Phys. Soc. London* **89**, 415 (1966).
77. Alikhanov, R. A., *J. Exptl. Theoret. Phys. U.S.S.R.* **45**, 812 (1963).
78. Hoge, H. J., *J. Res. Natl. Bur. Std.* **44**, 321 (1950).
79. Borovik-Romanov, A. L., Orlova, M. P., and Strelkov, G. P., *Dokl. Acad. Nauk SSSR* **99**, 699 (1954).
80. Jahnke, J., *J. Chem. Phys.* **47**, 336 (1967).
81. Anderson, A., and Leroi, G. E., *J. Chem. Phys.* **45**, 4359 (1966).

82. Ron, A., and Schnepp, O., quoted in reference 81.

83. Hatton, J., and Rollin, B. V., *Proc. Roy. Soc. London* **A199**, 222 (1949).

84. Reif, F., and Purcell, E. M., *Phys. Rev.* **91**, 631 (1953).

85. Smith, G. W., and Housley, R. M., *Phys. Rev.* **117**, 752 (1960).

86. Hass, W. P. A., Poulis, N. J., and Borleffs, J. J. W., *Physica* **27**, 1037 (1961).

87. James, H. M., and Raich, J. C., *Phys. Rev.* **162**, 649 (1967).

88. James, H. M., *Phys. Rev.* **167**, 862 (1968).

89a. Heine, V., and Weaire, D., *Phys. Rev.* **152**, 603 (1966).

89b. Weaire, D., *Proc. Phys. Soc. London* (2), **1**, 210 (1968).

89c. Heine, V., *Proc. Phys. Soc. London* (2), **1**, 222 (1968).

90a. Lee, M. J. G., *Proc. Roy. Soc. London* **A295**, 440 (1966).

90b. Lee, M. J. G., and Falicov, L. M., *Proc. Roy. Soc. London* **A304**, 1319 (1968).

91. Barron, T. H. K., and Domb, C., *Phil. Mag.* **45**, 654 (1954).

92. Barron, T. H. K., *Phil. Mag.* **46**, 720 (1955).

92a. Maradudin, A. A., Montroll, E. W., and Weiss, S. H., *Theory of Lattice Dynamics in the Harmonic Approximation*, Academic Press, New York, 1963.

93. Pauling, L., *Phys. Rev.* **36**, 430 (1930).

94. Harker, Y. D., and Brugger, R. M., *J. Chem. Phys.* **46**, 2201 (1967).

95. Dasannacharya, B. A., and Venkataraman, G., *Phys. Rev.* **156**, 196 (1967).

96. Venkataraman, G., Dasannacharya, B. A., and Rao, K. R., *Phys. Rev.* **161**, 133 (1967).

97. James, H. M., and Keenan, T. A., *J. Chem. Phys.* **31**, 12 (1959).

98. Cochran, W., *Proc. Roy. Soc.* **A253**, 260 (1959).

99a. Jansen, L., and de Wette, F., *Physica* **21**, 83 (1955).

99b. de Wette, F., *Physica* **22**, 644 (1956).

100. Nagai, O., and Nakamura, T., *Progr. Theoret. Physics* (*Kyoto*) **24**, 432 (1960).

101. Kohin, B. C., *J. Chem. Phys.* **33**, 882 (1960); Thesis, University of Maryland, 1960.

102. Hillier, I. H., and Rice, S. A., *J. Chem. Phys.* **46**, 3881 (1967).

103. Bostonjoglo, O., and Kleinschmidt, R., *Z. Naturforschung* **21a**, 2106 (1966).

104. Kittel, C., *Introduction to Solid State Physics*, 3rd Edition, John Wiley and Sons, New York, 1966, p. 185.

105. Long, H. M., and di Paolo, F. S., *Chem. Eng. Progress* **59**, 30 (1963).

106. Landau, L. D., and Lifshitz, E. M., *Statistical Physics*, Addison Wesley Publishing Company, Reading, Mass., 1958, p. 445.

107. Din, F., Goldman, K., and Monroe, A. G., *Proc. Intern. Cong. Refrig.* 9*th*, Paris, 1-003, 1955.

108. Hingsammer, J., and Lüscher, E., *Helvetica Physica Acta* **41**, 914 (1968).

109. Alder, B. J., and Paulson, R. H., *J. Chem. Phys.* **43**, 4172 (1965).

110. Jordan, T., Barrett, C. S., and Meyer, L., *J. Chem. Phys.*, in press.

111a. Jagodzinski, H., and Arnold, H., *Silicon Carbide*, Pergamon Press, 1960, pp. 136–146.
111b. Jagodzinski, *Neues Jahrb. Min.* **10**, 49 (1954).
111c. Jagodzinski, H., *Neues Jahrb. Min.* **10**, 209 (1954).
112. Hardy, W. N., Silvera, I. F., Klump, K. N., and Schnepp, O., *Phys. Rev. Letters* **21**, 291 (1968).
113. Greer, S. C., Meyer, L., and Barrett, C. S., *J. Chem. Phys.*, in press.
114. Hirschfelder, J. O., "Intermolecular forces", in *Advances in Chemical Physics*, **12**, John Wiley and Sons, New York, 1967, pp. 78, 168, 335.
115. Mucker, K. F., Harris, P. M., and White, D., *J. Chem. Phys.* **49**, 1922 (1968).
116. James, H. M., *Phys. Rev.* **167**, 862 (1968).

AUTHOR INDEX FOR VOLUME XVI

Entries in square brackets denote reference numbers (if any) cited in the preceding page numbers. Entries in **bold** type refer to authors of articles.

389

SUBJECT INDEX FOR VOLUME XVI

CUMULATIVE INDEX TO VOLUMES I-XVI

Authors of Articles

Numerals in bold type are volume numbers.

Titles of Articles